Automating Intelligence With Cloud-Native AI Tools

Minakshi
King Khalid University, Saudi Arabia

Madhulika Bhatia
Digital Healthcare Wales, USA

Gaurav Dubey
KIET Group of Institutions, India

Tanupriya Choudhury
University of Petroleum and Energy Studies, Dehradun, India

IGI Global
Scientific Publishing
Publishing Tomorrow's Research Today

Vice President of Editorial — Melissa Wagner
Director of Acquisitions — Mikaela Felty
Director of Book Development — Jocelynn Hessler
Production Manager — Mike Brehm
Cover Design — Jose Rosado

Published in the United States of America by
IGI Global Scientific Publishing
701 East Chocolate Avenue
Hershey, PA, 17033, USA
Tel: 717-533-8845 | Fax: 717-533-7115
Website: https://www.igi-global.com E-mail: cust@igi-global.com

Library of Congress Cataloging-in-Publication Data

Names: Minakshi (Computer engineer), 1980- editor | Bhatia, Madhulika, 1984- editor | Dubey, Gaurav editor | Choudhury, Tanupriya editor
Title: Automating intelligence with cloud-native AI tools / edited by: Minakshi, Madhulika Bhatia, Gaurav Dubey, Tanupriya Choudhury.
Description: Hershey PA : IGI Global Scientific Publishing, [2026] | Includes bibliographical references. | Summary: "This book explores how cloud computing and artificial intelligence converge to create intelligent, self-managing systems. Through case studies and practical frameworks, it demonstrates how integrating machine learning into scalable cloud architectures enables adaptive, efficient, and secure computing environments that drive digital transformation and next-generation innovation"-- Provided by publisher.
Identifiers: LCCN 2025054505 (print) | LCCN 2025054506 (ebook) | ISBN 9798337374420 hardcover | ISBN 9798337374437 paperback | ISBN 9798337374444 ebook
Subjects: LCSH: Artificial intelligence | Cloud computing
Classification: LCC Q335 .A887 2026 (print) | LCC Q335 (ebook)
LC record available at https://lccn.loc.gov/2025054505
LC ebook record available at https://lccn.loc.gov/2025054506

British Cataloguing in Publication Data
A Cataloguing in Publication record for this book is available from the British Library.

Table of Contents

Preface xxv

Chapter 1
Adaptive Learning Ecosystems: AI-Enabled Cloud Platforms for Employee Upskilling 1
Tanupriya Choudhury, School of Computer Science, University of Petroleum and Energy Studies, India
Mohit Yadav, JGBS, O.P. Jindal Global University, India
Narayanage Jayantha Dewasiri, Faculty of Management Studies, Sabaragamuwa University of Sri Lanka, Sri Lanka & Qasim Ibrahim School of Business, Villa College, Maldives
Minakshi, College of Computer Science, King Khalid University, Saudi Arabia

Chapter 2
Cloud AI of Strategic Human Resource Transformation 15
Bhumika Ray, JGBS, O.P. Jindal Global University, India
Shashi Kant, DMI-St. Euegne University, Zambia

Chapter 3
AI and the Human Touch: Building Trust in Cloud-Automated Business Ecosystems 31
Ali Najeeb, Qasim Ibrahim School of Business, Villa College, Maldives
Gunjan Mohan Sharma, JGBS, O.P. Jindal Global University, India
Priya Gaur, Centre for Online Education, GLA University, India

Chapter 4
Cloud-Driven Organizational Agility: AI as the Catalyst for Digital Transformation 45
Himanshi Rajora, JGBS, O.P. Jindal Global University, India
Azadeh Amoozegar, INTI International University, Malaysia
Dilshad Jeffery, Qasim Ibrahim School of Business, Villa College, Maldives

Chapter 5
AI Ethics in People Analytics: Balancing Automation, Privacy, and Fairness.... 57
Thi Minh Ngoc Luu, International School, Vietnam National University, Hanoi, Vietnam
Damith Gangodawilage, Qasim Ibrahim School of Business, Villa College, Maldives
Anugamini Priya Srivastava, Dhyaan Prasad Global University, Pune, India
Jeayaram Subramanian, JGBS, O.P. Jindal Global University, India

Chapter 6
AI-Augmented Leadership: Redefining Managerial Decision-Making in the Cloud Era 73
Damith Gangodawilage, Qasim Ibrahim School of Business, Villa College, Maldives
Anugamini Priya Srivastava, Dnyaan Prasad Global University, Pune, India
Jeayaram Subramanian, JGBS, O.P. Jindal Global University, India
Thi Minh Ngoc Luu, International School, Vietnam National University, Hanoi, Vietnam

Chapter 7
AI-Driven Sustainability: Automating ESG Reporting and Green Operations through the Cloud 87
Mohan Kumar, Faculty of Commerce and Management, SGT University, Gurugram, India
Sandeep Kumar Singh, JGBS, O.P. Jindal Global University, Haryana, India
Xuan-Hoa Nghiem, International School, Vietnam National University, Hanoi, Vietnam
Mananage Shanika Hansini Rathnasiri, Faculty of Management Studies, Sabaragamuwa University of Sri Lanka, Sri Lanka & Qasim Ibrahim School of Business, Villa College, Maldives

Chapter 8
AI-Powered Supply Chains: Cloud-Native Intelligence for Resilient and Adaptive Operations 99
Mohan Kumar, Faculty of Commerce and Management, SGT University, Gurugram, India
Xuan-Hoa Nghiem, International School, Vietnam National University, Hanoi, Vietnam
Sandeep Kumar Singh, JGBS, O.P. Jindal Global University, Haryana, India
Mananage Shanika Hansini Rathnasiri, Faculty of Management Studies, Sabaragamuwa University of Sri Lanka, Sri Lanka & Qasim Ibrahim School of Business, Villa College, Maldives

Chapter 9
Cross-Cultural Management in AI-Augmented Workplaces: Opportunities and Challenges 113
Himanshi Rajora, JGBS, O.P. Jindal Global University, India
Dilshad Jeffery, Qasim Ibrahim School of Business, Villa College, Maldives
Azadeh Amoozegar, INTI International University, Malaysia

Chapter 10
Shaping Entrepreneurial Creativity and Innovation Through Human-AI Cognition 127
Khushboo Jain, Shri Vaishnav Institute of Management and Science, Indore, India
Deepa Katiyal, Shri Vaishnav Institute of Management and Science, Indore, India
Shashank Mittal, JGBS, O.P. Jindal Global University, Sonipat, India
Mukhtaruddin Mukhtaruddin, Faculty of Economics, Universitas Sriwijaya, Palembang, Indonesia

Chapter 11
Infrastructural Intelligence as Sociotechnical Institution: Cloud-Native AI, Power, Governance, and Inequality in the Global South 139
Sneha Kadyan, Jindal Global Business School, O.P. Jindal Global University, Sonipat, India
Firdous Ahmad Malik, University of People, USA

Chapter 12
Flipped Learning Empowering Students Through Digital Innovation: Reverse Instruction for Enhanced Engagement and Deeper Understanding 155
Priti Gupta, Greater Noida Institute of Technology, Greater Noida, India
Richa Adlakha, Manav Rachna International Institute of Research and Studies, India
Ashish Grover, Manav Rachna International Institute of Research and Studies, India
Neha Chaudhary, Manav Rachna International Institute of Research and Studies, India

Chapter 13
Automating Corporate Strategy: How Cloud AI Transforms Managerial Intelligence 183
Himanshi Rajora, JGBS, O.P. Jindal Global University, India
Dilshad Jeffery, Qasim Ibrahim School of Business, Villa College, Maldives
Azadeh Amoozegar, INTI International University, Malaysia

Chapter 14
Cloud-Native AI for Operational Efficiency and Predictive Maintenance: Automating Intelligence in Practice 197
Swati Singh, Greater Noida Institute of Engineering and Technology, India
Kamaljit Kaur, Guru Nanak Dev University, Amritsar, India
Shivangi Tyagi, KIET Group of Institutions, Ghaziabad, India
Mayank Tyagi, KIET Group of Institutions, Ghaziabad, India

Chapter 15
Automating Intelligence: Cloud-Native AI, Generative Tools, and the Future of Research.. 217
Shimona Shriya, Xavier Institute of Management and Entrepreneurship, Bangalore, India
Mohit Verma, O.P. Jindal Global University, India
Aishwarya Vashishta, Rutgers University, USA

Chapter 16
The Cloud-Native Enterprise: Orchestrating Intelligent Automation Across Global Operations 235
Priya Gaur, Centre for Online Education, GLA University, India
Gunjan Mohan Sharma, JGBS, O.P. Jindal Global University, India
Ali Najeeb, Qasim Ibrahim School of Business, Villa College, Maldives

Chapter 17
Human-Centric AI: Reinventing Employee Experience Through Cloud-Native Intelligence 251
Isha Goley, Prestige Institute of Management and Research, Gwalior, India
Bhumika Ray, JGBS, O.P. Jindal Global University, India
Mukhtaruddin Mukhtaruddin, Faculty of Economics, Universitas Sriwijaya, Indonesia

Chapter 18
Intelligent Automation as a Strategic Lever: Harnessing Cloud-Native AI for Scalable Growth 267
Xuan-Hoa Nghiem, International School, Vietnam National University, Hanoi, Vietnam
Mananage Shanika Hansini Rathnasiri, Faculty of Management Studies, Sabaragamuwa University of Sri Lanka, Sri Lanka & Qasim Ibrahim School of Business, Villa College, Maldives
Sandeep Kumar Singh, JGBS, O.P. Jindal Global University, Haryana, India
Mohan Kumar, Faculty of Commerce and Management, SGT University, Gurugram, India
Manju Dagar, Faculty of Commerce and Management, SGT University, India

Chapter 19
Considering the Application of Artificial Intelligence in Society and Assessing the Future of AI 279
Shimona Shriya, Xavier Institute of Management and Entrepreneurship, Bangalore, India
Rituraj Singh, O.P. Jindal Global University, Haryana, India
Shashi Kant, DMI St. Eugene University, Zambia

Chapter 20
Governance in Cloud-Automated Enterprises: Frameworks for Responsible AI Deployment 295
Gunjan Mohan Sharma, JGBS, O.P. Jindal Global University, India
Ali Najeeb, Qasim Ibrahim School of Business, Villa College, Maldives
Priya Gaur, Centre for Online Education, GLA University, India

Chapter 21
Reimagining Talent Management: Cloud-Native AI Tools for Predictive HR Analytics 309
Mohit Yadav, JGBS, O.P. Jindal Global University, India
Narayanage Jayantha Dewasiri, Faculty of Management Studies, Sabaragamuwa University of Sri Lanka, Sri Lanka & Qasim Ibrahim School of Business, Villa College, Maldives
Minakshi, College of Computer Science, King Khalid University, Saudi Arabia
Tanupriya Choudhury, School of Computer Science, University of Petroleum and Energy Studies, Dehradun, India

Chapter 22
Risk Perception and AI-Driven Financial Decision-Making: A FinTech Perspective 325
Manish Dubey, Shri Vaishnav College of Arts and Commerce, Indore, India
Mansi Panwar, Devi Ahilya Vishwavidyalaya University in Indore, India
Rituraj Singh, O.P. Jindal Global University, Haryana, India
Shashi Kant, DMI St. Eugene University, Zambia

Chapter 23
Automating Intelligence in Healthcare: Cloud-Native Explainable AI for Brain Tumor Detection 349
Arun Kumar Rai, Greater Noida Institute of Engineering and Technology, India
Shraddha Kaushal, Greater Noida Institute of Engineering and Technology, India
Harsh Gour, Greater Noida Institute of Engineering and Technology, India
Sonali Jha, Greater Noida Institute of Engineering and Technology, India
Gazab Bhati, Greater Noida Institute of Engineering and Technology, India

Chapter 24
Emotion AI and Employee Experience: Cloud-Native Tools for Human-Centric Management 367
Jeayaram Subramanian, JGBS, O.P. Jindal Global University, India
Damith Gangodawilage, Qasim Ibrahim School of Business, Villa College, Maldives
Thi Minh Ngoc Luu, International School, Vietnam National University, Hanoi, Vietnam
Anugamini Priya Srivastava, Dnyaan Prasad Global University, Pune, India

Chapter 25
The Smart HR Cloud: Automating Workforce Planning and Skill Mapping 385
Mohit Yadav, JGBS, O.P. Jindal Global University, India
Minakshi, College of Computer Science, King Khalid University, Saudi Arabia
Tanupriya Choudhury, School of Computer Science, University of Petroleum and Energy Studies, India
Narayanage Jayantha Dewasiri, Faculty of Management Studies, Sabaragamuwa University of Sri Lanka, Sri Lanka & Qasim Ibrahim School of Business, Villa College, Maldives

Chapter 26
AI-Driven Personalization in Cloud-Powered Fashion: E-Commerce Customer Experience 395
Priti Gupta, Department of Electrical Engineering, Greater Noida Institute of Technology, Greater Noida, India
Neha Arora, Amity School of Fashion Technology, Amity University, Noida, India
Akanksha Kathuria, School of Business, Manav Rachna University, Faridabad, India
Neeraj Chopra, School of Business, Manav Rachna University, Faridabad, India

Compilation of References 415

About the Contributors 449

Index 459

Detailed Table of Contents

Preface xxv

Chapter 1
Adaptive Learning Ecosystems: AI-Enabled Cloud Platforms for Employee Upskilling 1
Tanupriya Choudhury, School of Computer Science, University of Petroleum and Energy Studies, India
Mohit Yadav, JGBS, O.P. Jindal Global University, India
Narayanage Jayantha Dewasiri, Faculty of Management Studies, Sabaragamuwa University of Sri Lanka, Sri Lanka & Qasim Ibrahim School of Business, Villa College, Maldives
Minakshi, College of Computer Science, King Khalid University, Saudi Arabia

The accelerating pace of technological change has created a critical skills gap, rendering traditional corporate training models obsolete. This chapter explores the transformative potential of adaptive learning ecosystems—AI-enabled, cloud-native platforms that create dynamic, personalized environments for continuous employee upskilling. It deconstructs the core architectural pillars of these systems: artificial intelligence for personalization, cloud infrastructure for scalability, and dynamic content for relevance. The analysis covers the mechanisms of adaptation, strategic benefits for organizational agility, and a practical implementation framework. It also addresses critical challenges, including data privacy and algorithmic bias. Ultimately, the chapter argues that investing in these ecosystems is a strategic imperative, enabling businesses to build a resilient, future-ready workforce and sustain a competitive advantage in the AI-augmented economy.

Chapter 2
Cloud AI of Strategic Human Resource Transformation 15
Bhumika Ray, JGBS, O.P. Jindal Global University, India
Shashi Kant, DMI-St. Euegne University, Zambia

The chapter explores how Human Resource Management is changing due to the use of cloud-native AI applications. It addresses how microservices, APIs, containerisation, serverless computing, and MLOps—elements that are discussed as cloud-native—can be used to build scalable and intelligent HR ecosystems. The conversation goes further to the application of AI in conventional HR activities, such as recruitment, workforce planning, performance management, learning and development, and employee experience. Also mentioned in the chapter is acute ethical, legal, and governance issues involved in AI implementation HR, especially given the sensitivity of employee data and the need to remain unbiased in the results of decisions. Case studies on how organisations in different industries use cloud-native AI to improve strategic impact, productivity, innovation, and agility culture are presented in reality. Finally, the chapter places cloud-native AI not as a technological revolution in itself but as a driving force that will transform the future, structure, and functionality of HR.

Chapter 3
AI and the Human Touch: Building Trust in Cloud-Automated Business Ecosystems....... 31
Ali Najeeb, Qasim Ibrahim School of Business, Villa College, Maldives
Gunjan Mohan Sharma, JGBS, O.P. Jindal Global University, India
Priya Gaur, Centre for Online Education, GLA University, India

The integration of artificial intelligence (AI) into cloud-automated business ecosystems promises unparalleled efficiency but poses a critical threat: the erosion of trust. This chapter argues that trust is not an obstacle to automation but its essential enabler. It diagnoses the roots of the trust deficit, including algorithmic opacity, data privacy concerns, and bias. The chapter then redefines the "human touch" not as mere sentiment, but as crucial roles in orchestrating, explaining, and overseeing AI. A practical framework for building trust is presented, founded on pillars of transparency, reliability, ethics, and human-centric design. Finally, the chapter explores the profound social, practical, ethical, and managerial implications organizations must navigate to foster a symbiotic future where technology amplifies, rather than replaces, human integrity.

Chapter 4
Cloud-Driven Organizational Agility: AI as the Catalyst for Digital Transformation....... 45
Himanshi Rajora, JGBS, O.P. Jindal Global University, India
Azadeh Amoozegar, INTI International University, Malaysia
Dilshad Jeffery, Qasim Ibrahim School of Business, Villa College, Maldives

This chapter examines the synergistic relationship between cloud computing and artificial intelligence (AI) as the core driver of modern digital transformation. It argues that while cloud computing provides the foundational infrastructure for organizational agility—enabling scalability, speed, and DevOps practices—it is the strategic application of AI that acts as the essential catalyst. AI transforms cloud capabilities into intelligent operations through predictive analytics, automation, and generative capabilities. The analysis moves from theoretical foundations to practical use cases across business functions, outlining the necessary human and organizational shifts. Crucially, the chapter concludes with a comprehensive discussion of the multi-faceted implications, addressing the managerial, practical, ethical, and social considerations leaders must navigate to harness this transformative potential responsibly and effectively.

Chapter 5
AI Ethics in People Analytics: Balancing Automation, Privacy, and Fairness.... 57
Thi Minh Ngoc Luu, International School, Vietnam National University, Hanoi, Vietnam
Damith Gangodawilage, Qasim Ibrahim School of Business, Villa College, Maldives
Anugamini Priya Srivastava, Dhyaan Prasad Global University, Pune, India
Jeayaram Subramanian, JGBS, O.P. Jindal Global University, India

The integration of Artificial Intelligence (AI) into people analytics presents a transformative opportunity for human resource management, yet it introduces critical ethical challenges. This chapter examines the central tension in balancing the efficiency of automation with the imperatives of employee privacy and algorithmic fairness. It argues that navigating this triad is a strategic necessity, not merely a technical concern. The analysis moves from foundational ethical principles to an operational framework for implementation, emphasizing governance, transparency, and human oversight. Finally, the chapter explores the profound social, practical, ethical, and managerial implications of AI-driven HR, concluding that a proactive, principled approach is essential to foster trust, mitigate risk, and build more humane and equitable workplaces.

Chapter 6
AI-Augmented Leadership: Redefining Managerial Decision-Making in the Cloud Era 73
Damith Gangodawilage, Qasim Ibrahim School of Business, Villa College, Maldives
Anugamini Priya Srivastava, Dnyaan Prasad Global University, Pune, India
Jeayaram Subramanian, JGBS, O.P. Jindal Global University, India
Thi Minh Ngoc Luu, International School, Vietnam National University, Hanoi, Vietnam

The convergence of artificial intelligence (AI) and cloud computing is fundamentally redefining managerial decision-making. This chapter explores the paradigm of AI-augmented leadership, where human intuition is synergistically combined with algorithmic intelligence to navigate an increasingly complex business environment.

It argues that the leader's role is evolving from a solitary decider to a designer of intelligent decision-making processes. The analysis deconstructs the foundational triad of AI, cloud infrastructure, and the modern leader's evolving skillset. It then presents a practical framework for augmented decision-making, spanning enhanced situational awareness, generative optioneering, and data-informed judgment. The chapter culminates in a critical examination of the profound managerial, ethical, social, and practical implications, asserting that while AI provides powerful capabilities, the ultimate responsibility for wise and ethical judgment remains a uniquely human imperative.

Chapter 7
AI-Driven Sustainability: Automating ESG Reporting and Green Operations through the Cloud 87
Mohan Kumar, Faculty of Commerce and Management, SGT University, Gurugram, India
Sandeep Kumar Singh, JGBS, O.P. Jindal Global University, Haryana, India
Xuan-Hoa Nghiem, International School, Vietnam National University, Hanoi, Vietnam
Mananage Shanika Hansini Rathnasiri, Faculty of Management Studies, Sabaragamuwa University of Sri Lanka, Sri Lanka & Qasim Ibrahim School of Business, Villa College, Maldives

The escalating demand for robust Environmental, Social, and Governance (ESG) reporting, coupled with the urgent need for sustainable operations, has exposed the severe limitations of traditional manual methods. This chapter explores the transformative synergy of Artificial Intelligence (AI) and Cloud Computing in automating and enhancing corporate sustainability. It demonstrates how this convergence automates the entire ESG reporting lifecycle-from intelligent data ingestion to audit-ready assurance-transforming it from a compliance burden into a strategic asset. Furthermore, it details how AI-driven analytics enable "green operations" by optimizing energy use, building smart supply chains, and facilitating proactive maintenance. Finally, the chapter addresses critical implementation considerations, including a strategic roadmap and the ethical imperatives required to navigate this technological shift responsibly and effectively.

Chapter 8
AI-Powered Supply Chains: Cloud-Native Intelligence for Resilient and Adaptive Operations 99
Mohan Kumar, Faculty of Commerce and Management, SGT University, Gurugram, India
Xuan-Hoa Nghiem, International School, Vietnam National University, Hanoi, Vietnam
Sandeep Kumar Singh, JGBS, O.P. Jindal Global University, Haryana, India
Mananage Shanika Hansini Rathnasiri, Faculty of Management Studies, Sabaragamuwa University of Sri Lanka, Sri Lanka & Qasim Ibrahim School of Business, Villa College, Maldives

Global supply chains face unprecedented volatility, rendering traditional, linear models obsolete. This chapter argues that resilience and adaptability are now critical competitive advantages, achievable through the integration of artificial intelligence (AI) and cloud-native architectures. It explores the evolution from legacy systems to cognitive networks, detailing how AI enables predictive demand sensing, autonomous planning, and proactive risk management. The cloud-native paradigm—built on microservices and containers—provides the essential scalable infrastructure for these AI capabilities. The chapter further outlines the core technology stack, implementation challenges, and key performance indicators for success. Finally, it envisions the future of hyper-automation and the autonomous supply chain, concluding that this technological fusion is indispensable for building the intelligent, adaptive enterprise capable of thriving in a complex global landscape.

Chapter 9
Cross-Cultural Management in AI-Augmented Workplaces: Opportunities and Challenges 113
Himanshi Rajora, JGBS, O.P. Jindal Global University, India
Dilshad Jeffery, Qasim Ibrahim School of Business, Villa College, Maldives
Azadeh Amoozegar, INTI International University, Malaysia

The contemporary global business environment is being simultaneously reshaped by two powerful forces: the deepening integration of cross-cultural teams and the rapid proliferation of artificial intelligence (AI). This chapter examines the intersection of these trends, exploring the opportunities and challenges of managing culturally diverse, AI-augmented workforces. We argue that AI is not culturally neutral but, when guided by cultural competence, can strengthen global collaboration, decision-making, and inclusion. Conversely, if implemented without regard for cultural nuances, AI risks amplifying biases, eroding trust, and creating new forms of digital exclusion. This chapter begins by establishing the theoretical foundations of cross-cultural management and AI augmentation. A framework for culturally-competent AI integration is proposed, focusing on managerial strategies and tool selection. The chapter concludes with a discussion of the social, ethical, and managerial implications, calling for a more inclusive and critically aware approach to AI-augmented workplaces.

Chapter 10
Shaping Entrepreneurial Creativity and Innovation Through Human-AI Cognition 127
Khushboo Jain, Shri Vaishnav Institute of Management and Science, Indore, India
Deepa Katiyal, Shri Vaishnav Institute of Management and Science, Indore, India
Shashank Mittal, JGBS, O.P. Jindal Global University, Sonipat, India
Mukhtaruddin Mukhtaruddin, Faculty of Economics, Universitas Sriwijaya, Palembang, Indonesia

Artificial Intelligence (AI) is reshaping entrepreneurial dynamics in modern times by fostering creativity, innovation and strategic decision-making. The current study examines the contribution of AI tools and technologies in the enhancement of entrepreneurial creativity and innovation. The data-driven insights provided by AI helps in automating cognitive processes and supporting idea generation. The qualitative and conceptual analysis of literature done using NVivo 14 software coding generated five themes for the study. Idea generative support of AI, cognitive intelligence, fostering innovation, human-AI's ethical alignment and entrepreneurial ecosystem readiness for AI. These themes helped identify the ways AI tools such as Machine Learning, natural language processing, generative AI models etc., support entrepreneurs in ideation, development, decision-making and implementation.

Chapter 11
Infrastructural Intelligence as Sociotechnical Institution: Cloud-Native AI, Power, Governance, and Inequality in the Global South 139
Sneha Kadyan, Jindal Global Business School, O.P. Jindal Global University, Sonipat, India
Firdous Ahmad Malik, University of People, USA

Cloud-native AI tools are impactful in the automation of cognition and organization decision making. This chapter is directed at advancing the sociological concept of infrastructural intelligence that explains how the cloud-based infrastructures of cognitive automation get intertwined, reshaping labor, governance, and global power. This chapter is anchored in the studies of sociotechnical systems, labor process theory, platform capitalism, datafication, and global inequality, and argues that the cloud-native AI represents a change in the infrastructure of automation: moving from isolated, algorithmic processes to extensive, codified, and worldwide systems of cognition. Evidence from logistics, finance, manufacturing, digital labor, and public governance is brought together to show how cloud-native AI is transforming work and at the same time sustaining geopolitical imbalances. The chapter presents a sociological perspective for understanding cloud-native AI and finally points out the ethical, environmental, and labor aspects that should be considered in future research.

Chapter 12
Flipped Learning Empowering Students Through Digital Innovation: Reverse Instruction for Enhanced Engagement and Deeper Understanding 155
Priti Gupta, Greater Noida Institute of Technology, Greater Noida, India
Richa Adlakha, Manav Rachna International Institute of Research and Studies, India
Ashish Grover, Manav Rachna International Institute of Research and Studies, India
Neha Chaudhary, Manav Rachna International Institute of Research and Studies, India

One major deviation from traditional teaching approaches is flipped learning, sometimes known as the inverted classroom paradigm. Under this method, the conventional wisdom—that of lecturing in class and assigning homework for solo study—is flipped. In-class time is instead devoted to discussions, active learning exercises, and hands-on activities while educational materials—texts, movies, or interactive modules—are distributed outside of the classroom. Fundamentally, flipped learning seeks to maximize in-person teaching time by means of interesting, dynamic activities promoting a better knowledge and application of ideas. There are several elements cooperating to establish a good learning environment. The core of the flipped learning model is the change from passive information absorption to active participation in the learning process.

Chapter 13
Automating Corporate Strategy: How Cloud AI Transforms Managerial Intelligence 183
Himanshi Rajora, JGBS, O.P. Jindal Global University, India
Dilshad Jeffery, Qasim Ibrahim School of Business, Villa College, Maldives
Azadeh Amoozegar, INTI International University, Malaysia

The contemporary corporate landscape, characterized by volatility and data saturation, renders traditional intuition-based strategic models inadequate. This chapter argues that Cloud Artificial Intelligence (AI) is catalyzing a paradigm shift, fundamentally transforming managerial intelligence. We demonstrate how Cloud AI automates and augments the entire strategic lifecycle—from environmental scanning and forecasting to planning and execution. By deconstructing core strategic functions and presenting detailed use cases, the chapter illustrates the practical application of machine learning, natural language processing, and optimization algorithms. Crucially, it provides a critical examination of the profound implications of this transition, analyzing the managerial, practical, ethical, and social consequences. The future of competitive advantage lies in a symbiotic human-AI collaboration, necessitating new competencies, ethical frameworks, and organizational structures for responsible and effective strategy formulation.

Chapter 14
Cloud-Native AI for Operational Efficiency and Predictive Maintenance: Automating Intelligence in Practice.. 197
Swati Singh, Greater Noida Institute of Engineering and Technology, India
Kamaljit Kaur, Guru Nanak Dev University, Amritsar, India
Shivangi Tyagi, KIET Group of Institutions, Ghaziabad, India
Mayank Tyagi, KIET Group of Institutions, Ghaziabad, India

Technology has changed from being a helpful tool to a key factor in growth and efficiency in today's dynamic business environment. For businesses looking to preserve their competitive edge, improve operational efficiency, and provide better customer experiences, digital transformation has become crucial. It's critical to create a thorough plan that takes into account both growth prospects and operational enhancements as businesses increasingly incorporate technology into their core operations. In the past, technology supported daily operations according to predetermined business needs, acting as a business enabler. Today, though, technology is far more important. These days, it influences business strategies, influencing everything from how businesses function to the development of new business models and techniques for engaging customers. Companies are securing their position in an increasingly digital marketplace by utilizing digital solutions to optimize operations and unlock growth potential.

Chapter 15
Automating Intelligence: Cloud-Native AI, Generative Tools, and the Future of Research... 217
Shimona Shriya, Xavier Institute of Management and Entrepreneurship, Bangalore, India
Mohit Verma, O.P. Jindal Global University, India
Aishwarya Vashishta, Rutgers University, USA

Defining a term before it becomes meaningful is vital in capturing the importance of any term. If a term does not have a meaning that people agree with, it is only a group of characters. The most important of all is to define an idiom. It is challenging as the term's meaning isn't clear from its constituent elements. The notion is most relevant in case of the technical terms which receive a significant press coverage in numerous ways and in different setups. Common apps, such as Google Assistant, Alexa, and Siri, have many people using artificial intelligence (AI) daily without even thinking about it. Productivity and creative apps such as ChatGPT, perplexity, and Gemini help us focus on the content rather than on how to get there. The media floods our entire social environment with so much information and disinformation that many people see AI as a kind of magic (which most certainly isn't). Therefore we start the chapter by defining what AI is, and what it isn't. We also discuss on how it connects to computers today.

Chapter 16
The Cloud-Native Enterprise: Orchestrating Intelligent Automation Across Global Operations 235
Priya Gaur, Centre for Online Education, GLA University, India
Gunjan Mohan Sharma, JGBS, O.P. Jindal Global University, India
Ali Najeeb, Qasim Ibrahim School of Business, Villa College, Maldives

The contemporary global business landscape demands a new operational paradigm centered on agility and intelligence. This chapter argues that the convergence of cloud-native architectures and intelligent automation (IA) is fundamental to this transformation. It deconstructs how cloud-native principles—microservices, containers, and DevOps—provide the essential, scalable foundation for orchestrating IA across complex, global operations. The analysis moves beyond a technical blueprint to explore a practical implementation playbook for transitioning from pilot to enterprise-wide scale. Crucially, the chapter dedicates significant focus to the profound implications of this shift, examining the social impact on the workforce, practical challenges like talent and cost, critical ethical considerations around algorithmic bias, and the evolving role of leadership. Success requires managing this socio-technical transformation to build resilient, responsible, and competitive enterprises.

Chapter 17
Human-Centric AI: Reinventing Employee Experience Through Cloud-Native Intelligence 251
Isha Goley, Prestige Institute of Management and Research, Gwalior, India
Bhumika Ray, JGBS, O.P. Jindal Global University, India
Mukhtaruddin Mukhtaruddin, Faculty of Economics, Universitas Sriwijaya, Indonesia

Cloud-native AI is not just a mere automation, but a combination of contextual knowledge, pattern recognition, behavioral analytics and predictive intelligence (Qin et al., 2025). These mechanisms interpret complicated employee cues, grasp dynamic shifts in motivation or well-being, and provide information that may be utilized to recommend individualized interventions (Segarra, and Williams, 2025). Workers are faced with new pressures like never before as global workplaces grow more digital and distributed, including increased skill transitions and/or the psychological strain of hybrid work (Choudhary et al., 2023). This poses an organizational need to develop HR processes that can help create resilience, engagement, and fulfillment. Human-centered AI will meet these essential needs by engraving compassion, personalization, and justice in the development of workplace technologies (Buhmann and Fieseler, 2022).

Chapter 18
Intelligent Automation as a Strategic Lever: Harnessing Cloud-Native AI for Scalable Growth 267
Xuan-Hoa Nghiem, International School, Vietnam National University, Hanoi, Vietnam
Mananage Shanika Hansini Rathnasiri, Faculty of Management Studies, Sabaragamuwa University of Sri Lanka, Sri Lanka & Qasim Ibrahim School of Business, Villa College, Maldives
Sandeep Kumar Singh, JGBS, O.P. Jindal Global University, Haryana, India
Mohan Kumar, Faculty of Commerce and Management, SGT University, Gurugram, India
Manju Dagar, Faculty of Commerce and Management, SGT University, India

Intelligent Automation (IA) represents a paradigm shift, evolving from routine task automation into a strategic lever for scalable growth. This chapter argues that the true potential of IA is unlocked when integrated with a cloud-native architecture. It deconstructs IA's core components-Robotic Process Automation, Machine Learning, Natural Language Processing, and Computer Vision-and demonstrates how their fusion creates cognitive, adaptive systems. The cloud-native model provides the essential fuel for this transformation, offering elastic scalability, access to sophisticated AI services, and accelerated deployment cycles. The chapter provides a comprehensive framework for formulating an IA strategy, a detailed implementation blueprint, and a discussion of critical challenges, including change management and ethical AI. Ultimately, it positions cloud-native IA as an imperative for building a resilient, agile, and future-proof enterprise.

Chapter 19
Considering the Application of Artificial Intelligence in Society and Assessing the Future of AI 279
Shimona Shriya, Xavier Institute of Management and Entrepreneurship, Bangalore, India
Rituraj Singh, O.P. Jindal Global University, Haryana, India
Shashi Kant, DMI St. Eugene University, Zambia

In this chapter the authors discuss on news applications of artificial intelligence (AI) in real world cases including industry and medical field with extended details. The chapter explores how AI can be easily used to automate the routine processes and improve the operational and functional efficiency of any department in any business sector. Regulations regarding AI is also discussed as one of the emerging concerns in the AI landscape. The chapter also defines different methods where there is a crossover between humans and AIs, be it direct or indirect contact.

Chapter 20
Governance in Cloud-Automated Enterprises: Frameworks for Responsible AI Deployment.. 295
Gunjan Mohan Sharma, JGBS, O.P. Jindal Global University, India
Ali Najeeb, Qasim Ibrahim School of Business, Villa College, Maldives
Priya Gaur, Centre for Online Education, GLA University, India

The integration of artificial intelligence (AI) with cloud computing is transforming enterprises through unprecedented automation and scalability. However, this convergence introduces significant risks—including algorithmic bias, opacity, and security vulnerabilities—that outpace traditional IT governance. This chapter argues that a new, proactive governance paradigm is essential for responsible AI deployment. It presents a comprehensive framework built on pillars of fairness, transparency, and accountability, and details how cloud-native technologies like Policy as Code can automate governance within MLOps pipelines. A step-by-step approach for operationalizing this framework is provided. The chapter concludes with a critical analysis of the broad social, ethical, practical, and managerial implications, positioning robust governance not as a barrier, but as a critical enabler of sustainable and trustworthy innovation in the cloud-automated enterprise.

Chapter 21
Reimagining Talent Management: Cloud-Native AI Tools for Predictive HR Analytics 309
Mohit Yadav, JGBS, O.P. Jindal Global University, India
Narayanage Jayantha Dewasiri, Faculty of Management Studies, Sabaragamuwa University of Sri Lanka, Sri Lanka & Qasim Ibrahim School of Business, Villa College, Maldives
Minakshi, College of Computer Science, King Khalid University, Saudi Arabia
Tanupriya Choudhury, School of Computer Science, University of Petroleum and Energy Studies, Dehradun, India

This chapter explores the paradigm shift in Human Resources (HR) from reactive, intuition-based practices to a proactive, data-driven function powered by cloud-native Artificial Intelligence (AI). It argues that the synergy between cloud computing's scalability and AI's predictive capabilities is fundamental to reimagining talent management. The chapter deconstructs the modern predictive HR tech stack and presents core use cases in acquisition, retention, development, and performance. It critically addresses the imperative of navigating ethical challenges—including bias, privacy, and transparency—and provides a framework for responsible implementation. Ultimately, this evolution empowers HR to become a strategic partner, using predictive analytics to augment human decision-making, mitigate workforce risks, and foster a more adaptive, resilient, and human-centric organization.

Chapter 22
Risk Perception and AI-Driven Financial Decision-Making: A FinTech Perspective 325
Manish Dubey, Shri Vaishnav College of Arts and Commerce, Indore, India
Mansi Panwar, Devi Ahilya Vishwavidyalaya University in Indore, India
Rituraj Singh, O.P. Jindal Global University, Haryana, India
Shashi Kant, DMI St. Eugene University, Zambia

Considering the changing future of digital finance, Artificial Intelligence (AI), cloud computing, and machine learning (ML) have transformed how financial decisions are made, credit-risks are assessed, and mechanisms of inclusion are applied. Irrespective of these improvements, user trust, perceived risk and behavioral biases remain influential to the usefulness and use of FinTech innovations. The proposed paper constructs a behavioral-technological conceptual model, which analyzes the mediating role of perceived risk between the relationship between AI-based FinTech adoption and the result of financial decisions, in which algorithmic trust acts as a moderator. The paper connects psychological and technological aspects of financial behavior by relying on various theoretical underpinnings including Prospect Theory,

Bounded Rationality, Signaling Theory, Expectancy-Violation Theory, Emotional Contagion Theory and the Algorithmic Bias-Trust Framework.

Chapter 23
Automating Intelligence in Healthcare: Cloud-Native Explainable AI for Brain Tumor Detection 349
Arun Kumar Rai, Greater Noida Institute of Engineering and Technology, India
Shraddha Kaushal, Greater Noida Institute of Engineering and Technology, India
Harsh Gour, Greater Noida Institute of Engineering and Technology, India
Sonali Jha, Greater Noida Institute of Engineering and Technology, India
Gazab Bhati, Greater Noida Institute of Engineering and Technology, India

A variety of brain neoplasms, including brain tumors, present serious health risks and diagnostic difficulties. They fall into two categories: primary tumors, which start in the brain, and secondary tumors, which spread to other parts of the body. The prevalence of brain tumors worldwide highlights the urgent need for accurate diagnostic instruments. Because brain tumors can cause a wide range of symptoms, from headaches to more serious neurological impairments, early and precise detection is essential to maximizing treatment results. Furthermore, a diagnostic tool that has high sensitivity and specificity is required due to the symptoms of brain tumors overlapping with those of other neurological disorders. Despite their effectiveness, traditional diagnostic techniques frequently involve invasive procedures or have limitations when it comes to detecting small or early-stage tumors.

Chapter 24
Emotion AI and Employee Experience: Cloud-Native Tools for Human-Centric Management 367
Jeayaram Subramanian, JGBS, O.P. Jindal Global University, India
Damith Gangodawilage, Qasim Ibrahim School of Business, Villa College, Maldives
Thi Minh Ngoc Luu, International School, Vietnam National University, Hanoi, Vietnam
Anugamini Priya Srivastava, Dnyaan Prasad Global University, Pune, India

The integration of Emotion AI (Affective Computing) into employee experience (EX) management represents a paradigm shift towards human-centric, data-driven leadership. This chapter explores how cloud-native tools enable organizations to move beyond lagging indicators like annual surveys by analyzing emotional data from digital communication platforms in real-time. It details use cases across the employee lifecycle, from enhancing onboarding to predicting burnout and fostering inclusive teams. However, this transformative potential is accompanied by significant implications. The chapter provides a critical examination of the ethical, social, and managerial challenges, including privacy, algorithmic bias, and the risk of emotional surveillance. It concludes that the responsible adoption of Emotion AI, governed by strong ethical frameworks, is crucial for creating empathetic, supportive, and productive workplaces.

Chapter 25
The Smart HR Cloud: Automating Workforce Planning and Skill Mapping 385
Mohit Yadav, JGBS, O.P. Jindal Global University, India
Minakshi, College of Computer Science, King Khalid University, Saudi Arabia
Tanupriya Choudhury, School of Computer Science, University of Petroleum and Energy Studies, India
Narayanage Jayantha Dewasiri, Faculty of Management Studies, Sabaragamuwa University of Sri Lanka, Sri Lanka & Qasim Ibrahim School of Business, Villa College, Maldives

The contemporary business landscape, characterized by rapid technological change and VUCA conditions, has rendered traditional, static HR processes obsolete. This chapter delineates the strategic imperative for organizations to adopt a "Smart HR Cloud"—an integrated, data-driven system for automating workforce planning and skill mapping. It explores the foundational elements, from building a dynamic skills inventory to leveraging AI for predictive analytics and gap identification. The chapter details how these insights translate into actionable strategies for talent development, internal mobility, and strategic hiring. Critically, it addresses the essential change management, ethical considerations, and practical implications of this transformation, providing a comprehensive blueprint for building an adaptive, future-proof organization grounded in a continuous skills-based strategy.

Chapter 26
AI-Driven Personalization in Cloud-Powered Fashion: E-Commerce Customer Experience 395
Priti Gupta, Department of Electrical Engineering, Greater Noida Institute of Technology, Greater Noida, India
Neha Arora, Amity School of Fashion Technology, Amity University, Noida, India
Akanksha Kathuria, School of Business, Manav Rachna University, Faridabad, India
Neeraj Chopra, School of Business, Manav Rachna University, Faridabad, India

Artificial intelligence (AI)-driven personalization has emerged as a transformative force in fashion e-commerce, redefining how consumers engage with online retail ecosystems and elevating customer experience (CX) through tailored interactions. Coupled with the scalability and flexibility of cloud-based infrastructure, modern fashion platforms can analyze extensive customer data in real time, optimize content delivery, and provide individualized shopping experiences that significantly enhance satisfaction, loyalty, and conversion metrics.

Compilation of References 415

About the Contributors 449

Index 459

Preface

The rapid evolution of cloud computing and artificial intelligence is redefining how modern computational systems are designed, deployed, and managed. As editors, we conceived *Automating Intelligence With Cloud-Native AI Tools: Harnessing Intelligent Clouds* in response to a growing need for a unified reference that captures this transformation—one that moves beyond isolated technologies to examine intelligent cloud ecosystems as integrated, adaptive, and self-managing environments.

This edited reference book explores the convergence of cloud computing, machine learning, and smart technologies to shape the future of computing. The chapters collectively examine how machine learning algorithms can be seamlessly embedded within scalable cloud architectures, enabling systems that automatically adapt, optimize resources, and respond intelligently to real-time demands. Rather than treating AI and cloud platforms as separate domains, the contributors emphasize their interdependence in building next-generation digital infrastructures.

A central focus of this volume is practical implementation. Readers will find in-depth discussions on deploying intelligent solutions across distributed and cloud-native environments, including edge computing integration, automated resource management, AI-driven security mechanisms, and predictive analytics frameworks. Through detailed case studies and architectural blueprints, the book bridges theory and practice, illustrating how intelligent cloud solutions are designed, implemented, and scaled in real-world contexts.

The objectives of this reference work are fivefold: to establish a comprehensive understanding of intelligent cloud ecosystems; to provide actionable strategies for integrating machine learning into scalable cloud architectures; to connect conceptual foundations with practical case studies; to empower organizations to lead digital transformation through adaptive, self-managing systems; and to prepare readers for emerging technologies that will define next-generation cloud infrastructures.

This book is intended for a diverse audience, including cloud architects, machine learning engineers, DevOps professionals, data scientists, and software developers, as well as researchers and practitioners seeking a holistic perspective on intelligent cloud systems. We hope this volume serves both as a foundational reference and a forward-looking guide, supporting readers as they design and deploy intelligent cloud solutions that deliver enhanced performance, efficiency, and innovation.

As editors, we extend our sincere gratitude to the contributing authors for their expertise and commitment, and to the readers who will carry these ideas forward into practice. We trust that this book will inspire new approaches to building intelligent, adaptive, and future-ready computing ecosystems.

ORGANIZATION OF THE BOOK

Chapter 1: Adaptive Learning Ecosystems: AI-Enabled Cloud Platforms for Employee Upskilling

In Chapter 1, the contributors set the foundation for understanding how AI-enabled, cloud-native learning ecosystems are redefining workforce development. From the editors' perspective, this chapter is pivotal in framing learning as a continuous, adaptive process rather than a static training function. It examines the architectural pillars of personalization, scalability, and dynamic content, while candidly addressing challenges such as data privacy and algorithmic bias. The chapter positions adaptive learning ecosystems as a strategic investment that enables organizations to cultivate resilient, future-ready talent in an era of relentless technological change.

Chapter 2: Cloud AI of Strategic Human Resource Transformation

Chapter 2 extends the discussion into the strategic reinvention of Human Resource Management through cloud-native AI. The editors highlight how this chapter systematically connects core cloud-native constructs—microservices, MLOps, serverless computing—with traditional HR functions such as recruitment, performance management, and learning. By grounding theory in real-world case studies, the chapter reframes HR not merely as an administrative function but as a data-driven, innovation-oriented strategic partner, while emphasizing ethical governance as central to sustainable transformation.

Chapter 3: AI and the Human Touch: Building Trust in Cloud-Automated Business Ecosystems

Trust emerges as the central theme of Chapter 3, which the editors view as a crucial counterbalance to purely technical narratives of automation. This chapter reframes the "human touch" as an essential governance and stewardship role within AI-driven ecosystems. By diagnosing sources of mistrust—opacity, bias, and privacy concerns—and proposing a practical trust-building framework, the chapter underscores that automation succeeds only when human oversight, ethics, and transparency are embedded by design.

Chapter 4: Cloud-Driven Organizational Agility: AI as the Catalyst for Digital Transformation

In Chapter 4, the editors emphasize the synergistic relationship between cloud computing and AI as the engine of organizational agility. The chapter convincingly argues that while cloud platforms provide speed and scalability, AI serves as the catalyst that converts infrastructure into intelligence. By integrating theory with cross-functional use cases and organizational considerations, this chapter offers leaders a holistic roadmap for navigating digital transformation responsibly and effectively.

Chapter 5: AI Ethics in People Analytics: Balancing Automation, Privacy, and Fairness

Chapter 5 addresses one of the most pressing concerns in AI adoption: ethics in people analytics. From the editors' viewpoint, this chapter is essential in articulating why privacy, fairness, and automation must be managed as an integrated triad. By moving from ethical principles to operational governance frameworks, the chapter reinforces the idea that responsible AI is not optional but foundational to trust, legitimacy, and long-term organizational success.

Chapter 6: AI-Augmented Leadership: Redefining Managerial Decision-Making in the Cloud Era

Chapter 6 reframes leadership itself in the context of AI and cloud convergence. The editors see this chapter as redefining the manager's role—from decision-maker to architect of intelligent decision systems. Through a structured framework for AI-augmented judgment, the chapter highlights how human intuition and algorithmic intelligence can coexist, while firmly asserting that ethical responsibility and accountability remain distinctly human obligations.

Chapter 7: AI-Driven Sustainability: Automating ESG Reporting and Green Operations through the Cloud

Chapter 7 situates AI and cloud computing at the heart of corporate sustainability efforts. The editors highlight how this chapter transforms ESG reporting from a compliance exercise into a strategic capability. By demonstrating automation across ESG lifecycles and optimization of green operations, the chapter provides both a technological roadmap and an ethical lens for embedding sustainability into core business strategy.

Chapter 8: AI-Powered Supply Chains: Cloud-Native Intelligence for Resilient and Adaptive Operations

In Chapter 8, the editors underscore the urgency of resilience in global supply chains. This chapter traces the evolution from linear, legacy models to cognitive, AI-driven networks enabled by cloud-native architectures. By combining technical stacks, performance metrics, and future-oriented visions, it positions AI-powered supply chains as indispensable to competitiveness in volatile global environments.

Chapter 9: Cross-Cultural Management in AI-Augmented Workplaces: Opportunities and Challenges

Chapter 9 brings a critical human and cultural dimension to AI adoption. From the editors' standpoint, this chapter is vital in challenging the assumption that AI is culturally neutral. By integrating cross-cultural management theory with AI implementation strategies, the chapter offers a nuanced framework for inclusive, globally sensitive AI-augmented workplaces.

Chapter 10: Shaping Entrepreneurial Creativity and Innovation Through Human-AI Cognition

Chapter 10 explores the intersection of AI and entrepreneurship, emphasizing creativity and innovation rather than automation alone. The editors note the chapter's contribution in synthesizing qualitative insights into how AI tools support ideation, cognition, and decision-making. It reinforces the theme of human–AI collaboration as a catalyst for entrepreneurial ecosystems.

Chapter 11: Infrastructural Intelligence as Sociotechnical Institution: Cloud-Native AI, Power, Governance, and Inequality in the Global South

Chapter 11 provides a critical sociological lens on cloud-native AI as infrastructural intelligence. The editors regard this chapter as essential for broadening the book's scope beyond organizational efficiency to questions of power, labor, and inequality. By situating AI within global political and economic structures, it challenges readers to confront the ethical and geopolitical implications of large-scale cognitive automation.

Chapter 12: Flipped Learning: Empowering Students Through Digital Innovation

In Chapter 12, the editors highlight flipped learning as a pedagogical innovation aligned with intelligent technologies. The chapter emphasizes active engagement, collaborative learning, and deeper understanding, illustrating how digital tools can enhance—not replace—meaningful in-person educational experiences.

Chapter 13: Automating Corporate Strategy: How Cloud AI Transforms Managerial Intelligence

Chapter 13 extends automation into the strategic core of organizations. The editors view this chapter as a compelling argument for cloud AI as a transformative force across the entire strategy lifecycle. By balancing technical use cases with ethical and managerial implications, it reinforces the necessity of human–AI symbiosis in strategic leadership.

Chapter 14: Cloud-Native AI for Operational Efficiency and Predictive Maintenance

Chapter 14 grounds the book in operational realities, illustrating how cloud-native AI drives efficiency, reliability, and growth. From the editors' perspective, this chapter highlights the shift from technology as a support function to technology as a strategic driver shaping business models and competitive positioning.

Chapter 15: Automating Intelligence: Cloud-Native AI, Generative Tools, and the Future of Research

Chapter 15 takes a reflective and foundational approach by clarifying what AI is—and is not—in contemporary discourse. The editors value this chapter for demystifying AI and situating generative tools within broader computational contexts, helping readers critically engage with both the promise and misconceptions surrounding AI in research and knowledge creation.

Chapter 16: The Cloud-Native Enterprise: Orchestrating Intelligent Automation Across Global Operations

In Chapter 16, the editors emphasize enterprise-wide orchestration of intelligent automation. By integrating technical architectures with socio-technical considerations, the chapter offers a pragmatic blueprint for scaling IA responsibly while addressing workforce, leadership, and ethical challenges.

Chapter 17: Human-Centric AI: Reinventing Employee Experience Through Cloud-Native Intelligence

Chapter 17 reinforces the book's human-centric theme by focusing on employee experience. The editors highlight how cloud-native AI can embed empathy, personalization, and fairness into HR processes, responding to the psychological and professional pressures of modern work environments.

Chapter 18: Intelligent Automation as a Strategic Lever: Harnessing Cloud-Native AI for Scalable Growth

Chapter 18 positions intelligent automation as a driver of scalable growth rather than cost reduction alone. From the editors' viewpoint, this chapter provides a cohesive strategic and technical framework that aligns automation initiatives with long-term organizational resilience and agility.

Chapter 19: Considering the Application of Artificial Intelligence in Society and Assessing the Future of AI

Chapter 19 broadens the lens to societal applications of AI across industries, healthcare, and regulation. The editors see this chapter as an integrative reflection on human–AI interaction, emphasizing the importance of governance and foresight as AI becomes embedded in everyday life.

Chapter 20: Governance in Cloud-Automated Enterprises: Frameworks for Responsible AI Deployment

Chapter 20 directly addresses governance as a foundational enabler of trustworthy AI. The editors underscore its contribution in translating abstract principles—fairness, transparency, accountability—into actionable, cloud-native governance mechanisms embedded within MLOps pipelines.

Chapter 21: Reimagining Talent Management: Cloud-Native AI Tools for Predictive HR Analytics

In Chapter 21, predictive analytics redefines talent management as a proactive, strategic function. The editors highlight how this chapter balances technological sophistication with ethical responsibility, positioning HR as a key driver of organizational adaptability and resilience.

Chapter 22: Risk Perception and AI-Driven Financial Decision-Making: A FinTech Perspective

Chapter 22 integrates behavioral finance with AI-driven decision systems. From the editors' perspective, this chapter is significant for demonstrating that trust, perceived risk, and human cognition remain central to FinTech adoption, even in highly automated financial environments.

Chapter 23: Automating Intelligence in Healthcare: Cloud-Native Explainable AI for Brain Tumor Detection

Chapter 23 exemplifies high-impact AI application in healthcare. The editors note the chapter's emphasis on accuracy, explainability, and early detection, reinforcing the critical role of cloud-native AI in improving diagnostic outcomes while acknowledging clinical and ethical constraints.

Chapter 24: Emotion AI and Employee Experience: Cloud-Native Tools for Human-Centric Management

Chapter 24 explores Emotion AI as a frontier in employee experience management. The editors value its balanced treatment of innovation and caution, highlighting how emotional analytics can enhance empathy and engagement when governed by strong ethical frameworks.

Chapter 25: The Smart HR Cloud: Automating Workforce Planning and Skill Mapping

Chapter 25 presents a comprehensive blueprint for skills-based workforce planning in volatile environments. From the editors' viewpoint, this chapter demonstrates how AI-driven skill intelligence can align talent development, mobility, and hiring with long-term strategic goals.

Chapter 26: AI-Driven Personalization in Cloud-Powered Fashion E-Commerce Customer Experience

The concluding chapter applies cloud-native AI to consumer-facing innovation in fashion e-commerce. The editors highlight how real-time personalization exemplifies the broader themes of the book—scalability, intelligence, and human-centered design—by transforming customer experience through data-driven insight and adaptive cloud platforms.

CONCLUSION

As editors, we conclude this reference work with a reaffirmation of its central premise: the future of computing is not defined solely by automation or scale, but by the intelligent, ethical, and human-centered integration of cloud-native artificial intelligence across organizational and societal domains. Throughout this volume, the contributing chapters collectively demonstrate that cloud computing and AI are no longer enabling technologies operating at the periphery of enterprise systems; they have become foundational infrastructures that shape how learning occurs, decisions are made, value is created, and responsibility is exercised.

The breadth of perspectives presented in this book—from adaptive learning ecosystems and AI-augmented leadership to intelligent supply chains, sustainable operations, healthcare innovation, and governance frameworks—highlights the truly interdisciplinary nature of automating intelligence. While the technical architectures of cloud-native AI provide scalability, speed, and efficiency, the chapters consistently underscore that the real differentiator lies in how these technologies are designed, governed, and aligned with human values. Trust, transparency, fairness, and accountability emerge not as constraints, but as prerequisites for sustainable innovation.

Equally important is the book's emphasis on transition rather than replacement. Across sectors, AI is shown to augment human judgment, creativity, and collaboration rather than render them obsolete. Whether in education, human resource management, entrepreneurship, finance, or healthcare, the contributors reinforce the notion that intelligent systems are most effective when they operate in partnership with human expertise, cultural awareness, and ethical stewardship.

This edited reference book is intended to serve both as a comprehensive scholarly resource and as a practical guide for practitioners, leaders, and policymakers navigating the complexities of digital transformation. By bridging theory with real-world implementation and by engaging critically with social, ethical, and governance implications, the volume prepares readers not only to adopt cloud-native AI tools, but to do so responsibly and strategically.

As we look ahead, it is clear that automating intelligence is an ongoing journey rather than a finite destination. The technologies explored in this book will continue to evolve, but the questions they raise—about power, equity, trust, and human purpose—will remain central. We hope this work inspires thoughtful innovation, informed leadership, and continued interdisciplinary dialogue, empowering readers to shape a future in which intelligent clouds enhance human potential and contribute meaningfully to a more adaptive, equitable, and resilient digital society.

Chapter 1
Adaptive Learning Ecosystems:
AI-Enabled Cloud Platforms for Employee Upskilling

Tanupriya Choudhury
https://orcid.org/0000-0002-9826-2759
School of Computer Science, University of Petroleum and Energy Studies, India

Mohit Yadav
https://orcid.org/0000-0002-9341-2527
JGBS, O.P. Jindal Global University, India

Narayanage Jayantha Dewasiri
https://orcid.org/0000-0002-5908-8890
Faculty of Management Studies, Sabaragamuwa University of Sri Lanka, Sri Lanka & Qasim Ibrahim School of Business, Villa College, Maldives

Minakshi
College of Computer Science, King Khalid University, Saudi Arabia

ABSTRACT

The accelerating pace of technological change has created a critical skills gap, rendering traditional corporate training models obsolete. This chapter explores the transformative potential of adaptive learning ecosystems—AI-enabled, cloud-native platforms that create dynamic, personalized environments for continuous employee upskilling. It deconstructs the core architectural pillars of these systems: artificial intelligence for personalization, cloud infrastructure for scalability, and dynamic content for relevance. The analysis covers the mechanisms of adaptation, strategic benefits for organizational agility, and a practical implementation framework. It

DOI: 10.4018/979-8-3373-7442-0.ch001

also addresses critical challenges, including data privacy and algorithmic bias. Ultimately, the chapter argues that investing in these ecosystems is a strategic imperative, enabling businesses to build a resilient, future-ready workforce and sustain a competitive advantage in the AI-augmented economy.

1. INTRODUCTION

The world economy is currently experiencing the Fourth Industrial Revolution, which represents a paradigm shift featuring the blurring of the boundaries between the physical, digital and biological environments (Schwab, 2016). This transformation, driven by artificial intelligence, robotics and internet of things (IoT) fundamentally changes the nature of work. The World Economic Forum (2023) estimates that by 2027, 44% of worker core skills will be disrupted, hence creating a huge mismatch between current workforce capabilities and competency requirements needed to succeed in the future. This “skills-gap tsunami” represents a current reality that stunts productivity, stifles innovation, and threatens the viability of organizations. Consequently, companies are no longer able to rely on hiring limited outside talent that will fill these gaps; instead, they have to cultivate talent from within, through systematic and scalable upskilling and reskilling programs.

Concurrently, traditional corporate learning and development (L&D) functions, which are usually centered on the Legacy Learning Management System (LMS), are insufficient to solve this challenge. The traditional LMS was created for a different time - mostly for compliance tracking, administering mandatory training, and hosting static monolithic courses (Schoox, 2022). This model is inherently passive, meaning learning must be sought by the employees and learning cannot be delivered in a proactive way. It works on a one-size-fits-all basis, not taking into account varying individual differences with regard to prior knowledge, pace of learning, career path aspirations, and individual modality preferences. The end result is often low engagement, poor knowledge retention and inability to translate learning to measurable improvements in performance on the job. As Bersin (2017) points out, the “overwhelmed employee” has little time or patience for learning that is not immediately relevant, accessible, and personalized.

Within this context of urgent need and systemic failure a new paradigm is emerging: the adaptive learning ecosystem. This concept is a fundamental shift from a centralized, course-centric training department to a distributed learner-centric environment. An ecosystem has been defined as a complex network of interrelated parts interacting so as to form a dynamic entirety (Boulding, 1979). An adaptive learning ecosystem is the solution that uses the power of AI and the scalability of cloud computing to create such a network for human development. It is a living system that is constantly learning data about its users, such as their skills, what they are trying to achieve, how well they are performing, and so on, and adapts in real time to create hyper-personalized learning experiences. Seamlessly integrated with the tools and workflows of the modern digital workplace, it makes learning a natural part of the workday rather than something that interrupts and intrudes on it. This chapter reviews the architecture, mechanics, benefits, and implementation of such AI-enabled cloud platforms, arguing that it is the cornerstone of a sustainable, 21st century workforce development strategy.

2. DECONSTRUCTING THE ADAPTIVE LEARNING ECOSYSTEM

The adaptive learning ecosystem is not a monolithic product but a complex technological architecture that consists of three interdependent core pillars. The synergy between these pillars allows the dynamic and responsive nature of the system, which is superior to the performance of any single component.

Core Pillar 1: Artificial Intelligence & Machine Learning. AI and machine learning (ML) are the central nervous system of the adaptive learning ecosystem that powers personalization and automation. At its heart is the personalization engine, which uses complex algorithms to build a multi-dimensional learner profile. This engine takes in data from explicit data (e.g. role, career interests declared in an HRIS), implicit data (e.g. content consumption patterns, amount of time spent on modules, search queries) and performance data (e.g. project outcomes, feedback from peers and managers, skills endorsements) (Kong, 2021). Continuous analysis of this data enables ML models to determine a learner's current proficiency levels on a broad range of skills, often with more accuracy than self-ratings. This rich profile allows the system to personalize the learning journey with high precision, both in terms of content ("what"), and of its sequencing and modality ("how").

Complementary to this engine of personalization are recommender systems that play the role of the discovery layer within this ecosystem. Just like algorithms that are used by Netflix or Amazon to recommend movies or products, these systems make use of collaborative filtering (identification of users with similar profiles and recommending what they found valuable) and content filtering (recommendation of items that are similar to the ones the user has engaged with) (Afzal, 2019). In a learning context, this means suggesting pertinent micro-courses, articles, internal experts, mentorship opportunities, or stretch assignments that are in line with inferred goals and skill gaps of the learner. Consequently, learning becomes not a passive activity of consumption, but an active journey of exploration and connection. Natural language processing (NLP), a subset of AI, adds another layer to this ecosystem by enabling intelligent chatbots that can answer learner questions 24/7, support users on the job and help guide them through the platform. NLP is also used to analyze open ended responses in assessments, grade written submissions, and scan vast repositories of internal and external content for tagging and categorizing learning assets, which can be done automatically to increase the discoverability of the content by the recommender engine.

Core Pillar 2: Cloud Native Platform Architecture. If AI is the brain, the cloud native platform acts as a circulatory system that offers critical infrastructure to achieve scale, connectivity, and resiliency. The scalability and access made possible by cloud computing is foundational. A cloud native platform can easily scale up and down the number of resources to accommodate a global workforce that can run in the thousands to hundreds of thousands without degrading performance. It allows learning to be seamlessly delivered to any device - laptop, tablet or smartphone - thus allowing learning to be delivered in what is known as "learning in the flow of work". This concept, promoted by Bersin (2019), brings learning into the daily working life and helps make it available at the point of need. This is in stark contrast to on-premise LMS solutions that are often inaccessible outside of corporate networks and are difficult to scale.

The power of the ecosystem is furthered due to its ability for data aggregation and integration. A modern adaptive learning platform is one that does not adopt the walled-garden concept; it is built as an open hub. Through application programming interfaces (APIs), it is able to bi-directionally integrate with a suite of enterprise systems. It extracts information from HRIS (e.g., Workday, SAP SuccessFactors) to know the organizational structure and roles, from performance management systems (e.g., Lattice, 15Five) in order to include goal progress and feedback, and from productivity tools like Slack, Microsoft Teams, Jira, or Salesforce. By analyzing activity in these tools, the platform can glean insights into the skills actually being used in real time (e.g. identifying an employee who successfully uses data analytics in Salesforce reports). This API-first design is critical because it allows organizations to develop a "best-of-breed" learning technology stack with specialized content providers, simulation parcels, and virtual lab environment integration without vendor lock-in.

Core Pillar 3 Dynamic Content & Microlearning. The third pillar relates to the actual substance of learning which AI curates and which is delivered in the cloud. An adaptive ecosystem is rich in a library of dynamic content. This includes traditional third-party courseware from vendors such as Coursera for Business or LinkedIn Learning as well as an enormous amount of internally generated material. These materials span from formally created training videos to informal "how-to" guides, recorded expert sessions, collected industry relevant articles and podcast episodes. The value of the ecosystem is that it is able to make this long tail of content discoverable and relevant.

Central to the learning experience is the idea of microlearning. Cognitive - science research shows that dividing information into small, focused chunks helps improve the retention of information and reduces cognitive load (Sweller, 2011). Microlearning modules, usually 3-10 minutes long, are ideal for the modern workforce with decreased attention spans and busy schedules. They allow employees to learn a particular skill or knowledge just-in-time and to implement it immediately to a current task, reinforcing learning and proving the value of learning. Finally, the ecosystem strongly promotes user generated content (UGC). By giving employees easy-to-use tools to create and share video explanations, document templates, or post-mortems, the platform can harness the collective intelligence of the organization. This promotes a culture of social learning and sharing of knowledge, thus turning every employee from a passive consumer to a potential contributor and a mentor, and growing the ecosystem organically.

3. THE ENGINE OF ADAPTATION: HOW AI PERSONALIZES THE LEARNING JOURNEY

The real value of the adaptive learning ecosystem comes from the sophisticated interplay of the AI components that create a unique personal and responsive learning experience for every learner. This process is continuous and multi-staged, and goes from diagnosis to delivery to prediction.

The journey starts with skills inference and gap analysis. Historically, skills assessment was a manual and infrequent process and tended to be based on self-reported surveys. AI automates and enhances this process by constantly scanning integrated data sources (e.g. job descriptions, completed projects, communications, even user-generated content) to build a dynamic "skills taxonomy" for each individual. For example, analyzing code commits in a repository like GitHub will enable the system to determine proficiency in Python; analyzing sales-call transcripts (with consent) will enable communication and negotiation skills to be assessed. This provides a data-driven and real-time portrait of an employee's current capabilities. The profile is then automatically benchmarked against skills which are required for the current role and aspirational roles of the individual. The AI is not only able to identify the gaps, but also quantifies the magnitude of the gaps and prioritizes them based on strategic importance of the gap to the business and career path of the individual, thereby creating a targeted and relevant development plan from the get-go.

On the basis of detected skills gaps, the AI creates adaptive learning pathways. Unlike a linear catalog where there is a pre-determined order of courses to be taken, an adaptive path is dynamic and non-linear. As a learner moves forward, performance is continuously evaluated by embedded quizzes, simulations and practical assignments using AI. When the learner demonstrates mastery in a timely fashion, the system is able to avoid redundant content or deliver more advanced content. Conversely, if a learner is having trouble with a specific topic, the system will automatically provide extra resources, different explanations, or simpler practice exercises until a sufficient level of proficiency is reached. This helps ensure that the learner is not bored or overwhelmed, but rather is in an optimal state of challenge and engagement - what is often called the zone of proximal development (Vygotsky, 1978). The path remains fluid; it changes with the learner; it is a living entity, changing with changing needs.

In the coming years, Predictive Analytics becomes the most strategic use of Artificial Intelligence within organizations. When analyzed across the entire organization, aggregated and anonymized data can be used in machine learning to identify macro-level trends, predict what skills will be on the rise due to automation, and which emerging competencies will be in high demand over the next six to 18 months. This approach encourages business strategic planning, training and development leaders to move from a reactive positive (responding to current needs) to a proactive positive (getting ready for future challenges). At the individual level, predictive analytics can help identify employees who are at high risk of skill obsolescence, enabling proactive interventions, such as reskilling programs, before roles become obsolete. It can also help to identify high-potential employees, who are ready for more challenging assignments or promotions, strengthening succession planning pipelines and improving talent retention.

To show this in action, let's take an example situation with two employees, Maria, a marketing manager, and Ben, a software engineer. Based on her project metrics and completed content, Maria has a high level of proficiency in traditional brand market marketing, but a gap in data-driven performance market marketing. The system automatically creates a learning path starting with a micro-course on Google Analytics basics, followed by a digital marketing mentor who is based in-house. After completion of a simulated A/B testing, the system will be able to identify mastery of the concept and suggest a higher level module in marketing mix modeling. For Ben, the system, combined with his Jira and Github data, highlights experience in Backend Development while increasing the demand for Cloud Security skills as projects are moved to AWS. His learning path features hands-on labs in a cloud sandbox, a recommended path for AWS Certified Security Specialty and articles on the latest cybersecurity threats. Both paths are distinct, productive, and directly connected with the performance of the job in the present and the career development in the future.

4. STRATEGIC BENEFITS: BEYOND INDIVIDUAL LEARNING TO ORGANIZATIONAL AGILITY

The benefits to the employee are significant and directly tied to engagement and retention. Having a platform where one's career aspirations are understood and where proactive development is offered is a powerful motivator. Employees feel valued and invested in, and sense a high degree of organizational commitment. The personalised, self-directed nature of the learning process helps them to take ownership of their career development away from a paternalistic model of the training process being given to them. On top of that, by delivering learning in the flow of work and centering around microlearning, the Ecosystem dramatically reduces time-to-competency. Employees can rapidly learn the specific skill required to solve an urgent problem or complete a project, resulting in immediate wins and a greater sense of efficiency and achievement.

For the manager, the ecosystem affords an unprecedented level of visibility and data-driven decision-making capabilities. Rather than having to guess what the capabilities of their team are, managers have access to a dynamic "skills dashboard" that paints a picture of the collective strengths and weaknesses of their team. This helps to more effectively staff projects as managers can easily see which team member has the skills (or is closest to getting the skills) needed to execute on a new project. IT is changing the performance and development discussion from basing the discussion on anecdotal evidence to being grounded in objective data. By co-creating development strategies with direct reports, managers can use the AI's recommendations as a springboard to have more meaningful and productive career conversations. It also makes succession planning easier because managers can easily identify and develop potential succession candidates for key roles within their staff based on a verified set of skills and learning agility.

These advantages are transformational for the organization. "This is the most direct benefit; the closing of the skills gap happens faster." By building talent from its own ranks, organizations will be less reliant on the expensive, competitive, and often reliable external talent market. They are more flexible in changing the skills of their workforce to suit new strategic directions such as a new market, or a new technology. This creates a culture of constant learning and innovation. If learning is integrated, accessible and rewarding, it becomes part of the DNA of the organization. "People are the most important asset of any organization, and a workforce that is highly adaptive, creative, and better equipped to innovate is a workforce that is learning continuously" (DeLoyte, 2020).

This directly affects Talent Retention and Attraction. According to research by the LinkedIn Learning, career advancement and development opportunities are some of the key factors driving employee retention in today's job market (LinkedIn

Learning, 2023). Lastly, the ecosystem can enable unprecedented strategic planning based on data. The aggregated and anonymized skill data across the organization provides a unique view into the human capital of the organization. Next, leadership can respond to key questions: Can we do what we said we would to accomplish our five-year plan? What are the riskiest departments for technological disruption? This intelligence can not only guide L&D investments, but also wider business strategies, such as mergers and acquisitions where the talent portfolio of the target company can prove a decisive factor.

5. IMPLEMENTING FRAMEWORK: DEVELOPING YOUR ECOSYSTEM

Implementing an adaptive learning ecosystem is a complex organizational task that needs careful planning, cross-functional collaboration, and a focus on change management. A Phased Approach is the Key to Success

Phase 1: Analysis and Objective Setting

This introductory phase brings the ecosystem in sync with business strategy. It starts by defining key business objectives. The question is: What are the critical business outcomes we need to achieve over the next one to three years, and what are the skills we need to be able to achieve them? This includes working closely with business leaders to understand their strategic priorities, whether they are focused on growing market share through digital sales, improving operational efficiency through automation, or enhancing customer experience. This top-down analysis needs to be supplemented by an analysis of the current L&D landscape (bottom-up approach) to give a clear picture of the current state of readiness. This includes cataloguing the available content, evaluating the functionality of the current LMS (if any) and determining the current learning culture. . A skills-gap analysis at an organizational level, using the methods described earlier, will provide a quantitative baseline against which progress can be measured.

Phase 2: Technology Selection and Integration

With a clear strategic understanding, the organization can proceed to the selection of the right technological platform. The technology selection must be driven by a stringent set of criteria. Factors to be considered include: sophistication and transparency of the AI and recommendation engines; API capabilities of the platform and demonstrated integrations with the company's core HR and productivity systems;

vendor's content strategy and partnerships; UX and mobile capability; vendor's security practices and data privacy; and finally vendor's vision and product roadmap. It is not a decision which should be made by the L&D function in isolation, but rather one in which IT, HR and information security teams must take an active part.

Once a platform has been selected, a phased integration plan of requirements is key. In general, the launch is a risky venture, and of course the more risky the better. A better way to go is to begin with a pilot program by choosing a department or business unit that has a definable skills shortfall and robust leadership commitment. The initial integration should prioritize integration of the most critical systems, for example a HRIS becomes the single source of user data, and a single content source is also an option.

Phase 3: Change Management and Creating a Learning Culture

Technology in and of itself is only an enabler; the ultimate success of the ecosystem requires people take it up and change their behaviors. This makes the change management phase the most important phase. It begins by gaining leadership buy-in. Outside of capitalizing the initiative, the leaders must actively promote the initiative by sharing their own learning journeys and acknowledging employees who engage with the platform. Communicating the what, why and how of the new ecosystem, along with the benefits to employees, personally and professionally, should be clearly and repeatedly communicated to all employees.

For organizations to maximize adoption, the focus should shift to incentives for participation and developing a learning culture. This means that learning is being recognised and rewarded as part of performance. This can be achieved by tying learning activities to the performance review and promotion criteria, developing a badge and certification system for mastery of a given skill, and publicly celebrating learning milestones. Finally, it is critical to upskill the L&D team itself. Their role is changing from content creator and program administrator to strategic curator, data analyst, and community manager. They must be able to interpret the analytics that the platform provides them, control the recommendations made by the AI, and promote the social and collaborative dimensions of the learning ecosystem.

6. NAVIGATING CHALLENGES AND ETHICAL CONSIDERATIONS

The potential of adaptive learning ecosystems is immense, but their implementation is not without significant challenges and ethical considerations that organizations must carefully consider and navigate.

Probably the most pressing issues are those of data privacy and security. These platforms need access to large volumes of sensitive employee data in order to work effectively. This begs the essential question: How much data tracking is too much? is the employee fully informed of his/her consent? What measures are taken to store, anonymize and protect the data from breaches? Organizations are required to comply with strict data protection regulations such as General Data Protection Regulation (GDPR) in Europe and California Consumer Privacy Act (CCPA) in California. There are clear policies, strong cybersecurity controls, and a dedication to data minimization - the collection of only the data necessary to fulfill the stated goal of employee development. An employee breach in this area can irreparably damage employee morale and the organization's reputation.

The second is that of algorithmic bias. Artificial Intelligence, or AI, and Machine Learning, or ML, algorithms are trained on data and if that data (historical data) reflects and magnifies human biases, the algorithms will inevitably learn to do the same (O'Neil, 2016). For instance, if past promotion data disproportionately rewards employees from a particular demographic or with a particular education level, the AI may disproportionately recommend high-potential programs or prestigious mentors to the same groups, thus perpetuating current inequities. To mitigate this, a multi-pronged approach will be necessary; from utilizing diverse and representative training data, to conducting regular audits on the AI's recommendations for biased outcomes, and human oversight to correct and inform the system. AI is not a tool to automate past discriminations; rather, it is a way to establish a more fair, meritocratic system of development.

Furthermore, organizations need to be cognizant to retain the human factor. Technology is to be a servant and not a replacement for the invaluable elements of human development. Though an artificial intelligence system can suggest a curriculum, it cannot replace empathy, contextualized suggestions or inspirational mentorship that comes from a human manager or coach. While AI can help facilitate connections, it cannot replace the trust and camaraderie that is built through informal social learning. The most effective ecosystems are ones that find a healthy balance: using AI to automate administrative tasks, personalise learning journeys and analyse data, freeing up human L&D and HR professionals, managers and coaches to focus on high touch, high impact: coaching, moderating discourse, and creating strategic learning experiences.

Finally, these numbers are difficult to measure and the ROI is a tricky proposition. That is, getting beyond the simple metrics of course completion rates and learner satisfaction scores (Level 1 of Kirkpatrick's model) is an important step. The real measure of an adaptive learning ecosystem is its impact on business performance. This requires that learning activities be connected to Level 3 (behavior change) and Level 4 (business results) outcomes (Kirkpatrick & Kirkpatrick, 2016). This can be done by mapping the data of the skills acquired against performance indicators such as sales data, project success rates, quality scores, and employee retention levels in specific teams. The data analytics necessary to make these causal links is complex, and the close collaboration between learning and development, human resources analytics and business units is essential to show the strategic value of the investment.

7. THE FUTURE HORIZON: EMERGING TRENDS

Adaptive learning ecosystems are only at the early stages of their development. Several new technologies are set to make them more immersive, smart and embedded into the fabric of work.

The most disruptive recent development is the emergence of generative AI. Unlike the analytical AI which is utilized by the current generation of recommender engines, the generative AI is able to generate new content out of nothing. In a learning context, this means a platform could in real time create personalized learning summaries, create practice scenarios and case studies that are specific to an employee's role and projects, or even simulate conversations with AI-powered role-playing characters for practicing difficult feedback or sales pitches. It could be a universal learning assistant, which, based on the company's entire knowledge base, could answer any question in a conversational manner. This will democratize content creation and offer a truly on-demand contextual learning experience.

Another frontier is in the cloud integration of immersive learning (VR/AR). Although virtual and augmented reality have been used for training for years, adoption has often been limited by expensive hardware as well as complex software deployments. Cloud streaming technologies are changing this landscape. Much like Netflix can now stream films, so too can VR training simulations be streamed to standard headsets (or even mobile phones in the case of AR) and be just as sophisticated. This makes immersive learning easier to scale and cost-effective, enabling employees to perform high-stakes/dangerous tasks (e.g. equipment repair, patient surgery, public speaking) in a safe, realistic environment anywhere in the world, with the data on their performance recorded and processed by the platform for feedback.

These technology changes are helping to enable a more general organizational movement toward skills-based organizations (SBOs). In an SBO, work is divided into projects and tasks based on required skills rather than inflexible job titles. Adaptive learning ecosystems are the force that facilitates this shift. They allow for real-time, verified skills data, needed to match employees to projects, create agile teams and identify internal talent for open positions. This transformation will have a powerful effect: it will release untapped talent throughout the organization, make the workforce more fluid, and provide employees with more flexible and satisfying career paths built on ability rather than job title.

Finally, the idea of lifelong learning record is becoming more popular. Driven by blockchain or other secure and verifiable technology, this would be a learner-owned, portable digital transcript representing every competency, micro-credential and achievement in that individual's career, across various employers and learning organizations. Corporate adaptive learning ecosystems would serve as a major input of verifiable data into this record to support individuals taking ownership of their career stories and making talent markets more transparent and efficient.

8. CONCLUSION: THE COMPETITIVE IMPERATIVE

The days of intermittent, standardized company training are gone. Artificial intelligence and digital transformation have driven up the pace of change in the global economy and there is no longer a state of stability in skills. In this environment, the ability of an organization to learn and adapt more quickly than its competitors is its only sustainable competitive advantage. Adaptive learning ecosystems based on the powerful combination of AI and cloud platforms are the most sophisticated and effective solution to this new reality.

These ecosystems are not just an incremental step up from the traditional LMS environment, they represent a paradigm shift. They make learning from a corporate-sponsored event into a continuous, personalized and integrated experience. They enable employees to take responsibility for their development, arm managers with actionable intelligence and enable organizations with the strategic agility to confront an uncertain future. While the journey to implementation is fraught with considerations that need careful planning, investment, and a thoughtful approach to ethical issues like data privacy and algorithmic bias, the potential reward - a resilient, innovative, and future-proof workforce - is immeasurable. As a result, the question for business leaders is not whether they should develop such an ecosystem but how fast they can get started. The future will belong to the fastest learners and adaptive learning ecosystems are the powerhouse of that learning.

REFERENCES

Afzal, S. (2019). *Recommender systems for technology enhanced learning*. Springer International Publishing.

Bersin, J. (2017). *The disruptive, integrated, and personalized learning systems are here*. Josh Bersin Academy. https://joshbersin.com/2017/12/the-disruptive-integrated-and-personalized-learning-systems-are-here/

Bersin, J. (2019). *Learning in the flow of life*. Josh Bersin Academy. https://joshbersin.com/2019/03/learning-in-the-flow-of-life/

Deloitte. (2020). *Global Human Capital Trends*. Deloitte Insights. https://www2.deloitte.com/us/en/insights/focus/human-capital-trends.html

Kirkpatrick, J. D., & Kirkpatrick, W. K. (2016). *Kirkpatrick's four levels of training evaluation*. Association for Talent Development.

Kong, X. (2021). AI in education: A systematic review of the literature. *Journal of Educational Technology Development and Exchange*, *14*(1), 1–20.

LinkedIn Learning. (2023). *Workplace Learning Report*. LinkedIn. https://learning.linkedin.com/resources/workplace-learning-report

O'Neil, C. (2016). *Weapons of math destruction: How big data increases inequality and threatens democracy*. Crown Publishing Group.

Schoox. (2022). *The LMS is dead, long live the LMS!* https://www.schoox.com/blog/the-lms-is-dead-long-live-the-lms/

Schwab, K. (2016). *The fourth industrial revolution*. World Economic Forum.

Sweller, J. (2011). Cognitive load theory. In Mestre, J. P., & Ross, B. H. (Eds.), *Psychology of learning and motivation* (Vol. 55, pp. 37–76). Academic Press.

Vygotsky, L. S. (1978). *Mind in society: The development of higher psychological processes*. Harvard University Press.

World Economic Forum. (2023). *Future of Jobs Report 2023*. https://www.weforum.org/reports/the-future-of-jobs-report-2023/

Chapter 2
Cloud AI of Strategic Human Resource Transformation

Bhumika Ray
https://orcid.org/0009-0000-5628-5859
JGBS, O.P. Jindal Global University, India

Shashi Kant
https://orcid.org/0000-0003-4722-5736
DMI-St. Euegne University, Zambia

ABSTRACT

The chapter explores how Human Resource Management is changing due to the use of cloud-native AI applications. It addresses how microservices, APIs, containerisation, serverless computing, and MLOps—elements that are discussed as cloud-native—can be used to build scalable and intelligent HR ecosystems. The conversation goes further to the application of AI in conventional HR activities, such as recruitment, workforce planning, performance management, learning and development, and employee experience. Also mentioned in the chapter is acute ethical, legal, and governance issues involved in AI implementation HR, especially given the sensitivity of employee data and the need to remain unbiased in the results of decisions. Case studies on how organisations in different industries use cloud-native AI to improve strategic impact, productivity, innovation, and agility culture are presented in reality. Finally, the chapter places cloud-native AI not as a technological revolution in itself but as a driving force that will transform the future, structure, and functionality of HR.

DOI: 10.4018/979-8-3373-7442-0.ch002

INTRODUCTION

Human Resource Management is passing an unprecedented stage of transformation due to the intensive growth of digital technologies and the increase in the strategic value of information-based decision-making (Chowdhury et al., 2024). With the digitisation of industry structures, organisations are increasingly becoming aware of the fact that the ability to manage, empower and strategically mobilise human capital is the most important source of sustainable competitive advantage, which replaces the long held product or process issues (Prikshat et al., 2023). The contemporary workforce, described through hybrid and remote work patterns, a global workforce, a consistent disruption of skills driven by the complexity of technologies, and increased pressure to include, be flexible, and care about the welfare of employees puts an increased strain on HR professionals. The traditional HR models, which are based on manual activity, decentralized sources of information, and decision-making silos coupled with subsequent analysis, are no longer feasible in this dynamic and unpredictable environment (Kim et al., 2024). The adoption of cloud-native artificial intelligence, in this case, is a paradigm shift in the way organisations envision and perform HR operations (Chowdhury et al., 2023). Cloud-native AI is an amalgamation of the scale, flexibility and power of cloud-computing with the predictive, analytic and automation features of artificial intelligence. Instead of a modular set of tools, cloud-native AI is a linked, self-training ecosystem that can manage, analyse and optimise the process of employees as a whole. Cloud-native AI can help HR to predict workforce trends and identify latent behavioural trends, as well as, build data-driven approaches to recruitment, development, and retention leveraging real-time data processing, intelligent automation, and deep analytics (Einola and Khoreva, 2022). As a result, cloud-native AI makes HR an operational and administrative traditionally functional unit into a strategic partner, which is organically incorporated in business growth and organisational development (Rabenu & Baruch, 2024). HR professionals no longer need to participate in paperwork, compliance, or transactional support, rather they are assigned strategic roles that involve talent anticipation, leadership growth, organisational culture impact, ability growth and sustaining the workforce. Cloud AI also enables the HR to provide evidence-based ideas to facilitate corporate governance, innovation, and long-term competitiveness (Bannya et al., 2022).

The chapter explores how Human Resource Management is changing due to the use of cloud-native AI applications. It addresses how microservices, APIs, containerisation, serverless computing, and MLOps - elements that are discussed as cloud-native - can be used to build scalable and intelligent HR ecosystems. The conversation goes further to the application of AI in the conventional HR activities, such as recruitment, workforce planning, performance management, learning and development, and employee experience. Also mentioned in the chapter is acute ethical, legal, and governance issues involved in AI implementation in HR, especially given the sensitivity of employee data and the need to remain unbiased in the results of decisions. Case studies on how organisations in different industries use cloud-native AI to improve strategic impact, productivity and innovation and agility culture are presented in reality. Finally, the chapter places cloud-native AI not as a technological revolution in itself, but as a driving force that will transform the future, structure, and functionality of Human Resource Management (Dineen et al., 2022).

Understanding the strategic potential of cloud-native artificial intelligence in human resource management would require the formation of concepts that establish the unique features of cloud-native systems as compared to traditional technological systems (Chowdhury et al., 2024). The HR information systems of the past were based on monolithic on-premise-based architectures that limited scalability, integration, and placed constraints on real-time data processing. Such monolithic structures were usually typified with rigid databases, uncoordinated modules, poor customisation, and heavy maintenance. Therefore, HR departments could not act promptly to business needs or integrate new technologies and provide individual employee experiences (Prikshat et al., 2023). Cloud presence systems overcome such constraints by adopting architectural values that promote decentralisation, agility, modularity and continuous enhancement. The smart ecosystem deployed on a cloud-native AI-based HR system incorporates distributed computing, microservices architecture, containerisation, API integration, and automated operation of the model, thus allowing an active response to new data, workforce trends, and organisational needs (Chowdhury et al., 2023).

The microservices solution is one of the key foundations of HR systems that are cloud-native. In this paradigm, the HR services of recruitment, onboarding, payroll, performance management, attendance monitoring, talent analytics, employee well-being, learning systems, and workforce planning services are autonomous and interoperable. Any microservice can be developed, deployed, changed or scaled without impacting the rest of the system. Such modularity enhances the pace of innovation, improves faster response to issues and makes HR technologies receptive to the ever-changing business environment. As an example, when an organisation aims to optimise its recruitment algorithms or implement a new learning platform, it will be able to do it without disturbing the rest of the HR processes, including payroll or attendance (Kim et al., 2024).

The API-based ecosystem is the connective tissue that binds the microservices and external platforms together. Practically, HR data are spread across a large number of systems, such as applicant tracking systems, human resource information systems, payroll engines, benefits administration applications, learning portals, collaboration platforms, and external labour-market databases, which can only work properly with API mediation. Absence of strong APIs will result in fragmentation of data and incomplete analytics. The AI models can create accurate workforce forecasts, streamline operations, and give a holistic perspective of employee paths through API integration and real-time flow of data (Chowdhury et al., 2023).

Even more technologies like Docker and Kubernetes, are based on containerisation, which makes applications, dependencies, and ancillary components portable and secure and can improve the performance of the systems. As a result, HR systems are able to expand automatically when demand is high. As an illustration, organisations will see spikes in the system use when undertaking appraisal processes, recruiting new employees, conducting compliance audits, or training their employees. The HR systems are deployed automatically and performance is ensured and cost of operation optimised under a container orchestration (Prikshat et al., 2023). MLOps operationalisation is the pillar of the deployment and management of machine-learning models. The HR processes create a stream of data on issue of worker interaction, performance measurements, engagement rates and learning and communication patterns at the workplace. Such dynamic sources of data require the use of AI models. MLOps automated various phases of the model life cycle, including training, validation, deployment, monitoring, and retraining; therefore, enabling predictive models, e.g., attrition risk analytics or performance forecasting, to be accurate, fair, and aligned with organisational goals (Kim et al., 2024).

In line with this, cloud-native AI suggests the use of a technological backbone that fosters acceleration, durability, and continuous intelligence. These attributes are unavoidable in modern HR environments that are typified by fast changes in employee needs and desires, regular organisational shifts, and the need to be flexible in the global labor market (Rabenu and Baruch, 2024). The use of cloud-native artificial intelligence has transformed every aspect of the HR profession, making it possible to transition to intelligent, automated, and strategic systems rather than the manual, routine, and reactive practices (Chowdhury et al., 2024). These tools improve the efficiency of the operations, enhance accuracy, create transparency, relieve the administrative burdens, and provide predictive information that directly affects the organisational strategy. They have the biggest influence on recruitment, workforce planning, performance assessment, learning and development and the overall employee experience.

The process of recruitment is one of the most complicated and time-consuming HR processes that require a lot of resources to source, screen, interview, and select candidates. These processes are revolutionised using cloud-native AI, where automation, intelligence, and foresight are combined into the talent-acquisition pipeline (Dineen et al., 2022). Natural language processing can also be used to increase the resume parsing by interpreting the context, work history and skill relevance. In contrast to the keyword-based systems that tend to ignore the subtle career trajectories,

AI-powered semantic search finds transferable skills, implicit competencies, and job fits, thus eliminating chances to miss qualified applicants and improving the accuracy of the matching between candidates and jobs (Chowdhury et al., 2023).

The use of AI in conducting interviews adds a new layer to objective appraisal. Video interview tools assess the speech patterns, communication style, engagement, emotional cues and behavioural cues; these details form an extra degree of analysis that complements human judgement and does not eliminate it. Cloud-based interview intelligence solutions can bring more informed and fair hiring because they allow an objective, bias-free evaluation based on data (Einola and Khoreva, 2022). Another important aspect of AI-based recruitment is bias detection. Past data is frequently biased and biases that are inherited and result in unequal results across groups. Cloud-native AI systems are built with equity checks, bias-detection systems, and remedial algorithms so that they may promote fair recruitment practices; they are capable of detecting a potential inequity, emphasize discriminatory behaviors, and implement policies to adhere to diversity and inclusivity goal (Rabenu and Baruch, 2024). In general, cloud-native AI saves time-to-hires, better engages candidates, increases quality of hiring, and frees the HR team to concentrate on strategic work, as opposed to screening activities (Kim et al., 2024).

The increasing instability of the global labour markets, the disruption caused by technologies in a very short period of time and the changing organisational needs have made workforce planning to be a very important HR task. Cloud-native AI enhances workforce planning by offering the ability to make powerful predictive analytics choices that deliver comprehensive knowledge of staffing requirements, talent gaps, turnover risk, and productivity trends (Chowdhury et al., 2023). Attrition prediction models examine the behaviours of the employees, the performance metrics, the engagement rating, the compensation, internal mobility, manager rating, workload, and external labour trends to predict the likelihood of a given employee quitting their job. Such forecasts allow the HR departments to act proactively and develop individual retention plans, such as specialised development programmes, incentives, progression requests, or work conditions changes (Prikshat et al., 2023).

The demand forecasting of manpower focuses on combining past trends on workforce management, growth of the business, market dynamics, and demands to forecast future needs of staffing. Such predictive insights give organisations an opportunity to plan recruitment cycles strategically, reduce labour shortages, and control costs associated with the workforce (Kim et al., 2024). The workforce optimisation programs take into account the skills of employees, project needs, leadership presence, team interaction, and employee performance trends to assign the resources effectively, eliminating the occurrence of a scenario where one team is overworked and another is underutilised (Rabenu & Baruch, 2024). Compensation benchmarking tools also offer access to cloud-based labour-market data in real time, which allows organisations to have a competitive pay structure, align compensation policies with industry rates, and, thus, create employee satisfaction and minimise the risk of losing the top talents to the competitors (Chowdhury et al., 2024).

The old model of the annual performance review is quickly becoming irrelevant in the contemporary organisations. The workers are now requiring feedback regularly, instantaneous acknowledgements, individual growth, and evidence-based assessments. Cloud-native AI is a crucial characteristic that is able to refreeze the concept of performance management by incorporating various data sources, including project results, team working activities, goal achievement, peer ratings, communication habits, and leadership metrics, into complete judgments (Einola and Khoreva, 2022). Performance systems powered by AI acknowledge behavioural patterns, emphasise the performance trends, and pinpoint strengths and weaknesses which will provide managers with more objective, transparent, and evidence-based feedback. Nudges, which are automated, prompt employees to remain focused on objectives, get tasks done, and participate in development (Kim et al., 2024).

The performance systems based on the use of OKRs are also aligned with communication tools and project platforms, allowing the organization of automated updates and the monitoring of goal achievement in real-time. Azure Machine Learning, AWS SageMaker, and Google Vertex AI are cloud services that offer the processing power required to handle large performance datasets with thousands of employees (Chowdhury et al., 2023). Implementation of ethics is also necessary in performance management system. Cloud-native AI uses explainable AI methods whereby the architecture of generating performance scores is explained. Demographic imbalances in assessments are identified through fairness audits to make sure that performance systems powered by AI are not opaque, unfair, or unaccountable (Rabenu and Baruch, 2024).

However, learning and growth has emerged as a strategic issue among organisations that are faced by rapid technological upheaval and skills gaps. Cloud-native AI changes L&D and develops personalised, adaptive, and data-driven learning ecosystems that coordinate employee development with organisational objectives (Prikshat et al., 2023). The AI-driven competency tests are used to assess the skills of the employees based on the performance records, history of projects, learning behaviour, and career goals. Such evaluations produce dynamic skill profiles that allow HR departments to define the gaps and formulate specific development plans (Bannya et al., 2022).

Recommendation systems pair employees with value learning modules depending on skills needs, job roles and career paths. These engines are cloud-based and use internal repositories, web-based learning, as well as professional accreditation (Chowdhury et al., 2024). AI-based learning nudges will motivate employees to finish their courses, thus decreasing the dropout rates. Learning impact is measured and skills developed with automated analytics to assess the payback of training programmes. As a result of this, the L&D no longer exists in standardised or compliance-focused modules but the ecosystem that develops capabilities, contributing to ever-growing (Kim et al., 2024).

Employee experience is a success factor in organisations that has been attributed to engagement, retention, performance and culture. Cloud-native AI gives organisations the effective tools to comprehend the feelings and motivations of the employees, their needs and the difficulties they face in the work place (Einola and Khoreva, 2022). Sentiment analysis engines can analyze feedback on employees gathered through surveys, emails, in-house social networks, chat messages, and collaboration tools. These insights would allow organisations to notice the development of morale concerns, define the risk of burnout, and get awareness of the sentiment trends in the workplace (Chowdhury et al., 2024). Pulse survey websites are used to gather feedback real-time, and thus the leadership team of a business can assess the perception of employees and resolve any issue in a short time. Employment Expereince platforms examine feedback patterns, identify areas of stress, and suggest changes to policies, interventions on the part of managers, or support resources (Rabenu and Baruch, 2024).

Cloud-native AI encourages a culture of constant interaction, which promotes the psychological safety, belonging, and welfare throughout the organisation (Bannya et al., 2022).

CLOUD-NATIVE HR AI SYSTEMS TECHNICAL ARCHITECTURE.

Cloud-native HR AI is a new technology with a technical architecture that marks one of the deepest technological changes in the Human Resource Management history (Chowdhury et al., 2024). Increasingly ineffective to serve the agility, global scalability, analytical insights, and continuous intelligence demands of modern organizations, traditional HR systems, often laden with legacy infrastructures, monolithic software architectures and siloed databases, end up being less and less effective (Prikshat et al., 2023). The cloud-native AI systems of HR are able to surmount these shortcomings by embracing the architectural principles that focus on modularity, automating, flexible, and real-time responsiveness (Kim et al., 2024). They combine AI and machine learning as the main elements of HR work and allow optimising it continually and orchestrating the work processes in the entire employee lifecycle (Rabenu and Baruch, 2024).

These systems are anchored on a microservices-based architecture. Microservices also replace the traditional systems where all HR functionalities are integrated into one software suite with separate and distinct services that can operate independently. Any one of the services, including candidate sourcing, interview scheduling, onboarding workflow management, compensation analytics and learning personalization, can be developed, launched, and expanded alone. This autonomy allows organisations to add new features quickly, modify the established workflows without interruption, and implement new technologies without the need to revamp the entire systems (Chowdhury et al., 2023). To the HR teams, this would mean increased flexibility since functions can also be adjusted to changes in labour regulations, or organisational restructuring or even the changing approach to the workforce.

These microservices interact via API-based ecosystems which connect the whole HR digital environment into one network. The APIs allow the exchange of data in real-time between the HRIS, ATS engines, payroll engines, enterprise communication tools, learning management systems, and performance management applications. As a candidate leaves the interview screening stage and enters the job-offer negotiation stage, APIs facilitate that data is smoothly transferred to onboarding processes, salary-structuring financial systems, access-provisioning IT processes and job-role-specific training modules. This interrelationship will drive out unnecessary data entry, enhance data consistency, and contribute to an employee life cycle between pre-hire and departure (Einola and Khoreva, 2022).

Event-driven architecture is another principle that is central to the cloud-native HR AI systems. Every day, hundreds of thousands of events happen in HR processes: a job applicant makes an application, a worker finishes a training course, a performance evaluation is accepted, or an engagement score drops below a point. Event-driven systems sense these indicators and autonomously perform workflows defined by default. As an example, an employee performance rating of less than expected on collaboration skills can automatically trigger the system to forge customised training material, allege the manager, change internal dashboards, and set specific development objectives. Event-based triggers can be prompted when an employee displays signs of burnout in terms of sentiment analysis of emails, collaboration data, and survey responses. This architectural paradigm is able to make HRM a proactive instead of a reactive administrative role and a strategic partner that is intelligence-driven (Kim et al., 2024).

This automation is further improved in serverless computing when the HR systems can perform tasks without having to operate underlying servers or infrastructure. Serverless functions are only activated upon certain triggers. As an example, hundreds of resumes can be processed within a few seconds by a serverless function when the company is actively recruiting and then remain idle (and free of charge) when the hiring process declines. Serverless functions may be used to create performance summaries, analyse competency matrices and calculate salary increments in real time, without the use of permanent compute resources, during annual appraisal cycles. This elasticity allows the operations of the HR to be maintained on a high level of efficiency, cost-optimised, and dynamically scaled during the times when the business process is highly active, i.e., when mergers, compliance audit, or mass hiring efforts are involved (Chowdhury et al., 2024).

The data architecture is located beneath any automated workflow. Cloud-native HR AI systems are based on the robust data pipelines that absorb, cleanse, amalgamate, and modify data that comes in through different HR and enterprise applications. These pipelines combine structured data, including demographic data, payroll data, attendance data, performance data, and training completion data, with unstructured data, including interview transcripts, feedback data, email sentiment data, voice recordings, and collaboration metadata. The current cloud data lakes or lakehouses represent a centralised storage that puts this versatile data in its pure form so that AI models can uncover more insights. Snowflake, BigQuery, Azure Synapse and Databricks are tools that can be used to process data in large amounts and provide real-time analysis at the organisational levels (Prikshat et al., 2023).

After centralising data, cloud-based machine learning processes become actualised using MLOps pipelines. The HR models do not remain the same, the organisational cultures, team dynamics, labour markets and expectations of employees change constantly. MLOps makes sure that AI models like attrition predictors, talent-matching algorithms, performance-scoring engines and workforce-planning forecasters continue to be relevant, free of bias and aligned with the business goals. MLOps automates model training, model validation, deployment, model monitoring, model drift detection, and retraining. When the pattern of applicants starts to make a recruitment screening model to favour certain profile, the drift detection will raise the alarm, initiate the retraining process and uphold the standards of fairness at all times (Kim et al., 2024). MLOps incorporate responsibility and transparency elements of the model lifecycle by incorporating version control, automated testing, and governance checkpoints (Rabenu & Baruch, 2024).

Containerisation and orchestration are also of crucial importance. Docker containers isolate microservices and machine-learning models into separate units that can be configured to run in line with development, testing, and production environments. The Kubernetes coordinates these containers, scales, load balancing, and resource distribution. The capacity to automatically scale the capacity of the system also ensures that HR applications can be responsive and highly available in global organisations with thousands of employees with spikes in workload or geographic distribution (Chowdhury et al., 2023).

Due to the sensitivity of HR data and associated ethical risks of automated decision-making, cloud-native architectures have specific security, fairness, explainability, and compliance layers. The information of employees and job applicants is safeguarded through data encryption, multi-factor authentication systems, anonymisation systems, and role-based access control systems. Fairness dashboards track the behaviour of models in demographic groups, and thus AI predictions will not unjustly disadvantage women, minorities, older workers, and individuals with under-represented backgrounds (Einola and Khoreva, 2022). Explainable AI processes enable HR departments to have the interpretation of AI recommendations and create trust among employees and guarantee congruence with organisational values. The transparency and ethical governance are also supported with the help of regular audits and compliance checks (Rabenu & Baruch, 2024).

The architecture eventually leads towards the current employee-experience platforms that combine conversational AI, mobile applications, and smart digital assistants into daily working processes. Conversational interfaces in tools like Microsoft Teams, Slack or workplace intranets allow employees to request leave, view their payslips, seek advice, have trainings or customised developmental advice. These interfaces make HR resources more democratic and provide employee support, which can be accessed anytime (Bannya et al., 2022).

The HR is transformed by cloud-native AI systems, which introduce inter-linked concepts of architecture, microservices, APIs, event-driven automation, serverless computing, data lakes, MLOps, containerisation, and ethical governance, to create a more dynamic and intelligence-driven ecosystem of HR that can deliver future-proof workforces (Chowdhury et al., 2024).

LEGAL, ETHICAL AND GOVERNANCE.

The emergence of cloud-native AI in Human Resource Management is associated with massive opportunities as well as massive ethical, legal and governance liabilities (Rabenu and Baruch, 2024). HR is an area, in which the choices made directly affect the employment, health, career growth and pay of the people. The decisions made by AI related to HR concerns can have significant implications on the lives of people unlike other organisational functions such as the operational efficiency or behaviour of the customer. Thus, the application of AI to HR requires strict guidelines of fairness, transparency, privacy, and accountability (Einola and Khoreva, 2022).

Among the fundamental ethical factors is that of data protection. Employees provide organisations with information of a sensitive nature in personal terms, including demographic details, performance records, behavioural trends, benefits records derived through medical records and financial records and in some cases even biometric records. In cloud-native architectures, this data can be usually stored in distributed systems, but on international borders. It is necessary to comply with the regulations like the General Data Protection Regulation (GDPR), California Consumer Privacy Act (CCPA), and new AI governance legislation such as the EU Artificial Intelligence Act. They require legal data processing, direct consent of the employees, reduction of the scope of the data collection process, limiting the scope of the purpose, encrypting and anonymising data, and implementing strict access policies. Breachs put organisations at risk of legal sanctions and destroy the confidence of the employees (Chowdhury et al., 2024).

The other significant problem is algorithmic bias. The development of AI models that are based on historical workforce data can also inadvertently reproduce or increase any current inequities. In case historical hiring has been biased, AI models could be programmed to follow the same patterns. Discrimination may be more covert, such as using resume-screening algorithms to focus on certain educational establishments or performance-grading metrics accidentally developing a correlation between communication style and leadership ability. It is the responsibility of organisations to undertake stringent fairness checks and compare the model results in terms of gender, ethnicity, age, disability status, and other categories that are protected. Machine-learning methods that are fair, bias-reduction algorithms, and debiasing methods are necessary to avoid discriminatory results (Dineen et al., 2022).

Explainability and transparency are also important. Employees also need to know how AI is involved in making decisions that impact on them. Black-box models particularly in high stakes, like hiring or promotion, can give the impression of transparency and unfairness. Explainable AI techniques, such as interpretable model architecture or post-hoc explanation systems, can make the HR teams explain the rationale behind AI suggestions. Such openness fosters trust in employees, ethical responsibilities, and helps in adhering to legal requirements that are demanding more and more AI explainability (Kim et al., 2024).

The governance frameworks make sure that the AI operations are aligned to organisational values and ethical principles. Most organisations are forming AI ethics committees, which include HR leaders, data scientists, legal experts, ethicists, and employees representatives. Such committees determine risk, AI deployment approvals, fairness audit, and responsible innovation. There are well-established accountability designs that outline the roles of model training, governance of data, decision-making, and reporting. The communication strategies with employees are also critical because they will have awareness of how the AI is implemented, what information is gathered, and how it affects them (Prikshat et al., 2023).

Finally, AI technologies need to be governed using ethical and legal approaches. As cloud-native HR AI systems become increasingly complex, analysing feelings, anticipating turnover, assessing teamwork trends, and suggesting career paths, organisations will have to constantly review governance systems to ensure fairness, respect, and human rights (Bannya et al., 2022).

CASE STUDIES

One of the world's largest technology companies used cloud-based, all-purpose artificial intelligence in its recruitment lifecycle process to resolve the problem of scalability and efficiency (Chowdhury et al., 2023). Before the implementation of AI-based procedures, the company had long turnaround periods, and was screening applicants subjectively, and providing inconsistent experiences to applicants. A switch to a cloud-native applicant tracking system based on semantic search, automated matching, and video interview analytics made a strong difference and hastened the hiring pipeline significantly. The time spent screening resumes reduced by almost 65 percent, with cloud-based natural language processing models to screen applications and rank applications based on skills, experience, and job fit. Video interview assessments used affective features and psychometric features, which allowed hiring managers to focus on good candidates. Scalability of the cloud in real time helped to cope with the surges of applications in the campus recruitment drives. In addition, the fairness dashboards were used to track the demographic trend of candidates throughout the candidate stages to ensure that the rules of equal-opportunity are followed (Einola and Khoreva, 2022). As a result, the organization got better quality hires, increased the productivity of the recruiter and better candidate engagement.

A global production company was faced with the continued issue of employee dropout especially frontline employees. The conventional exit-interviews did not provide much information and the human-resources interventions were reflective but not proactive. A provision of the cloud-native workforce analytics platform, which combines its HRIS, payroll, and shift-scheduling system, allowed the enterprise to create predictive attrition models that determined the high-risk employees that were prone to leaving (Kim et al., 2024). The overtime patterns, shift volatility, manager quality, performance trends, and engagement-survey results were the variables that were included in these models. Active dashboards and automated notifications helped HRs to promote specialized retention efforts, such as adjusting work hours, offering recognition programmes, offering skill-enhancement prospects, and handling managerial challenges. The company registered a voluntary turnover of 18 per cent reduction and a better allocation of resources at the manufacturing plants within a period of one year. Scalability of the system also enabled HR departments in other areas to have access to standardised analytics and bring about consistency in decision-making across the world (Prikshat et al., 2023).

One of the global corporations has implemented a large-scale change in its approach to leadership development with the help of cloud-native AI services. The earlier training programmes were generic, time wastive and in many cases, not relevant to real skills requirements. The organisation established individualised leadership development journeys through incorporating AI-driven learning platforms that are able to analyse the existing competency gaps, workplace behaviours, and career trajectories (Bannya et al., 2022). Recommendation engines that were designed to be cloud-native proposed training content based on personal strength, aspirations and performance profiles. The behavioural analytics have revealed the emergent leaders in terms of pattern of collaboration, involvement in project and informal influence network. Adaptive learning systems lent content challenge and speed keeping up with employee progress. As a result, leadership preparedness was on a path to success, internal promotions increased, and the organisation developed a better succession channel (Rabenu & Baruch, 2024).

DIFFICULTIES AND FUTURE PROJECTIONS.

Although cloud-native AI has a significant potential, organisations face a number of challenges with its implementation. The swift development of AI technologies demands that HR professionals should acquire digital literacy, data-analysis skills and technological proficiency (Kim et al., 2024). Most HR practitioners have to shift their functions to strategic analysts and AI partners. This transition requires significant changes in terms of investment into training and change management (Chowdhury et al., 2024).

Trust and explainability are still significant impediments. When employees think that AI-based assessment or forecasts may misunderstand their behaviour or underestimate their input, they often fear it. To build confidence, it is necessary to establish open communication, inform the staff about the use of AI and its role, and incorporate human control into the high-stakes decisions (Einola and Khoreva, 2022).

One more difficulty is the dilemma of augmenting the workforce and automation. Although AI can be used to automate repetitive jobs, organisations need to find a fine balance between considerate social implications of job replacement. Human-AI collaboration is the future of HR as AI can offer information and automation, and humans can offer empathy, judgement, and the contextual rationale (Rabenu and Baruch, 2024).

In the future, HR AI will be more hyper-personalised, predictive, and smart workforce structures. Skills-based organisations will turn to AI by finding new skills, anticipating reskilling, and defining flexible career development (Prikshat et al., 2023). It will enable proactive employee wellness strategies, using real-time sentiment analysis, stress detection, and well-being analytics (Bannya et al., 2022). Artificial intelligent helpers will appear as a regular employee companion, who will guide the employee through tasks, learning courses and performance targets (Chowdhury et al., 2023).

CONCLUSION

Cloud-native AI is a radical change in human resource management, putting HR out of the scope of administrative functions and making it a strategic driver that defines the successfulness of organisations. Cloud-native AI reinvents recruitment, workforce planning, performance management, learning and development, and employee experience by combining automation, predictive analytics, continuous learning, and real-time intelligence. These systems provide never before seen agility, scalability, and fairness with the ability to support evidence-based decision-making at the employee lifecycle. What is more important is that cloud-native AI is a remodelling of the strategic identity of HR. HR will no longer be a service but an architect of organisational capability, building workforces ready to meet the demands of the future, with the capability to respond to quick technology and economic changes. The interface of cloud computing, machine learning, and human-capital strategy has a strong basis of sustainable competitive advantage. The role of HR in an increasingly complex world will be to transform organisations in terms of creating cultures of innovation, resilience, and constant growth due to the introduction of this transformation.

REFERENCES

Bannya, A., Bainbridge, H., & Chan-Serafin, S. (2022). Hr practices and work relationships: A 20 year review of relational hrm research. *Human Resource Management, 62*(4), 391–412. DOI: 10.1002/hrm.22151

Chowdhury, S., Budhwar, P., & Wood, G. (2024). Generative artificial intelligence in business: Towards a strategic human resource management framework. *British Journal of Management, 35*(4), 1680–1691. DOI: 10.1111/1467-8551.12824

Chowdhury, S., Dey, P., Joel-Edgar, S., Bhattacharya, S., Rodríguez-Espíndola, O., Abadie, A., & Truong, L. (2023). Unlocking the value of artificial intelligence in human resource management through ai capability framework. *Human Resource Management Review, 33*(1), 100899. DOI: 10.1016/j.hrmr.2022.100899

Dineen, B., Yu, K., & Stevenson-Street, J. (2022). Recruitment in personnel psychology and beyond: Where we've been working, and where we might work next. *Personnel Psychology, 76*(2), 617–650. DOI: 10.1111/peps.12548

Einola, K., & Khoreva, V. (2022). Best friend or broken tool? exploring the co-existence of humans and artificial intelligence in the workplace ecosystem. *Human Resource Management, 62*(1), 117–135. DOI: 10.1002/hrm.22147

Kim, S., Khoreva, V., & Vaiman, V. (2024). Strategic human resource management in the era of algorithmic technologies: Key insights and future research agenda. *Human Resource Management, 64*(2), 447–464. DOI: 10.1002/hrm.22268

Prikshat, V., Malik, A., & Budhwar, P. (2023). Ai-augmented hrm: Antecedents, assimilation and multilevel consequences. *Human Resource Management Review, 33*(1), 100860. DOI: 10.1016/j.hrmr.2021.100860

Rabenu, E., & Baruch, Y. (2024). Cyborging hrm theory: From evolution to revolution – the challenges and trajectories of ai for the future role of hrm. *Personnel Review, 54*(1), 174–198. DOI: 10.1108/pr-02-2024-0111

Chapter 3
AI and the Human Touch: Building Trust in Cloud-Automated Business Ecosystems

Ali Najeeb
https://orcid.org/0000-0003-4663-9347
Qasim Ibrahim School of Business, Villa College, Maldives

Gunjan Mohan Sharma
https://orcid.org/0000-0003-2272-7385
JGBS, O.P. Jindal Global University, India

Priya Gaur
Centre for Online Education, GLA University, India

ABSTRACT

The integration of artificial intelligence (AI) into cloud-automated business ecosystems promises unparalleled efficiency but poses a critical threat: the erosion of trust. This chapter argues that trust is not an obstacle to automation but its essential enabler. It diagnoses the roots of the trust deficit, including algorithmic opacity, data privacy concerns, and bias. The chapter then redefines the "human touch" not as mere sentiment, but as crucial roles in orchestrating, explaining, and overseeing AI. A practical framework for building trust is presented, founded on pillars of transparency, reliability, ethics, and human-centric design. Finally, the chapter explores the profound social, practical, ethical, and managerial implications organizations must navigate to foster a symbiotic future where technology amplifies, rather than replaces, human integrity.

DOI: 10.4018/979-8-3373-7442-0.ch003

INTRODUCTION: PARADOX OF AUTOMATION AND TRUST

Today's business landscape finds itself in the midst of a big transformation, fueled by the pairing of artificial intelligence (AI) and the cloud. This pairing has given rise to automated business ecosystems in the clouds—changing, interconnected networks in which intelligent algorithms control such things as supply chain operations, customer relationships, financial trading, and human resources (Iansiti & Lakhani, 2020). Their promise: unmatched efficiency, scalability, and knowledge built on data. But the rapid advance of automation brings with it a paradoxical flaw: while systems become smarter and more autonomous, the very building block of all successful commercial exchanges—trust—becomes ever more indispensable and at the same time ever more fragile. A customer who receives a spot-on, individualized product recommendation at exactly the right moment from an AI feels a sense of comprehension, while another customer presented with an indecipherable and unyielding denial by a loan-eligibility algorithm feels alone and suspicious. This dichotomy lies at the heart of the modern digital business.

This chapter explores the hypothesis that in racing to the future of automation, organizations risk sacrificing the human dimension, which is the very heart of trust. Arguably, the main argument here is that trust should not be considered a minor, tertiary matter or a barrier to automation; rather, it's the ultimate facilitator of such processes. Where trust disappears, employee buy-in to AI-based technology will likely be slow and obstinate, customer loyalty will be short-lived, and the full potential of these sophisticated systems will go untapped (Følstad et al., 2021). This chapter's goal is to navigate this complex terrain. We begin by pinpointing the exact factors driving the trust deficit in automated systems, then reconceptualizing the importance of the "human touch" in an age of artificial intelligence. Next, we articulate a practical, multi-faced framework for actively building and maintaining trust. Finally, and arguably most important, we delve into the substantial social, practical, ethical, and managerial issues that organizations must grapple with to thrive in this new epoch. The shift is from seeing AI simply as an efficiency instrument to seeing it as part of a larger, human-centric system whose performance is measured in revenue only in part and in reliability and integrity in full.

A LACK OF TRUST IN AUTOMATED SYSTEMS

First comes understanding where this distrust comes from. Unease among humans with respect to automated systems is no exaggeration; it is always a reasonable response to real or potential dangers of convoluted and opaque technology. This lack of trust stems from several interconnected issues that, if left unattended, could seriously hamper the incorporation of AI in business environments.

One of the root sources of skepticism comes from the "black box" problem inherent to many advanced AI systems, in particular deep learning neural networks. Such systems have the capability to make incredibly accurate decisions—like diagnosing a medical condition or predicting machinery failures—without providing an account in human-understandable terms of the grounds upon which they base their decisions (Castelvecchi, 2016). In the eyes of stakeholders such as doctors, patients, or ratings officials, an unexplained result is deemed to lack credibility. This failure to provide openness goes against an important axiom of trust: the ability to understand and verify the reason behind a choice. In the event such an AI program denies a mortgage application or flags a transaction as fraudulent, the person so affected has a right to an explanation, which many such systems have not technically advanced to the point of supporting (Goodman & Flaxman, 2017). This lack of openness not only inspires irritation but also makes it more difficult to identify and correct errors, and which in turn leads to suspicion that the system works reliably or rather at random.

Closely related to the black box puzzle is the global debate surrounding data protection and security. Central to the nature of cloud-automated systems, these have an insatiable appetite to ingest data. This data—both personal identifiable and advanced behavioral data—becomes the raw material on which the AI algorithms feed. But data collection and processing in distant, off-site third-party clouds create legitimate concerns surrounding access and usage and the effectiveness of safeguards to prevent breach (Zuboff, 2019). Highly publicized data scandals have given public profile to the potential misuse perils such as inappropriate surveillance, manipulation of consumer behavior, or sale of personal data to faceless third partics. Lying at the root of the very technology that supports advanced AI—the cloud—is the concentration of risk, which in turn becomes an attractive target for cyberviolence. Faith in the output of AI then becomes reliant on faith in the overall data processing ecosystem upon which it depends, a faith that all too frequently is wanting.

One of the most socially harmful sources of mistrust comes from the problems of algorithmic unfairness and prejudice. Intelligence systems base their learning on past data sets, and where such data reflects societal or historical biases, the AI is likely to absorb and even amplify these biases (O'Neil, 2016). There have been numerous studies to produce empirical proof of this process: e.g., hiring algorithms that lower the ranking of resumes with the term "women's," such as in "women's chess club," or face-recognition software with significantly higher inaccuracies among non-whites and women (Buolamwini & Gebru, 2018). Such events are not merely technical errors; they represent a severe failure of equity that could perpetuate systemic inequality. When automated processes make prejudiced decisions on questions of job fitness, credit worthiness, or even sentences in criminal proceedings, they erode confidence not only in the particular application, but in the institutions that apply them. In most cases, the widespread belief, and often the reality, is that the system itself works on essentially unequal principles.

In the end, there is a legitimate fear of obsolescence among both workers and customers. With workers, the fear of job loss facilitated by artificial intelligence becomes a considerable source of discomfort and resistance. This concern could either arise in terms of reluctant ness to utilize new AI technology or even sabotage, especially if workers believe that knowledge of such technology will ultimately make them obsolete (Brougham & Haar, 2018). With customers, the relationship could come in the form of a feeling of dehumanization. Having to go through endless, infuriating automated phone systems or chatbots that cannot understand complex issues produces the impression that the company no longer holds its customer base in sufficient value to provide human interaction. This "faceless" interaction reduces relational bonds that are fundamental in creating long-term customer loyalty. In both cases, the adoption of AI in a non-introspective consideration of the human context creates a trust deficit that inevitably compromises its purpose.

REIMAGINING THE "HUMAN TOUCH" IN THE ARTIFCIAL INTELLIGENCE ERA

Against these significant challenges, it might come to mind that the solution is to ensure human intervention in all decisions. However, this kind of solution is impractical and unnecessary. A more advanced and efficient solution is to reconceptualize the human function from being interventionist— a controller— to being indirect— a coordinator, supervisor, and adviser. The importance of the "human touch" in the age of artificial intelligence does not mean copying machine functions; it means introducing the abilities that machines by definition lack.

The human aspect goes beyond the simple correlation of the interpretation of empathy with feelings at the rudimentary level. In the context of business, it translates into the dimension of context-awareness, ethics, and creative solutioning to problems. Whereas an AI program can assess a customer's history of purchases, only a human may understand the context in which a delayed payment—like job loss or a medical emergency—has been categorized by the AI program as risk. A human individual recognizes cultural nuance, irony, and implicit mandates that go unnoticed by even the most advanced algorithms in natural language processing. This distinctive ability of a human being allows the act of applying discernment and wisdom to the raw data and risk assessments generated by AI. It represents the ability to ask "what if" and "why" questions that go beyond the operational bounds of deterministic or statistical models (Davenport & Kirby, 2016). Therefore, the human aspect serves in the role of the unifying force that puts into context the results generated by AI in the more complex and real world of human experience and societal expectations.

Within this newly articulated framework, several pivotal human roles arise that are indispensable for the establishment of reliable AI ecosystems. The first role is that of The Trainer. AI models do not possess innate intelligence; rather, they require education. Humans are tasked with the curation of training datasets, the identification and mitigation of biases present within those datasets, and the refinement of model parameters to ensure alignment with organizational values and objectives. An inadequately trained AI functions similarly to a poorly educated employee; it will demonstrate subpar performance and has the potential to inflict substantial harm. The second role is that of The Explainer. This position entails the translation of the intricate, technical outputs generated by AI into coherent and actionable explanations for stakeholders, including customers, regulators, or colleagues. Explainers serve to bridge the divide between the AI's "black box" and the human demand for clarity and rationale, thereby meeting the essential need for transparency (Ribeiro et al., 2016).

Thirdly, and of utmost importance, is the function of The Sustainer. Artificial intelligence systems are not fixed entities; they are subject to "model drift," a phenomenon whereby their effectiveness diminishes over time as the real-world data they process evolves. The Sustainer's role encompasses the ongoing surveillance, upkeep, and validation of AI systems after they have been deployed. This individual monitors for declines in performance, the emergence of biases, and unforeseen edge cases, guaranteeing that the system remains resilient, equitable, and pertinent long after its initial introduction. Lastly, we have The Decider. This position is designated for scenarios characterized by high stakes, ambiguity, or ethical complexities where mere algorithmic computation falls short. The Decider manages exceptions, modifies AI recommendations when contextual factors necessitate, and assumes ultimate accountability for outcomes. This role encapsulates the principle of human responsibility, ensuring that a moral agent is always present to address the ramifications of an AI-influenced decision.

A FRAMEWORK TO DEVELOP TRUSTWORTHY AI ECOSYSTEMS

Building trust constitutes an ongoing and progressive process and not a single certification. In an attempt to systematically treat the foundations of distrust while capitalizing on the newly established human dimension, organizations could embrace a framework relying on four interconnected principles: Transparency, Reliability, Ethics, and Human-Centric Design.

Pillar 1: Explainability and Transparency in Artificial Intelligence (XAI): Trust cannot exist in solitude; it requires some level of transparency. This pillar's goal is to improve the explainability of the AI's operations and reasoning processes to relevant stakeholders. This involves the use of methods in the field of Explainable Artificial Intelligence (XAI), which strives to create models that are intrinsically interpretable or to provide post-hoc explanations for black-box models (Adadi & Berrada, 2018). A credit-scoring AI, for instance, could give a simple summary: "Application denied due to high debt-to-income ratio (70%) and short credit history (2 years)." This kind of explanation is much more reliable than simply being given a "denied" signal. Furthermore, transparency not only refers to the algorithm itself but also to data that has been used and the aims in which it's being used. Institutions should maintain inclusive documentation, commonly known as "model cards" or "fact sheets," which summarize the AI's proposed application, performance characteristics, and known limitations (Mitchell et al., 2019). By defining the workings of the AI, organizations show concern for end-users and introduce a foundation of liability.

Pillar 2: Performance and Reliability: Trust is built on a track record of steady and competent performance. A system of artificial intelligence that regularly goes down, performs slowly, or produces unpredictable results will lose trust, no matter what it says about transparency or ethical considerations. This aspect essentially deals with the engineering and operational rigor that goes into the AI. It requires a scalable, secure, and reliable cloud infrastructure that guarantees high performance and availability. It also requires thorough testing in controlled environments before deployment and continued scrutiny in production to uncover and correct flaws related to model drift or deterioration. Creation of clear, quantifiable performance specifications (like accuracy, latency, and fairness measures) and transparent tracking of them is also required. Reliability also encompasses security; stakeholders need to have confidence that the system has resilience to cybersecurity threats that could compromise its function or integrity of its data. Overall, this pillar confirms that the AI is not only well-intentioned but also essentially competent and reliable.

Pillar 3: Ethics, Fairness & Accountability: This pillar expands on technical expertise to address the ethical dimension of AI deployment. It involves establishing a formal governance structure, such as an AI Ethics Board with members from diverse backgrounds including technical, legal, ethical, and business to coordinate AI systems building and deployment (Field et al., 2020). One of the noteworthy activities under this pillar includes conducting regular bias audits, using statistics to detect and prevent discriminatory outcomes beforehand in multiple population groups (Raji et al., 2020). Aside from that, organizations also need to set clear channels of human accountability. In the event that an AI system causes loss or comes up with life-critical failure, it ought to become obvious who among the developers, operators, business unit head, or CEO ought to accept the blame. This creates a "responsibility sink" that prevents leaving the AI to accept the blame for human negligence. Establishing and maintaining a clear set of ethics principles of AI usage—for instance, putting human well-being, fairness, and contestability at the center—helps to provide the entire organization with a sense of direction.

Pillar 4: Human-Centric Design and User Experience (UX): Trust is built in the end by designing good interactions. This pillar focuses on building artificial intelligence systems that extend and augment human users, rather than replacing or alienating them. The user interface should be clear and provide users with a clear understanding of the functions of the AI, its capabilities, and, no less important, its limitations. A guiding principle of human-centric design is the maintenance of user control and agency. This means always offering a clear and convenient pathway to human intervention, such as a easily visible "escalate to a human agent" button in a chatbot interface or an uncomplicated process to appeal an automated decision (Shneiderman, 2020). In addition, the design should adhere to inclusivity, considering the varied needs of users with dissimilar abilities, technical expertise, and cultural settings. Where users see themselves in control, understood, and capable of requesting human intervention where needed, their likelihood to trust and engage with the automated system significantly rises.

EMPIRICAL RESEARCH ON TRUST

To illustrate the practical application of this framework, two possible yet hypothetical case studies could be contrasted that depict the successful incorporation of the human factor in automated systems in the cloud.

Case Study A: The Trustworthy FinTech Lender - "LoanLogic" LoanLogic is an online loan platform that utilises an advanced artificial intelligence system to process loan applications in real-time. Acknowledging the pervasiveness of suspicion about automated decisions regarding credit, the firm constructed its model on foundations of transparency and the human touch. After notification of a decision, prospects get more than a thumbs-up or thumbs-down. Instead, the website presents an open, pictorial interpretation based on its explainable AI (XAI) framework, which emphasises the top three determinants that drove the decision (e.g., "Strong Credit History (Duration: 10 years)," "Moderate Debt-to-Income Ratio (45%)," "Stable Employment (2 years with current employer)"). In case of denial, the system also presents useful and actionable recommendations, like "Reducing your credit card load by $2,000 will significantly improve your score." What's more, each decision notification email and website page prominently also includes a link that states:

"Discuss this decision with a loan specialist." This feature brings together the capabilities of both The Explainer and The Decider into the process flow. AI processes the majority of straightforward cases in an efficient manner, with human professionals covering tricky, non-conventional, or emotive cases, with the understanding and fine judgment that the AI is unable to provide. This combined approach enables LoanLogic to scale up its process while building up a reputation for fairness and customer satisfaction.

Case Study B: The Empathetic Healthcare Chatbot - "HealthBuddy" HealthBuddy works as an artificial intelligence-based symptom checker and patient triage system deployed by a leading healthcare provider. Its main challenge is to attain a fine equilibrium between efficiency in operations and the indispensable need to create trust and security in a healthcare setting. HealthBuddy's design embodies superior dependability and ethical control. It uses a conservative, clinically accredited algorithm to assess risk. Each interaction begins with a clear disclaimer on its capabilities: "I am an AI assistant and cannot provide a medical diagnosis. In case of a medical emergency, please call emergency services immediately." The system has strong safeguards in place; if a patient states symptoms that potentially imply a severe health issue (like chest pain or stroke symptoms), the AI instantly stops its automated questioning and provides clear instructions to go to the doctor immediately while simultaneously informing a human nurse to follow up. In less severe cases, it is able to schedule appointments or provide general information. However, a group of human nurses (The Sustainers and Deciders) continuously monitors the interaction and suggestions provided by the AI, thoroughly checking a random set and all high-risk flagged cases to ensure suitability and precision. This constant human review ensures that the system remains reliable and adheres to the highest level of ethically correct conduct, following the motto "first, do no harm" and hence creating deep trust among patients and healthcare professionals alike.

IMPLICATIONS: NAVIGATING THE EMERGING LANDSCAPE

Incorporation of artificial intelligence in business systems constitutes more than a technological change; it's a socio-technical change that permeates all aspects of societal existence and organizational structure. To those who aspire to advance their organizations ethically into the future, comprehension and foreseen consequences of these modifications are urgent.

Social Implications

The extensive implementation of automation driven by artificial intelligence is significantly altering the structural dynamics of society. It necessitates a reevaluation of the social contract among individuals, corporations, and governmental entities. There is a notable shift in public expectations, with an increasing call for corporate accountability that transcends mere profit generation to encompass ethical data management, fairness in algorithms, and transparency (Martin, 2019). A considerable concern is the aggravation of the digital divide. This gap will evolve beyond mere technological accessibility to encompass the capacity to comprehend, engage with, and derive benefits from AI systems. Such a scenario could lead to the emergence of a bifurcated society: one comprising an "AI-empowered" class adept at utilizing these tools for progress, and another categorized as "AI-marginalized," which faces persistent disadvantages due to opaque automated systems that are beyond their understanding or capability to contest (Zuboff, 2019). The implications for community and employment are similarly substantial. With the automation of routine tasks, both the nature of work and wealth distribution undergo significant transformation. This situation prompts critical societal inquiries regarding the necessity for innovative social safety nets, such as universal basic income, alongside a shared obligation for comprehensive retraining and lifelong education programs to facilitate individuals' adaptation to the evolving economy.

Practical Applications

From a practical point of view, organizations face a plethora of implementation challenges. There is a considerable and growing requirement for new capabilities. The workforce of the future will require "bilingual" experts who have both deep knowledge in their professional domains (like marketing, finance, and law) and the ability to communicate with data and artificial intelligence systems (Brynjolfsson & McAfee, 2014). This calls for an overall redesign of both corporate training programs and undergraduate/graduate school curricula. Furthermore, system integration poses a primary barrier. Integrating new, intelligent cloud-based services with legacy on-premise systems and existing workflows poses a sophisticated technical and cultural barrier that could potentially derail AI efforts. Arguably the least addressed practical implication is the need to continuously monitor and maintain. Unlike traditional software, AI models have a inherent self-degrading nature. They have the potential to decline in effectiveness with changing data patterns, which is known as "model drift." The result is that organizations need to invest and plan for the ongoing cost of model retraining, adjustment, and supervision in terms of AI systems that are not static capital expenditures but organisms that require constant maintenance and attention.

Ethical Considerations

These challenges raise fundamental questions about moral agency and liability and in cases where an autonomous vehicle suffers an accident or where a clinical AI recommends a treatment that is harmful. Whom do we hold morally and legally liable: the designer, the operator, the owner, or even the AI itself? Current laws and regulations prove insufficient to cover this aspect, and we urgently need new protections and liability frameworks (Boddington, 2017). The value-alignment problem is central. How can we ensure that sophisticated, autonomous systems make choices that mirror the complex and sometimes conflicting nature of human values? It's notoriously hard to induct ethics into machines owing to the frequently implicit and contextual nature of values. In addition to this, even the principle of informed consent becomes ever more antiquated. In a world of complex AI ecologies, it's impossible for users to know all the possible uses and processes of data that's being used. This scenario calls forth the creation of new modes of consent, potentially based on contextual integrity or continuous, contextual permissioning, that truly respect user autonomy in the age of the digital.

Managerial Implications

For managers and leaders, the revolution in AI requires a radical transformation in their own roles and capabilities. Leadership will need to acquire new capabilities to lead teams of human and artificial intelligent workers. This necessitates expertise in developing teamwork between team members and AI systems, navigating the change and stress surrounding AI deployment, and executing strategic decisions predicated on AI-based insights (Brynjolfsson & McAfee, 2014). Success depends on leveraging a cultural transformation in the enterprise. Leaders will need to create a culture of psychological safety where workers will feel comfortable questioning the output of AI, reporting potential biases in AI output, and recommending improvements without fear of retaliation. A safe haven where AI failure can be discussed and learned from is important to improve continuously. From a governance angle, managers need to create new risk and governance structures specifically customized to the unique set of risks in AI, such as reputational harm due to biased results, operational failure due to model drift, and legal liability. Lastly, leaders need to consider spending on trust-building measures such as ethics boards, bias review audits, explainability technology, and employee education as centerpieces of strategic investments and not side-of-the-road cost of compliance to achieve long-term brand integrity, customer loyalty, and sustainable advantage.

CONCLUSION

The passage through the challenges and opportunities of cloud-automated business ecosystems culminates in an inevitable conclusion: in the age of AI, trust will be the ultimate advantage that matters in competition. Most successful and durable organizations will no longer be those with superior algorithms in a vacuum, but those that become virtuosos in the subtle and evolving harmony between the sheer scale and computational might of cloud AI and unique human capabilities of wisdom, empathy, and ethical thinking. This chapter has shown such a direction, from diagnosing the sources of mistrust to deploying an actionable framework based on transparency, dependability, ethics, and human-centered design.

The business leaders of today have a clear and pressing call to action: they need to deliberately and consciously design their strategies of automation with trust and the human touch at the core, non-negotiable pillars. This necessitates breaking out of tight ROI and efficiency-based thinking and adopting a richer scorecard that encompasses fairness, explainability, and user empowerment. This is a deep responsibility. What we create today in systems terms aren't tools; these become the foundations upon which the future business and, indeed, societal landscape will be built. By choosing to create systems that complement rather than supplant, that explain and illuminate rather than conceal and mystify, and that empower rather than command, we have the potential to unlock the power of AI to create a future that's both more efficient and more humane, more just and more trustworthy. The vision is a future in which technology doesn't erode our humanity, it elevates it, creating intelligent systems that aren't intelligent in processing alone, but intelligent in purpose.

REFERENCES

Adadi, A., & Berrada, M. (2018). Peeking inside the black-box: A survey on explainable artificial intelligence (XAI). *IEEE Access : Practical Innovations, Open Solutions*, *6*, 52138–52160.

Boddington, P. (2017). *Towards a code of ethics for artificial intelligence*. Springer.

Brougham, D., & Haar, J. (2018). Smart technology, artificial intelligence, robotics, and algorithms (STARA): Employees' perceptions of our future workplace. *Journal of Management & Organization*, *24*(2), 239–257.

Brynjolfsson, E., & McAfee, A. (2014). *The second machine age: Work, progress, and prosperity in a time of brilliant technologies*. WW Norton & Company.

Buolamwini, J., & Gebru, T. (2018, January). Gender shades: Intersectional accuracy disparities in commercial gender classification. In *Conference on fairness, accountability and transparency* (pp. 77-91). PMLR.

Castelvecchi, D. (2016). Can we open the black box of AI? *Nature*, *538*(7623), 20–23. PMID: 27708329

Davenport, T. H., & Kirby, J. (2016). *Only humans need apply: Winners and losers in the age of smart machines*. HarperCollins.

Field, J., Achten, N., Hilligoss, H., Nagy, A., & Srikumar, M. (2020). *Principled artificial intelligence: Mapping consensus in ethical and rights-based approaches to principles for AI*. Society Research Publication. Berkman Klein Center for Internet.

Følstad, A., Araujo, T., Law, E. L. C., Brandtzaeg, P. B., Papadopoulos, S., Reis, L., & Luger, E. (2021). Future directions for chatbot research: An interdisciplinary research agenda. *Computing*, *103*(12), 2915–2942.

Goodman, B., & Flaxman, S. (2017). European Union regulations on algorithmic decision-making and a "right to explanation". *AI Magazine*, *38*(3), 50–57.

Iansiti, M., & Lakhani, K. R. (2020). *Competing in the age of AI: Strategy and leadership when algorithms and networks run the world*. Harvard Business Press.

Martin, K. (2019). Ethical implications and accountability of algorithms. *Journal of Business Ethics*, *160*(4), 835–850.

Mitchell, M., Wu, S., Zaldivar, A., Barnes, P., Vasserman, L., Hutchinson, B., & Gebru, T. (2019, January). Model cards for model reporting. In *Proceedings of the conference on fairness, accountability, and transparency* (pp. 220-229).

O'Neil, C. (2016). *Weapons of math destruction: How big data increases inequality and threatens democracy*. Crown Publishing Group.

Raji, I. D., Smart, A., White, R. N., Mitchell, M., Gebru, T., Hutchinson, B., & Barnes, P. (2020, January). Closing the AI accountability gap: Defining an end-to-end framework for internal algorithmic auditing. In *Proceedings of the 2020 conference on fairness, accountability, and transparency* (pp. 33-44).

Ribeiro, M. T., Singh, S., & Guestrin, C. (2016, August). " Why should I trust you?" Explaining the predictions of any classifier. In *Proceedings of the 22nd ACM SIGKDD international conference on knowledge discovery and data mining* (pp. 1135-1144).

Shneiderman, B. (2020). Human-centered artificial intelligence: Reliable, safe & trustworthy. *International Journal of Human-Computer Interaction, 36*(6), 495–504.

Zuboff, S. (2019). *The age of surveillance capitalism: The fight for a human future at the new frontier of power*. PublicAffairs.

Chapter 4
Cloud-Driven Organizational Agility:
AI as the Catalyst for Digital Transformation

Himanshi Rajora
https://orcid.org/0000-0001-5686-3030
JGBS, O.P. Jindal Global University, India

Azadeh Amoozegar
https://orcid.org/0000-0003-4988-5113
INTI International University, Malaysia

Dilshad Jeffery
Qasim Ibrahim School of Business, Villa College, Maldives

ABSTRACT

This chapter examines the synergistic relationship between cloud computing and artificial intelligence (AI) as the core driver of modern digital transformation. It argues that while cloud computing provides the foundational infrastructure for organizational agility—enabling scalability, speed, and DevOps practices—it is the strategic application of AI that acts as the essential catalyst. AI transforms cloud capabilities into intelligent operations through predictive analytics, automation, and generative capabilities. The analysis moves from theoretical foundations to practical use cases across business functions, outlining the necessary human and organizational shifts. Crucially, the chapter concludes with a comprehensive discussion of the multi-faceted implications, addressing the managerial, practical, ethical, and social considerations leaders must navigate to harness this transformative potential responsibly and effectively.

DOI: 10.4018/979-8-3373-7442-0.ch004

INTRODUCTION

The modern business environment is defined by the condition of constant, accelerating change, sometimes referred to as VUCA, which means the Volatility, Uncertainty, Complexity, and Ambiguity. Within this context, adaptability, innovation, and quick response or agility of the organization, which is an organizational agility, has ceased to be a competitive advantage but rather a survival and success factor. Organizations over decades have been running on a legacy infrastructure that was, despite its stability, causing data silos, had a long procurement process to acquire new hardware, and had inherently reactive business models. The use of cloud computing has toppled this paradigm because it promised an end to capital-intensive and inflexible IT estates and instead an operational-expenditure-based set of services that could be accessed on-demand (Marquez, 2020). The cloud was the engine behind agility providing more scalability and speed like never before. But, this does not make it a winning race car it needs a sophisticated fuel and a talented driver to realize its potential. It is at this point where Artificial Intelligence (AI) makes an appearance in the story. AI is the force that is catalytic and transferring the potential of the cloud in agility to actualized, smarter action, digital transformation is no longer about operational efficiency but basic business model innovation and the creation of markets (Iansiti and Lakhani, 2020).

The chapter assumes that, even though cloud computing is the mandatory and non-negotiable engine that allows the current organizational agility, it is the strategic combination and implementation of AI that is the real impetus and, therefore, the move that transforms organizations into simply responsive structures and into proactively adaptive and presciently sighted structures. The combination of the computational capacity of the cloud with data storage options and the analytical and predictive capacity of AI helps to form a circle of innovation and creating value. In this chapter, the concept of a cloud-driven agility will be dismantled defining its basic components and the restrictions. It will then explore the detailed processes of how AI enhances these cloud abilities acting as evidence of this symbiotic relationship to give tangible examples of how these capabilities are used and across which business functions. Having realized the fact that technological implementation cannot be complete without human and organizational alignment the chapter will inculcate the much needed cultural and skill changes that are vital. Last but definitely not least, it will offer a thorough discussion of the multi-layered implications- managerial, practical, ethical, and social- that leaders need to work in order to utilize this to create a very strong mix in a responsible and effective manner. The trip we describe is not only to accept new technologies but to radically remodel the organization in a world where the intelligence of humans is enriched with artificial intelligence, which will be running on the vibrant foundation of the cloud.

SECTION I: THE FOUNDATION - DECONSTRUCTING CLOUD-DRIVEN AGILITY

It is necessary to define the base on which AI builds, which is cloud-driven agility, before analysing the catalytic role of AI. Firstly, the cloud migration business case was mainly financial as it was dedicated to the Capital Expenditure (CapEx) to Operational Expenditure (OpEx) switching, which relieved organizations of heavy initial investments in data centers and physical server room (Marston, Li, Bandyopadhyay, Zhang, and Ghalsasi, 2011). Nevertheless, the most disruptive value of the cloud has been found to be its position as an overall agility platform. Agility, used in the organizational context, is the ability to perceive and act in response to changes of the external market, sellers needs, and competitive pressures swiftly, accurately, and with the least amount of disturbance (Tallon, Queiroz, Coltman, and Sharma, 2019). The cloud makes this possible by a number of interrelated pillars that make the foundation of the contemporary digital operations.

The first pillar is elasticity and scalability. Cloud resources can be brought up or brought down automatically and immediately unlike on-premises infrastructure which has to be provisioned to reach peak loads and will sit idle during troughs. This implies that an e-commerce site can easily manage a thousand times more traffic in a seasonal sale with no loss in performance and will only pay the amount of compute power and storage that it is using at the time. The underlying principle behind this elasticity is that it enables organizations to test out new services or venture into new markets without the presence of the prohibitive infrastructure costs that would have been an obstacle in the past (Cito, Leitner, Fritz, and Gall, 2017). Speed and accelerated time-to-market is the second pillar. Cloud platforms offer self service portals and enormous lists of ready-to-use services, allowing development teams to build all the application environments, including development and testing, staging and production, within minutes and not months. This is the compression of cycle time by a dramatic margin that drives the current software development techniques.

This brings to the third pillar of DevOps and continuous innovation. DevOps practices make the cloud their natural habitat, and these practices dissolve the old borders between the development and operations teams. With the help of Infrastructure as Code (IaC) and Continuous Integration/Continuous Deployment (CI/CD) pipelines deployed to the cloud, organizations may automate the entire software delivery process, including the process of writing the code and bringing it to the production (Ebert, Gallardo, Hernantes, and Serrano, 2016). This creates an environment of constant experimentation where minor yet rapid updates can be issued on a frequent basis and feedback can be quickly integrated. The agile responsiveness to customer needs is characterized by this loop. Nevertheless, even with such powerful capabilities, it is still common to have a substantial agility gap. A perfectly elastic,

DevOps-enabled cloud environment may be fine but an organization has to deal with the huge volumes of data that may enter the environment on a daily basis. It is capable of launching software in a relatively short time but might not be aware of what will actually be well received by the customers. It is capable of scaling its operations though it might fail to forecast future demands accurately. The cloud offers the supercharged, intelligent car, yet it does not have inbuilt intelligence to help choose the most effective path or foresee the presence of roadblocks. This disconnect between the possibility and smart implementation is precisely where Artificial intelligence inserts itself as the catalyst that can not be done without.

SECTION II: THE CATALYST - HOW AI SUPERCHARGES CLOUD CAPABILITIES

Cloud computing and AI are symbiotic to each other. The cloud offers the necessary substrate the enormous, scalable computing capabilities and enormous, centralized data stores, necessary to educate and execute complicated AI models. In its turn, AI gives the brains of analysis, that is, the ability to make this raw infrastructure an active and intelligent collaborator of business processes. Such a synergy establishes a positive feedback loop: the more information is stored in the cloud, the more powerful and more accurate AI models are created, which, in turn, generate more valuable insights and automation, which attracts more information and application cases to the cloud (Huang and Rust, 2021). This catalytic role expresses itself in various critical forms and the first one is the fundamental transformation of data storage to intelligent insight.

The cloud serves as a giant data warehouse, where structured and unstructured data of the enterprise, between various sources, ERP and CRM systems, IoT sensors, social media feeds, and transaction logs, are combined into data lakes. Nonetheless, the presence of a reservoir of data with limited means of purification, analysis, and interpretation is of little use. This is the core value of AI. It is a branch of AI called Machine Learning (ML) algorithms that could process petabytes of such data and find hidden patterns, correlations, and trends that human analysts fail to see (Provost and Fawcett, 2013). As an example, whereas the cloud data warehouse could store years of sales data, an ML model could process it with weather data, economic indicators, and social media sentiment to predict the future demand with a high level of accuracy. This shifts the organization to the descriptive analytics (what happened) to predictive and prescriptive analytics (what will happen and what should we do about it).

The supercharging effect of AI is provided by a set of capabilities, which are now easily accessible in the form of cloud services. The most influential are probably

Machine Learning (ML) and Predictive Analytics, which allow applications in areas including fraud detection in financial transactions and predictive maintenance in manufacturing, where sensors on equipment feed algorithms to predict equipment failures before they happen, reducing downtime (Tao, Qi, Liu, and Kusiak, 2018). Natural Language Processing (NLP) enables the machine to comprehend, perform interpretation and generate human language. NLP drives the smart virtual assistants and chatbots in the cloud which process customer care queries 24/7 yet it can also be used to analyze legal documents, summarize patient medical histories, and predict the opinions of millions of online reviews (Hirschberg and Manning, 2015). Computer Vision provides software with the capacity to perceive and acquire the meaning of images and videos. Implemented on cloud systems, it allows controlling the quality of the items produced on assembly lines by detecting microscopic errors, handles inventory with automated shelf scans, and even improves shopping experiences with cashierless stores (Amazon Go, 2024). Recent launch of Generative AI is a leap in this catalytic relationship. Such models as GPT-4 and its descendants are able to create new content, such as code and marketing text, synthetic data and business simulations, directly in clouds, which leads to a significant acceleration of the creative and development process (Bubeck et al., 2023).

The formalization of this democratization of advanced AI is the idea of AI-as-a-Service (AIaaS). Vision, speech, language, and decision-making models and APIs provided by cloud providers such as Amazon Web Services (AWS), Google Cloud Platform (GCP), and Microsoft Azure can be pre-trained. It implies that a retail business can incorporate the advanced product recommendation systems into the e-commerce platform, or a healthcare facility can roll out the medical imaging analysis systems without having to hire a team of data scientists with PhDs. The cloud removes the complexity of the AI models, and powerful intelligence is a consumable service that can be integrated into the core functionality of any application, which is a giant force multiplier to the agility and innovation of organizations (Benlian, Klumpe, and Hinz, 2020).

SECTION III: THE TRANSFORMATION IN ACTION - USE CASES ACROSS BUSINESS FUNCTIONS

In practice, the theoretical synergy of cloud and AI is most compelling and it is transforming the fundamental functions of business. This is not the revolution of evolution; it is a radical redefining of the way work is performed, choices are made and value is created. In Product Development and R&D, the cloud is used to give the collaborative platform and computational power of complex simulations, and AI serves as a co-creator. To illustrate, in the car and aerospace sectors, AI in the form of generative design algorithms can be executed on cloud supercomputers to sample a given set of potential design solutions to a part, optimizing the weight, strength, and material usage under a set of constraints, which would otherwise take human engineers impossibly long (Shi, 2022). In addition, AI-powered analysis of user behavior data, gathered by cloud-based applications, offers new, real-time, granular feedback on what features are used and what are not used, enabling product teams to pivot and innovate faster than ever before, with a much closer fit to customers.

One of the most promising areas of AI-cloud synergy is probably the Operations and Supply Chain management. The cloud brings together ERP system, warehouse management system, GPS trackers, and IoT sensors data on the worldwide supply chain. This data is then fed through AI models to generate a simulation and analysis of an entire operation in real time to form a so-called digital twin the dynamic and virtual model (Tao et al., 2018). This allows logistics to be highly efficient with dynamic routing that re-calculates routes of fleets depending on live traffic, weather and demand data. It not only allows predictive maintenance as noted but also to demand forecasting that automatically initiates procurement and inventory replenishment producing a self-optimizing supply chain that reduces stockouts and excess inventory as well as responds resiliently to disruptions.

Another area that is on the limelight is the reinvention of Customer Experience (CX). CRM is a type of cloud-based Customer Relationship Management systems like Salesforce, which is used to store all interactions with customers. The 360-degree customer profiles and micro-segment audiences are made with AI algorithms overlaid on top of this data to provide hyper-personalized marketing campaigns, product recommendations, and promotion deals (Kumar et al., 2019). NLP-driven intelligent virtual assistants with a scalable cloud to support them in the field of providing instant, context-sensitive support will solve typical problems without human intervention and release human agents to address more complex and empathetic questions. This enhances a seamless, proactive, and intensely personalized customer experience, which generates loyalty and generates revenue.

Lastly, the workforce is also being severely affected by the transformation. AI is no longer a tool to automate routine jobs but a so-called co-pilot or a collaborator that complements the human intelligence. Software development has been dramatically reduced with the use of cloud-based systems such as GitHub Copilot, which recommends lines of code and complete functions to the developer. Knowledge work AI assistants can draw conclusions about long documents, write emails, investigate intricate spreadsheets, and create the first draft of reports, using the enormous amount of information stored in the corporate cloud (Dellermann, Calma, Lipusch, Weber, Weigel, and Ebel, 2019). In the talent management field, AI can scan thousands of resumes in cloud databases to find the most promising candidates and analyze video interviews to assess communication abilities and even recommend internal mobility opportunities to employees based on their skills and career goals. This addition enables human workers to concentrate on the higher-order work that involves creativity, strategic thinking, and emotional intelligence.

SECTION IV: THE HUMAN & ORGANIZATIONAL SHIFT

The technological upgrade of AI in the cloud-driven organization does not simply imply the technological change; it requires the radical human and organizational change. The implementation of the most advanced AI tools will bring inadequate outcomes, or even will not be utilized at all, unless the culture and capabilities of the workforce are oriented towards this new paradigm. There is the initial and the most urgent need to develop an AI-ready culture. This means changing the culture that is likely to be threatened by AI to one that sees it as an effective augmentation and empowerment instrument. The culture of this kind is based on the premises of data literacy, in which employees on all levels learn to comprehend data-driven insights and AI-generated suggestions (Davenport and Ronanki, 2018). It also involves the creation of a spirit of experimentation and psychological safety, whereby the teams feel empowered to experiment AI-based hypotheses and learn through failures without fear of reprimand. The key element being trust, the employees will have to trust the outputs of the AI, and this aspect will require the leadership to be transparent on how these systems will operate and their perceived benefits.

This cultural change should be backed up by an equal change in capabilities, which will result in the development of new roles and capabilities of the AI-enhanced organization. Although the number of data scientists and ML engineers will kecp increasing, the need in the number of translator-like occupations will skyrocket, i.e., specialists able to mediate between technical AI teams and business decision-makers (Shankar, 2018). Moreover, with the increased use of Generative AI, such skills as the art of writing good instructions to AI models will be useful in the majority of work processes. It will also be critical that positions that concentrate on the ethical aspects of AI emerge, including AI ethicists and fairness auditors, who will make sure that AI systems are used ethically. In the case of the rest of the workforce, constant learning and upskilling programs are necessary to train individuals to collaborate successfully with AI, focusing on the most human skills that machines cannot readily duplicate.

Finally, such transformation reinvents leadership and strategy. The function of a leader changes as the main decision-maker with his/her experience and intuition to the organizer of human-AI cooperation. The leaders should be capable of asking the right questions that AI will respond to, interpretations of its outputs in the wider business context, and judgment decisions that take into consideration data-driven insights and human values (Ransbotham, Kiron, Gerbert, and Reeves, 2017). Tactically, leadership should change focus on strategic planning of roadmaps of 3-5 years to a clear vision and direction, with adaptive and emergent strategy allowed by the fast, AI-based insights produced on the cloud. The company loses a strict hierarchy and more a flowing, feeling network, with intelligence decentralized and agility the management style.

SECTION V: IMPLICATIONS OF THE AI-CLOUD SYNERGY

There are deep issues and implications of the mighty collision between cloud and AI. An intelligent and prudent strategy involves having a sober perspective of the multi-faceted consequences that cut across managerial, practical, ethical and social aspects.

A. **Managerial Implications:** The AI-cloud era is a new strategic challenge to managers. Among the most acute ones are the talent war and skills gaps. The market is highly competitive towards people who have knowledge in AI, data science, and cloud architecture, which increases wages and retention is a challenge. This requires both an aggressive approach of external recruitment and a heavy investment on internal upskilling and reskilling initiatives to develop internal capability (Brynjolfsson and McAfee, 2017). In addition, there is a need to have new investment and ROI models. The value of AI initiatives may be in improved innovation, better customer satisfaction, or strategic positioning, which is more difficult to quantify, and justification of AI initiatives can often not depend on the traditional and linear calculations of ROI. The new metrics that managers need to come up with reveal the agility and strategic value created. The management of strategic partners is also complicated because the organizations are deeply intertwined with their cloud providers (hyperscalers). This poses threats of vendor lock-in, where the cost and complexity of switching to a different platform is prohibitive, which may limit the bargaining power and flexibility. Lastly, the organizational structure itself needs to change, sometimes necessitating the formation of cross-functional teams of data, AI, and business savvy, like MLOps teams, to handle the end-to-end lifecycle of AI products.
B. **Practical Implications:** At the practical level, organizations encounter great challenges as far as technical issues are concerned. Technical debt and MLOps is a significant challenge. In contrast to conventional software, AI models may degrade their performance as they continue to experience the real world (this is called model drift). To make AI systems accurate and reliable, it demands a strong MLOps practice, i.e., a collection of data and model versioning, contin-

uous monitoring, and automatic retraining, deployed on cloud systems (Sculley, Holt, Golovin, Davydov, Phillips, Ebner, and Dennison, 2015). The complexity of the integration between new AI services and the legacy on-premises systems and several cloud environments might pose a substantial operational overhead and data consistency challenge. In terms of security, the AI-cloud stack adds new attack surfaces. Opponents may also engage in data poisoning (adding malicious data to training data) or model stealing (asking a publicly available API to reverse-engineer a private model). As a result, security systems will have to change in order to incorporate AI-specific threat modeling and protection systems.

C. **Ethical Implications:** This has serious ethical implications of mass adoption of AI that should be strictly governed. The most popular issue is perhaps algorithmic bias and fairness. When an AI model is trained with historical information that is biased (e.g., in the employment sector, lending, policing etc.), it will learn, enhance, and scale those biases (Mehrabi, Morstatter, Saxena, Lerman, and Galstyan, 2021). The principle of fairness is both technical and ethical. The black box problem and the necessity of transparency and Explainable AI (XAI) are closely related. Most complicated ML models, especially deep learning networks, are opaque and it is hard to reason why a given decision was arrived at. Such inexplicability is an impediment to trust and a grave problem in traditional business like finance and health care. Privacy and consent are also essential; the large amounts of data required to train AI can include personal data, and it is not clear how this data is gathered and processed and anonymized in accordance with the law such as the GDPR. In the end, this gives rise to the issue of responsibility and regulation: who is to blame in an accident when an AI system that drives a self-driving car crashes or an AI hiring tool discriminates against someone; is it the developer, the user, or the company that deployed it? These questions require establishment of clear governance structures.

D. **Social Implications:** The future of the society will be determined by the long-term social implications of the AI-cloud synergy. The future of employment and job replacement is a point of much apprehension. Although AI will generate new employment opportunities, it will certainly automate most of the routine and some of the non-routine cognitive work, leaving a significant portion of the labor force jobless. This requires an educational reform and colossal reskilling efforts at the societal-level (World Economic Forum, 2023). Another threat is the further expansion of a digital divide. Giant companies with huge resources are able to spend billions on cloud and AI, which could lead to the formation of AI monopolies, and faster economic concentration, but other small businesses and developing countries would face challenges in following suit. The emergence of Generative AI increases the issue of trust and misinformation in society. When demonstrated on a massive scale, the power to produce highly realistic and yet entirely synthetic text, audio, and video (deepfakes) jeopardizes the trust in the media, institutions, and even the discourse, and is a fundamental problem to the democratic process (Chesney and Citron, 2019). Lastly, we cannot disregard the environmental impact. Large AI models are also very costly to train, consuming a lot of computational power, which creates a big carbon footprint through the data centers that are energy intensive. The quest of the technological agility should be coexistent with the desire to adhere to sustainable and energy-efficient computer practices.

CONCLUSION: MAKING THE WAY AHEAD.

The expedition in the realm of cloud-based agility, which is propelled by AI, is both thrilling and terrifying in its potential of transformation. This chapter has suggested that cloud computing offers the necessary, scalable, and flexible infrastructure, the contemporary nervous system of organizations, which is required to respond quickly to change. Nonetheless, it is Artificial Intelligence that performs the catalytic role of the brain which infuses this nervous system with the ability to predict, optimize and generate, thus providing a new level of organizational agility and intelligence. This collaboration between the two forms a positive feedback loop of information, understanding and execution that is transforming all business operations, including R&D and operations, as well as customer experience and workforce management.

However, what is inevitable is that this technological promise is intertwined with deep human and organizational issues. Technology alone is not a sure way of succeeding in this new paradigm. It demands a conscious development of an AI-receptive culture, a strategic commitment to new skills, and a re-invention of leadership in a new era of human-AI partnership. Most importantly, it requires a sober and active involvement of the far-reaching consequences of this power. The above-mentioned managerial, practical, ethical, and social considerations are not peripheral to the responsible and sustainable usage of such technologies. The end-point should be responsible innovation - using the potential of the cloud and AI not just to make a profit or to be efficient, but to build organizations that are stronger, more sensitive to human needs, and more equal in their influence. The agile future state is the one of constant learning and change, in which the organizations become intelligent systems that use the best of human and artificial intelligence to navigate the world which is becoming more and more complex. The way ahead is not a mere technical change of direction, but an overall organizational change that is both able and conscience.

REFERENCES

Benlian, A., Klumpe, J., & Hinz, O. (2020). Mitigating the intrusive effects of smart home assistants by using anthropomorphic design features: A multimethod investigation. *Information Systems Journal*, *30*(6), 1010–1042.

Brynjolfsson, E., & McAfee, A. (2017). *Machine, platform, crowd: Harnessing our digital future*. W. W. Norton & Company.

Bubeck, S., Chandrasekaran, V., Eldan, R., Gehrke, J., Horvitz, E., Kamar, E., . . . Zhang, Y. (2023). Sparks of artificial general intelligence: Early experiments with GPT-4. *arXiv preprint arXiv:2303.12712*.

Chesney, R., & Citron, D. (2019). Deep fakes: A looming challenge for privacy, democracy, and national security. *Lawfare Research Paper Series, 1*(1), 1-25.

Cito, J., Leitner, P., Fritz, T., & Gall, H. C. (2017). The making of cloud applications: An empirical study on software development for the cloud. In *Proceedings of the 2017 11th Joint Meeting on Foundations of Software Engineering* (pp. 728-739).

Davenport, T. H., & Ronanki, R. (2018). Artificial intelligence for the real world. *Harvard Business Review*, *96*(1), 108–116.

Dellermann, D., Calma, A., Lipusch, N., Weber, T., Weigel, S., & Ebel, P. (2019). The future of human-AI collaboration: A taxonomy of design knowledge for hybrid intelligence systems. In *Proceedings of the 52nd Hawaii International Conference on System Sciences*.

Ebert, C., Gallardo, G., Hernantes, J., & Serrano, N. (2016). DevOps. *IEEE Software*, *33*(3), 94–100.

Hirschberg, J., & Manning, C. D. (2015). Advances in natural language processing. *Science*, *349*(6245), 261–266. PMID: 26185244

Huang, M. H., & Rust, R. T. (2021). A strategic framework for artificial intelligence in marketing. *Journal of the Academy of Marketing Science*, *49*(1), 30–50.

Iansiti, M., & Lakhani, K. R. (2020). *Competing in the age of AI: Strategy and leadership when algorithms and networks run the world*. Harvard Business Review Press.

Kumar, V., Rajan, B., Venkatesan, R., & Lecinski, J. (2019). Understanding the role of artificial intelligence in personalized engagement marketing. *California Management Review*, *61*(4), 135–155.

Márquez, J. (2020). *Cloud Computing: Theory and Practice*. Morgan Kaufmann.

Marston, S., Li, Z., Bandyopadhyay, S., Zhang, J., & Ghalsasi, A. (2011). Cloud computing—The business perspective. *Decision Support Systems, 51*(1), 176–189.

Mehrabi, N., Morstatter, F., Saxena, N., Lerman, K., & Galstyan, A. (2021). A survey on bias and fairness in machine learning. *ACM Computing Surveys, 54*(6), 1–35.

Provost, F., & Fawcett, T. (2013). Data science and its relationship to big data and data-driven decision making. *Big Data, 1*(1), 51–59. PMID: 27447038

Ransbotham, S., Kiron, D., Gerbert, P., & Reeves, M. (2017). Reshaping business with artificial intelligence: Closing the gap between ambition and action. *MIT Sloan Management Review, 59*(1), 1–17.

Sculley, D., Holt, G., Golovin, D., Davydov, E., Phillips, T., Ebner, D., . . . Dennison, D. (2015). Hidden technical debt in machine learning systems. In *Advances in Neural Information Processing Systems* (pp. 2503-2511).

Shankar, V. (2018). How artificial intelligence is transforming marketing. *GfK Marketing Intelligence Review, 10*(2), 34–39.

Shi, Z. (2022). Generative Design: A Paradigm Shift in the Design Process. *Proceedings of the 2022 International Conference on Engineering Design.*

Tallon, P. P., Queiroz, M., Coltman, T., & Sharma, R. (2019). Information technology and the search for organizational agility: A systematic review with future research possibilities. *The Journal of Strategic Information Systems, 28*(2), 218–237.

Tao, F., Qi, Q., Liu, A., & Kusiak, A. (2018). Data-driven smart manufacturing. *Journal of Manufacturing Systems, 48*, 157–169.

World Economic Forum. (2023). *Future of Jobs Report 2023*. World Economic Forum.

Chapter 5

AI Ethics in People Analytics:

Balancing Automation, Privacy, and Fairness

Thi Minh Ngoc Luu
https://orcid.org/0000-0002-5972-7752
International School, Vietnam National University, Hanoi, Vietnam

Damith Gangodawilage
https://orcid.org/0000-0001-7968-6284
Qasim Ibrahim School of Business, Villa College, Maldives

Anugamini Priya Srivastava
https://orcid.org/0000-0003-0617-2711
Dhyaan Prasad Global University, Pune, India

Jeayaram Subramanian
JGBS, O.P. Jindal Global University, India

ABSTRACT

The integration of Artificial Intelligence (AI) into people analytics presents a transformative opportunity for human resource management, yet it introduces critical ethical challenges. This chapter examines the central tension in balancing the efficiency of automation with the imperatives of employee privacy and algorithmic fairness. It argues that navigating this triad is a strategic necessity, not merely a technical concern. The analysis moves from foundational ethical principles to an operational framework for implementation, emphasizing governance, transparency, and human oversight. Finally, the chapter explores the profound social, practical,

DOI: 10.4018/979-8-3373-7442-0.ch005

ethical, and managerial implications of AI-driven HR, concluding that a proactive, principled approach is essential to foster trust, mitigate risk, and build more humane and equitable workplaces.

CHAPTER INTRODUCTION

The adoption of the Artificial Intelligence (AI) in human resource management and specifically in the field of people analytics is a paradigm shift that has the potential to transform the modern workplace. Advocates also argue that AI-controlled applications have the potential to unlock new possibilities of organizational productivity, improve objective decision-making, and tailor the experience of employees (Koch, 2021). The pledge is that of a data-based utopia where attrition is anticipated before it takes place, talent is perfectly aligned with locations, and biases are eliminated in the employment procedures. But fast adoption has emerged with a set of ethical issues that are very serious and complicated. The algorithms designed to discover high performers may unintentionally reproduce and even enhance historical biases in the society thus turning the HR systems into the engines of inequality (Raghavan et al., 2020). Automation creates the danger of dehumanizing processes that require human intervention and turning people who are complex entities into data points. This is the main tension that is directly addressed in this chapter, and the author examines a complex and often perilous phenomenon of balancing the overwhelming appeal of automation with the sacred rights to privacy and the primary moral requirement of justice. We believe that it is not only a technical issue of navigating this triad, but a key strategic, ethical and leadership requirement of contemporary organizations. To wade us through this exploration, the chapter is organized in the following way: we introduce the underlining principles of people analytics and the ethics on which this practice should be built; we take an in-depth look at the triad of tension, exploring the promises and dangers of automation, the sanctity of privacy, and the multi-faceted exploration of fairness; we then translate these principles into the practical operational plan that we will put into operation; and we then open our perspectives to observe the deep social, practical, ethical, and managerial implications of this technological change.

PART I: THE PRINCIPLE OF AI ETHICS IN PEOPLE ANALYTICS.

1.1. What Is People Analytics? (A Brief Refresher)

People analytics, or HR analytics, or talent analytics, is the methodology of managing human capital in organizations, based on data. This practice is not new but its development has been transformational. It is no longer single-level descriptive reporting, answering the question what happened? to primitive turnover rates and time-to-hire to the sophisticated predictive and prescriptive analytics. Predictive analytics uses machine learning and statistical models to make future predictions, namely which employees are highly likely to leave the company, or what candidates will be successful (Marler and ⁻Boudreau, 2017). Prescriptive analytics goes a step beyond, recommending an intervention that can change the results predicted; an example is being able to recommend a certain training program or managerial practice to keep a valuable employee. People analytics have broad and wide areas of application. In hiring, AI-enabled applications filter resumes, conduct video interviews and evaluate paralinguistic and linguistic features, as well as rank candidates. In performance management, project completion, email metadata, and collaboration patterns are analyzed using algorithms to give them a so-called productivity score. Natural language processing (NLP) also measures sentiment in employee engagement through the open-ended survey or internal communication tool. In addition, people analytics is being used more to analyse skills gaps, plan the workforce and personalise learning and development paths (Davenport et al., 2019). Such an overwhelming evolution requires a solid ethical system, as the consequences of such data-driven decision-making include the careers, livelihoods, and health of people.

1.2. The Ethical Imperative: Why Moving Fast and Breaking Things Is Not an Effective Approach With People

Silicon Valley titanium rule that one would move very fast and break things is highly inappropriate to the people analytics sphere. The main explanation is the fact that the information at hand is unique. Employee data are one of the most secret types of personal data, not only the demographics, but also the intimate aspects of the professional, and, more and more, personal life: social networks, communication patterns, health and wellness indicators, and, finally, psychological inclinations (Mohlmann and ⁻ Zalmanson, 2017). Decisions that are informed by employee data, unlike consumer preference data, have immediate and direct impacts on the economic security of an individual, the trajectory of his/her career and self-worth. An imperfectly designed product recommendation algorithm can help miss a sale, whereas an imperfect algorithm evaluation in a people analytics system can help miss a promotion, a biased performance evaluation, or even cause wrongful termination, therefore causing direct harm to the individual and his or her family. The ethical imperative is also promoted by the high power distance between the employment relationship. Just because there is an informed consent of the data collection and use even does not mean that attaining true consent is impossible as the lack of consent can be perceived as non-cooperation or in fact cannot be structurally achieved when the employment is conditional upon complying (Martin & Freeman, 2003). Organizational risks are also as risky as they are in the moral failure and far deeper in the legal, financial and reputational losses. Algorithms related lawsuits, regulatory penalties in laws like the GDPR, and reputation damage through public outcry can strangle the name of an organization and its capacity to recruit the best talent. Therefore, a proactive, principled, and cautious stance is not only morally right but also is one of the elements of risk management that is prudent.

1.3. Basic Ethical Principles: An Evaluative Framework

In order to negotiate this complicated terrain, organizations need to implement a definite ethical system. A number of overlapping collections of principles have been put forward but they generally lead to a set of values that can be used to help guide the creation and implementation of AI in people analytics (Jobin et al., 2019). These principles are based on bioethics, philosophy and offer a strong basis. Beneficence refers to the imperative to do good, i.e., AI systems must be programmed with the key objective of benefiting both the organization and its personnel, i.e. by making job matches or improving employee welfare. Its adjacency, non-maleficence (do no harm), is a very important injunction to act proactively to recognize, reduce, and avoid possible harms, such as discrimination, infringement of privacy, and psychological stress. Autonomy is about the importance of the human agency and self-determination of the employees. This would in practice mean meaningful transparency, consent, and the right to appeal to algorithms decisions, which is necessary to make sure that employees are not passive members of a technological system but participants. Justice requires the distribution of the benefits and burdens of AI systems without favoring the groups of people that are protected and making sure that the groups are treated equally. This principle compels an open challenge of algorithmic bias and fairness. Lastly, explainability is a feasible requirement to make some of these principles work. It demands that there be comprehensibility of the decisions or recommendations made by AI systems to a human listener or reader, which simplifies accountability and makes sure that there is the right to challenge, as well as the trust that is necessary to be adopted by the organization (Adabi and Berrada, 2018). These principles combined make a critical checklist that every people analytics program must be measured against.

PART II: THE TRIAD OF TENSION: STRIKING A BALANCE BETWEEN THE CORE CHALLENGES.

2.1. The Lure of Automation: Productivity vs. Workforce Supervision

The initial reason behind the implementation of artificial intelligence in human resources is that it promises automation, which theoretically provides unemployment efficiency, scalability, and objectivity to personnel operations as such. This can also be achieved by having the automated systems perform repetitive and high-volume processes, such as resume screening, interview scheduling, and onboarding documentation, so that the HR practitioners can dedicate more time to more strategic and human-focused tasks, which include leadership development, cultural enhancement, and managing complicated employee relations (Tambe et al., 2019). Advocates argue that algorithms driven by data help reduce the amount of noise and subjectivity inherent in the decision-making process by humans, which can produce more consistent and fairly imposed results due to the same set of selection criteria being used.

Nevertheless, there are always significant risks that are related to the quest of efficiency. The key threat is an overdependence on opaque models by organizations, that is, complex algorithms, the internal decision logic of which is not comprehensible to the developers themselves. When managers or HR professionals are not capable of understanding the logic behind the results of the algorithm, this encourages the automation bias, which is a cognitive bias that is well-documented, to prefer automated systems to the use of human judgments despite the contextual evidence that may indicate the opposite (Cummings, 2004). Therefore, poor recommendations can be blindly embraced. In addition to this, the wholesale automation of HR practices poses a risk of dehumanizing the work environment, which transforms employees into a succession of data points and metrics that destroy trust, psychological safety, and the social fabric upon which the cohesion of organizations is built. Sidelining of managerial judgment, which is a privilege of integrating nuance, context and empathy, is a dangerous move.

The introduction to totally computer-controlled resume-screening mechanisms is a good illustration of such tension. Although operationally effective, these systems have proven to be disadvantageous to candidates who have non-traditional career paths, gaps in employment due to caring duties, or alumni of less competitive colleges or universities, thus perpetrating digital opportunity disadvantages. The best tradeoff here is therefore not to substitute human judgment but to improve it in terms of a human-in-the-loop system, where automation is used to handle the scale and to compute the volumes of data but people give the final evaluative and essential context and ethical evaluation.

2.2. In the Digital Workplace: The Sanctity of Privacy: Data Collection

Inviolability of personal privacy is a cornerstone that is deeply threatened by the informationality of AI in people analytics. Employee privacy is an ever more complex and disputed concept in the modern work environments that are digital in nature. Companies can now have the technological capability to accumulate an enormous amount of data that is not limited to conventional HR databases, but rather passive monitoring of email communications (e.g. response times, network routes), badge-in/out times, GPS tracking of company vehicles, keystroke patterns, and even the linguistic tone and mood of communications and video feeds (Ball, 2021).

The moral issue lies in the separation between, on the one hand, the legitimate business interests in the form of the need to ensure security, oversee productivity, and examine team collaboration, and on the other, inappropriate surveillance that creates an effect of panopticon, which in turn triggers the stress of employees and their self-censorship and the development of a culture of distrust among employees. Another fundamental tenet of data ethics is informed consent, though this is very controversial in employment-related situations. The natural power disparity can force the employees to consent to the extensive surveillance in threat of negative consequences, and thus their consent will be rather a coerced one than a voluntary one.

Ethical practice requires, therefore, compliance with such principles as data minimization, which implies gathering of data only that are strictly required by a certain, legitimate reason, and purpose limitation, which does not allow using data towards other, unrelated purposes without providing new consent (Mantelero, 2014).

There is a very interesting case study that demonstrates these issues: the employee well-being monitoring with the use of data. On the one hand, it is true that when an organization wants to recognize burnout by considering work patterns (e.g., emails received at late hours, the density of calendars), the practice can reach a very important ethical line. The process will turn what otherwise was a friendly effort into a repressive surveillance mechanism where employees perceived as vulnerable can be monitored intrusively where such surveillance can be used against them. Finally, privacy protection, therefore, is the creation of transparent, clear and consentual limits, which accept employees as individuals, not resource pools.

2.3. The Pursuit of Fairness: Reducing Bias and Securing Equity

Fairness is, perhaps, the most examined and technically complicated ethical AI people analytics challenge. The very idea of removing human bias in recruitment and personnel management is faced with the fact that the AI models are trained on the data that have been produced by the historically biased societies and organizations. Therefore, an algorithm that has been trained on historical hiring data of a company will learn and recreate the biases that exist in the input, including a tendency to favor a person of a particular gender, race, or education level (Barocas and Selbst, 2016).

Fairness is not a monolithic concept and researchers both in computer science and ethics have proposed a variety of fairness measures often mathematically incompatible with each other. An example is that the equality of opportunities can be considered by asking that the rates of selection should be the same among the demographic groups, but equality of opportunity would be the same true-positive rates, or the probability of the qualifying candidates receiving the selection, per group. All these conditions cannot usually be simultaneously met within the operational contexts, which has necessitated organizations to take explicit, value-based decisions, as to which definition of fairness to foreground.

The system can be biased in several points: the historical data layer, the model design (e.g., the target variable is being defined based on the subjective assessment of a manager, as in the case of defining high performer), and the analysis of model results. Technical strategies have been developed in order to avert these risks. Preprocessing methods purify and normalize training data to eliminate biases before training them through models. In-processing techniques include fairness constraints as part of the objective function of an algorithm. The outputs of the algorithm are modified with post-processing techniques to facilitate equal results of the predictions within groups (Mehrabi et al., 2021).

Technology however is not enough. The most important mitigation measure is organizational: different, multidisciplinary teams, such as data scientists, HR professionals, ethicists, and representatives of employees, should be involved in the development, testing, and validation of these systems. An illustrative case in point is the trial recruitment utility created by Amazon, which was discarded due to systematically lowering resumes with the word women in them (e.g., women captain of chess club), and punishing those who went to colleges with an exclusive female enrollment. The algorithm was trained on 10 years of postings, mostly of male applicants thus maintaining the gender bias at scale. This example emphasizes the fact that, without active oversight and active governments, AI technologies will not destroy bias, but will only automate it, exacerbating existing inequalities.

2.4. The Interaction: The Collision and Compromise of Automation, Privacy, and Fairness

The most important point to note is that the three pillars of automation, privacy, and fairness are not independent, and they are in a dynamic and at times tense relationship, where the promotion of one can worsen the other. This presents difficult trade-offs which leaders have to negotiate. As an example, the quest to achieve fairness tends to need additional information. The system must possess the knowledge of the demographic attributes of people to identify and rectify bias in a given demographic group. Nevertheless, gathering this type of sensitive information (e.g. about race, gender, disability) is in direct opposition to the principle of minimization of data and poses a serious privacy threat. The workers can be reasonably unwilling to give such information, and the legal regulations such as the GDPR leave strict requirements concerning the manner in which it is processed. On the other hand, unfairness can emerge by a strict following of privacy by minimizing data. When an organization is unwilling to gather any demographic information due to the privacy reasons, then it will be technically infeasible to audit the systems to identify cases of disparate impact and this will result in a blind spot of fairness. Likewise, the automation motivation may compromise privacy and fairness. The most efficient, the most opaque (black boxes) and the most data hungry systems are also the most opaque, posing a twofold risk of diminished explainability and greater surveillance. As an example, an automated video interview analysis system is the most automated system that necessitates a gatherance of highly sensitive biometric information (facial expressions, tone of voice) and its internal functionality is commonly proprietary and un-explainable, which makes it hard to audit to achieve fairness or do not infringe upon privacy. All these tensions cannot be solved in a simple formula. The way ahead consists of a candid and effective process in which these trade-offs are clearly recognized, discussed by a wide governance body, and reported to stakeholders. Achieving a perfect equilibrium is not the objective but rather a deliberate, justifiable, and morally rationalized balance that will in conformity with the fundamental values of the organization and its responsibility of care to its workforce.

PART III: PRINCIPLES TO PRACTICE: AN OPERATIONAL FRAMEWORK

3.1. Setting the Governance: The Purpose of an AI Ethics Board/Committee

To translate the ethical principles into the regular practice, it will be necessary to implement a formal governance. This would best be organized as an cross-functional AI Ethics Board/Committee. This must not be an organization that consists of technologists or HR experts only; the variety of point of view is its strength. An effective board would comprise personnel of the HR leadership, to offer domain knowledge about people processes; Legal and Compliance, to offer a view of compliance with relevant regulations and laws; Data Science and IT, to comprehend the technical viability and limitation of the models; Ethics or Corporate Social Responsibility, to offer a philosophical and values-based approach; and most importantly, Employee Representatives or members of worker councils/unions, to have the employee voice in the center-stage (Field et al., 2020). The mandate of this committee is versatile. It has the responsibility of creating, establishing, and upholding the ethical charter of people analytics of the organization. It acts as an inspecting agency of high risk projects prior to their implementation against the laid down principles. It also serves as an appeal board to employees who want to appeal an algorithmic decision to offer an important accountability and redress mechanism. In so institutionalizing ethics, organizations can go beyond ad-hoc and post-hoc justifications, and entrench ethical consideration into the lifecycle of adopting technology.

3.2. The Ethical Artificial Intelligence Lifecycle: Design, Development, and Deployment

Ethical people analytics is not a single audit, but an ongoing process that is incorporated in the lifecycle of the AI system. Such lifecycle can be subtracted into three major sections with their ethical checkpoints. The most poignant one is The Design Phase which is aimed at establishing the ethical direction. It starts with Ethical Risk Assessment, which is an official procedure to determine the possible harms in terms of fairness, privacy, and human autonomy prior to writing even a single line of code. It is also this stage where the purpose of the project should be defined and made conciliated with human values, and the operational metrics of fairness that will be employed in this context (e.g. we will provide demographic parity on this hiring tool) should be explicitly defined. Technical mitigation takes place at the Development Phase. It will include strict and continuous bias testing contrary to the established measures of fairness, such as a fairness-conscious machine learning.

It also presupposes the emphasis on Explainable AI (XAI), which implies using such methods as LIME, SHAP to generate explainable interpretations of the model results, which would make its reasoning understandable to the HR professionals and managers that will utilize it (Adabi and Berrada, 2018). Deployment Phase is concerned with responsible implementation. This includes setting up the human-in-the-loop protocols that clarify the roles division of labor between the algorithm and the human decision-maker, and when a recommendation is advisory or binding. It also involves in-depth change and training of managers and HRBPs on how to interpret and utilize the insights of the tool in an ethical rather than a deterministic manner. Lastly, the system has to be constantly monitored in the production since biases may have changed over the years and its effect on the real-life results is to be monitored in order to make sure the system is actually meeting its ethical and business commitments.

3.3. Openness and Interaction: Networking With Workers

Ethical people analytics are all about trust, which can be earned on a condition of radical transparency and outreach communication almost purely. The existence of secrecy leads to suspicion and fear, which is fatal to any analytics initiative. Organizations then have to dedicate themselves to transparent, open, and continuous communication regarding how they use AI. One of the effective means to this effect is to develop an "Algorithms Bill of Rights" to employees. This paper would explicitly state the rights that employees would have in an algorithmic workplace, including the right to be informed that an algorithm is being applied in a decision that involves them, the right to a meaningful explanation of how the decision is made, the right to challenge and to appeal algorithmic decisions, and the right to the presence of a human in a decision that will greatly impact the employee (Kim and Schessler, 2022). In addition to this, organizations are expected to offer simple language records on their vital people analytics systems. Some basic questions should be answered in this documentation: What data is being collected? How is it being used? What is the top-level reasoning of the model (e.g. The model finds skills gaps based on training completion records and project tags, not by spying on your personal emails). Who will be responsible to the system? And how can an employee be able to get more information or make an appeal? Demystifying the technology and informing the employees about the facts and rights can help the organization to change the model of surveillance and control to the model of partnership and empowerment, in which analytics is regarded as the tool of mutual benefit, and not a tool of managerial control.

PART IV: IMPLICATIONS AND THE PATH FORWARD

4.1. Implications

The mass implementation of AI in people analytics has such in-depth consequences that linguistically surpass the HR department, transforming the social, practical, ethical, and managerial landscape of companies.

Social Implications: On a macro level, the aspect of algorithmic management can radically change the psychological contract (unwritten set of expectations between an employer and an employee). The old contract with its loyalty and long-term security already fades away, but AI threatens to transform it into the contract of optimized performance in the form of measurable and quantifiable performance criteria, turning a human value into an optimized output. This may result in a kind of digital version of Taylorism where work cannot be relentlessly measured and standardized through algorithms, creativity, spontaneity, and informal social interactions are essential to innovation and culture are stifled (Kellogg et al., 2020). The surveillance of the workplace has the potential to foster the culture of anxiety and self-censorship, in which employees are always ready to change their behavior to better fit the algorithm and to ultimately undermine well-being and genuine co-operation. Another potential effect of these technologies is the emergence of a new digital divide among the workforce, with the jobs that are not complex enough to be automated and those that scan every action and streamline it, potentially worsening the social and economic inequality existing in the workplace.

Practical Implications: At the practical level, the implementation of ethical AI requires a major change of organizational abilities and resources. The HR function is in urgent need of new competencies. HR professionals should become data literate with a high level of awareness of ethics to inform its usage. At the same time, data scientists in this field should be trained in ethics and sociology in order to see the human aspect of the task. Such upskilling needs massive training and development. Also, establishing a strong ethical governance does not come free. It involves monetary investment on auditing tools, time dedication of the members of the ethics board and even longer time to roll out new technologies since they are subjected to ethical scrutiny. The HR technology vendor environment will also change, where a rising cost will be placed on vendors who can offer an ethics-by-design experience, including integrated bias audits, explainability interfaces, and effective data privacy as these will be core points of differentiation in the market.

Ethical Implications: Ethical implications leave a profound philosophical reckoning on the part of organizational leaders. They bring issues as to whether the different actors are morally responsible or not. Who should finally be held responsible in case an AI system harms somebody: the data scientist who designed the model, the HR leader who ordered one, the vendor who sold it, or the CEO who approved its usage? This is the main ethical issue in the form of a responsibility gap. In addition, there exists an inherent conflict between utilitarian ethics (maintenance of the highest good to the highest number, e.g. the efficiency of the overall hiring process) and deontological ethics (e.g. the right to a fair process and privacy that every person has with a company, although this may hinder hiring). People analytics systems tend to be run on a utilitarian ideology, yet the utilitarian philosophy can be used to justify serious damages to minority groups or individuals. The leaders will thus have to wrestle with such conflicting and undefined ethical actions and establish where their organization draws red lines on such, knowing that certain values such as human dignity are indeed non-negotiable even in the sake of efficiency.

Managerial Implications: To the people managers, the emergence of algorithmic management is a total redefinition of their business. Data-driven systems are supplementing the traditional model of the manager as the main assessor and the lead guide of work and are replacing it in certain situations. This will involve the need to move away to being a boss to being a coach that is backed by data. The managers should also be taught to visualize the AI-based insights not as a set of truth but as a foundation that should be supplemented with their personal contextual knowledge, empathy, and leadership intuition. This makes them play a role of interpreting the why behind an algorithmic recommendation, offer the human context the data does not have and engage employees in hard conversations, based on that information. This involves a lot of emotional intelligence and critical thinking. In addition, there should be clear accountability structures where the managers are made aware that they are the ones who make the final decisions with regard to people in their teams and not the algorithm. The algorithm is a means; the manager is the leader.

4.2. The Future of Ethical Person Analytics

People analytics is in a critical crossroads. Two major external forces that will have a significant impact on the path it takes are regulation and professionalization. The AI Act of the European Union is an unprecedented law that will develop the HR management systems into a category of high-risk AI, which is exposed to stringent demands of risk assessment, data management, transparency, and human control (European Commission, 2021). The wave of regulation will most probably have an international effect, as it will compel organizations across the world to raise the standards of their ethics. At the same time, we will witness the emergence of the third-party AI auditing market, where independent companies will evaluate the systems of people analytics on bias, fairness, and compliance, like the financial auditors studying company books. This will be a much needed outside control of corporate behavior. The final destination should not be a transformation in compliance, which is reactive, i.e. doing the bare minimum to escape legal issues but proactive ethics, where ethical considerations are used as a competitive edge and they become part of a trust culture that will help attract and retain the best talent. Companies that lead in this regard will eliminate risk as well as create more resilient, innovative, and humane working environments.

CHAPTER CONCLUSION

To sum up, AI in people analytics is an opportunity of significant depth to improve decision-making and staff experience in an organization. Nevertheless, this change is closely connected to the high ethical risks that focus on the trio of automation, privacy, and fairness. As this chapter has pointed out, these conflicting demands cannot be resolved through a technical method of data scientists because it is a strategic, ethical, and leadership issue that should be addressed through a participatory, organisation-wide movement. It requires the underpinning of fundamental ethical values, functional working structure based upon sound governance and clear communication, and the comprehensive knowledge of the over-arching social, practical, ethical and managerial repercussions. The way ahead is not to renounce technology, but to use it wisely, carefully, with an undying dedication to human values. In this way, organizations can obtain the power of data and automation not to manage and downsize their staff, but to empower, develop, and dignify each worker, which eventually results in more efficient working environments, but also in more just and human ones.

REFERENCES

Adadi, A., & Berrada, M. (2018). Peeking inside the black-box: A survey on explainable artificial intelligence (XAI). *IEEE Access : Practical Innovations, Open Solutions, 6*, 52138–52160.

Ball, K. (2021). Electronic monitoring and surveillance in the workplace: Literature review and policy recommendations. *Journal of Business Ethics, 170*(4), 609–625.

Barocas, S., & Selbst, A. D. (2016). Big data's disparate impact. *California Law Review, 104*, 671–732.

Cummings, M. L. (2004). Automation bias in intelligent time critical decision support systems. *American Institute of Aeronautics and Astronautics*, 1-6.

Davenport, T. H., Harris, J., & Shapiro, J. (2019). Competing on talent analytics. *Harvard Business Review, 88*(10), 52–58.

European Commission. (2021). *Proposal for a regulation of the European Parliament and of the Council laying down harmonised rules on artificial intelligence (Artificial Intelligence Act).* COM(2021) 206 final.

Field, J., Achten, N., Hilligoss, H., Nagy, A., & Srikumar, M. (2020). Principled artificial intelligence: A map of ethical and rights-based approaches. *Berkman Klein Center for Internet & Society Research Publication*, 2020-1.

Jobin, A., Ienca, M., & Vayena, E. (2019). The global landscape of AI ethics guidelines. *Nature Machine Intelligence, 1*(9), 389–399.

Kellogg, K. C., Valentine, M. A., & Christin, A. (2020). Algorithms and the future of work: A research agenda. In *The Future of Work* (pp. 1–40). Emerald Publishing Limited.

Kim, P. T., & Schessler, M. (2022). The algorithmic bill of rights: A framework for governing automated employment decisions. *Yale Law & Policy Review, 40*(2), 345–410.

Koch, M. (2021). The digital transformation of HR: A review of the literature. *German Journal of Human Resource Management, 35*(1), 3–28.

Mantelero, A. (2014). The future of consumer data protection in the EU: Re-thinking the "notice and consent" paradigm in the new era of predictive analytics. *Computer Law & Security Review, 30*(6), 643–660.

Marler, J. H., & Boudreau, J. W. (2017). An evidence-based review of HR Analytics. *International Journal of Human Resource Management*, *28*(1), 3–26.

Martin, K., & Freeman, R. E. (2003). Some problems with employee monitoring. *Journal of Business Ethics*, *43*(4), 353–361.

Mehrabi, N., Morstatter, F., Saxena, N., Lerman, K., & Galstyan, A. (2021). A survey on bias and fairness in machine learning. *ACM Computing Surveys*, *54*(6), 1–35.

Möhlmann, M., & Zalmanson, L. (2017). Hands on the wheel: Navigating algorithmic management and Uber drivers' autonomy. In *Proceedings of the International Conference on Information Systems (ICIS)*.

Raghavan, M., Barocas, S., Kleinberg, J., & Levy, K. (2020). Mitigating bias in algorithmic hiring: Evaluating claims and practices. In *Proceedings of the 2020 Conference on Fairness, Accountability, and Transparency* (pp. 469-481).

Tambe, P., Cappelli, P., & Yakubovich, V. (2019). Artificial intelligence in human resources management: Challenges and a path forward. *California Management Review*, *61*(4), 15–42.

Chapter 6

AI-Augmented Leadership:
Redefining Managerial Decision-Making in the Cloud Era

Damith Gangodawilage
https://orcid.org/0000-0001-7968-6284
Qasim Ibrahim School of Business, Villa College, Maldives

Anugamini Priya Srivastava
https://orcid.org/0000-0003-0617-2711
Dnyaan Prasad Global University, Pune, India

Jeayaram Subramanian
JGBS, O.P. Jindal Global University, India

Thi Minh Ngoc Luu
https://orcid.org/0000-0002-5972-7752
International School, Vietnam National University, Hanoi, Vietnam

ABSTRACT

The convergence of artificial intelligence (AI) and cloud computing is fundamentally redefining managerial decision-making. This chapter explores the paradigm of AI-augmented leadership, where human intuition is synergistically combined with algorithmic intelligence to navigate an increasingly complex business environment. It argues that the leader's role is evolving from a solitary decider to a designer of intelligent decision-making processes. The analysis deconstructs the foundational triad of AI, cloud infrastructure, and the modern leader's evolving skillset. It then presents a practical framework for augmented decision-making, spanning enhanced situational awareness, generative optioneering, and data-informed judgment. The

DOI: 10.4018/979-8-3373-7442-0.ch006

chapter culminates in a critical examination of the profound managerial, ethical, social, and practical implications, asserting that while AI provides powerful capabilities, the ultimate responsibility for wise and ethical judgment remains a uniquely human imperative.

INTRODUCTION

The modern business environment has been defined by its speed and complexity that makes the traditional management decision making process more and more ineffective. Leaders become overwhelmed with information, unstable market trends and are forced to make strategic decisions quickly with long-term implications. Consider a top management ten years ago when a top executive had depended on quarterly reports and gut instinct to sanction a significant capital investment. Compare that with a modern leader who prior to the same decision can use artificial intelligence (AI) to model the performance of the investment in hundreds of economic regimes around the globe, real-time vulnerabilities of the supply chain, and competitor responses all doable using the scalable computing resources in the cloud (Kumar, Dagar, et al., 2025; Nguyen et al., 2025). This paradigm shift is not regarding the obsolescence of the manager, rather the transformation of the role into something more powerful the AI-augmented leader. The thesis of this chapter is that the essential role of leadership in the 21st century is being redefined fundamentally in the form of a symbiotic relationship between human intuition and artificial intelligence, which in its operational form is made operationally possible by the ubiquitous architecture of cloud computing. The AI augmentation is not just about automation of routine but an increase in human cognition, the potential of the leader to have more insightful information, a wider range of perspectives and a clearer vision of what is possible (Davenport and Ronanki, 2018). The cloud is the engine of this revolution, making access to strong computational resources available to everyone where previously it was the preserve of tech giants. The first step in this chapter will be to dismantle the triad of AI, cloud and modern leadership. It will then outline a sensible model of AI-enhanced decision-making, and then conclude by critical analysis of the deep-seated managerial, ethical, social, and pragmatic implications that this new epoch will force us to face. The path forward is not technological, but rather a re-evaluation of the very way wisdom is developed and used at the top of our organizations.

PART I: THE TRIAD OF THE FOUNDATIONS

1.1. Deconstructing "AI-Augmentation"

There should be a breakout of the simplistic and dystopian discourse about machines replacing human managers to get to AI-augmented leadership. Augmentation is inspired by the historical tradition of human-computer interaction where the technology is not used to replace human abilities, but to enhance them instead. In this regard, AI augmentation means the strategic implementation of artificial intelligence to assist the leaders with their thinking and decision-making. It is a collaboration in which machines do the computationally complex task of data processing, pattern recognition, and predictive modeling, therefore allowing human leaders to concentrate on more complex tasks, including strategic framing, ethical consideration, emotional intelligence, and creative synthesis (Raisch and Krakowski, 2021)(Kumar, Nain, et al., 2025; Mohan Kumar, Yadav & Sahoo, 2025). It is a crucial difference with automation. Whereas automation considers the elimination of human labor in structured and repetitive processes, including invoice processing or the generation of standard reports, augmentation is concerned with working in partnership with intelligence to solve non-structured and complex problems. As a leader, this is making use of an array of AI capabilities. Natural Language Processing (NLP) can scan through every single customer review, employee survey, and news article to narrow down on the emerging trends and subtle sentiments that a human can hardly even grasp in large quantities. Predictive Analytics are able to predict the sales patterns, who is most likely to leave the company or how much the company will lose upon entering a new market, as well as do it with a level of accuracy and speed that is otherwise unfamiliar to traditional analysis. Generative AI, more recently, has become an effective optioneering technology, able to write the strategic plans, or come up with new product concepts, or various, in-depth scenarios of a potential business project (Shrestha et al., 2019). In the process of this, the leader, instead of being a passive consumer of products, is in-the-loop and an active adjudicator. The leader offers the context that is critical, challenges the assumptions behind the AI models, uses his ethical and cultural judgment, and makes the final decision. The AI gives the map and the potential paths, the leader is the one guiding and determining the direction to go, and avoiding unexpected challenges.

1.2. The Cloud-computing Engine That Cannot Be Unplugged

Without the practical force of cloud computing, the theoretical potential of AI-augmented leadership would be mostly ineffective. The cloud is the uncompromising technological substrate, which allows implementing AI in management in large scale and massively. Until the cloud age, the implementation of advanced AI involved huge initial capital expenditure in on-premise computing devices, dedicated IT personnel, and long implementation times, in effect making it only accessible to the largest companies of all. This access has been democratized by the cloud because of its basic model of on-demand and pay-as-you-go computing services (Mell and Grance, 2011)(Kumar et al., 2023; Kumar et al., 2025). To the AI-augmented leader, this will be a number of vital benefits. One, it gives virtually unlimited as well as elastic scalability. Thousands of servers can be spinned by a leader in a few hours to perform a complicated simulation of a corporation merger and spun down, only paying the compute time. This flexibility can not be achieved with on-premise, fixed infrastructure. Secondly, the big three cloud providers, Amazon Web Services (AWS), Microsoft Azure, and Google Cloud Platform (GCP), have an ecosystem of managed and pre-built AI and machine learning services. The manager does not require a PhD in data science to utilize such tools as Azure Machine Learning, Google Vertex AI, or Amazon SageMaker, they all have high-level interfaces to build, train, and deploy models, which makes the entry barrier considerably low. Also, the cloud plays a key role in dismantling the infamous data silos that have been a bane to big organizations. The cloud allows the consolidation of dissimilar data sources, including CRM and ERP systems, IoT sensor and social media feeds, into one cognizant source of truth through a single, centralized, scalable data lake or data warehouse. The lifeblood of an effective AI is this data democratization without big, diverse, and clean datasets, AI models are unreliable and biased. The cloud is not a utility but a strategic enabler, which enables leaders to stop making decisions based on fragmented information and do so based on a holistic and data-rich perspective on their whole operation (Iansiti and Lakhani, 2020).

1.3. The Changing Image of the Leader of the 21st Century

The emergence of AI and the cloud makes it necessary to evolve the very identity of the leader and its skill set as well. The command and control manager who valued having a monopoly on information and made top down decisions is becoming a relic. The AI-enhanced organization has a floating information and broadly available information, and value is generated by being agile and working together. That is why the role of the leader is no longer to be a single, independent person who decides, but to be a designer of decision-making processes and a coach to empower his or her team (Knight, 2023). The emerging portrait of the leader requires a resettlement of the core competencies. Although technical competence is not falling out of favor, it is no longer as important today compared to data literacy, the capacity to understand, interpret and critically challenge data and algorithm outputs. To a certain extent, a leader should be capable of probing an AI system: What data did this model learn?

What are the possible biases of it? What are the confidence intervals of this prediction? It is just as dangerous to blindly accept the recommendation of an AI as to ignore it. Critical thinking is of utmost importance in conjunction with data literacy. The leader is the bullshit detector of the AI age, and he/she can detect spurious correlations, causality, and when an output of a model fails to comply with business reality, or human values. Also, systems thinking is crucial to identify the second and third-order impacts of decisions in a complicated, unified business world. Nevertheless, in the world where algorithms are becoming a dominant force, the so-called soft skills are being re-assessed radically. Such qualities as empathy, moral reasoning, communication, cultural intelligence become the finally distinguishing factors of human leadership (Bersin, 2020). These are functions that the AI with all its might can not convincingly imitate. The AI has the ability to process the engagement surveys but the human leader will only be able to engage in a sympathetic discussion that can be used to regain confidence. A human being can make a decision about the more complex ethical environment of a complex strategic decision that an AI can identify, only after they have flagged a potentially unethical transaction based on rules. Therefore, an AI-enhanced leader is a digital-era professional: he or she is digital savvy, analytically keen and deeply human.

PART II: THE AI-ENRICHED DECISION-MAKING FRAMEWORK

2.1. Phase 1: Increased Situational Awareness

The initial stage of AI-enhanced decision-making process shifts the leader into the situation of increased situational awarcness that changes the external and internal image of the organizational setting. On the outside, leaders have been used to using market research reports, which tend to be backward-oriented and slow to generate. Environmental scanning instruments based on AI can now scan and track the information ecosystem around the globe in real-time. These systems can scan news stories, social media feeds, patent applications, and government data to inform about the emergence of new rivals, the existence of minor swings in consumer behavior, regulatory changes, and send early alerts of geopolitical risks that might interfere with supply chains using Natural Language Processing and machine learning (Kumar et al., 2019). This gives a dynamic, alive map of the competitive environment. On the inside, AI enhances diagnostic skills of a leader. Leaders can use AI to constantly analyze the data on operations instead of waiting until monthly financial reports or annual surveys of employees are created. Machine learning algorithms are capable of detecting the inefficiencies on the production lines, anticipating the machine breakdown, and making real-time adjustments in the logistics network. Regarding the human capital level, AI applications can conduct sentiment analysis on internal communication and feedback systems to understand organizational morale, uncover areas of the organization with a high concentration of toxic culture, or determine areas of the workforce with skills shortages. They will also be able to understand trends in data in order to see who the most high-performing employees are most likely to leave, giving them a chance to retaliate. A combination of both outside and inside intelligence forms an all-inclusive situational dashboard that is available to the leader. It is not any longer about the increased quantity of data but rather the correct insights, contextualized and formatted in such a manner as to signal out the most crucial signals within the noise. This stage radically alters the position of strategic decision-making process as reactive problem-solving to proactive opportunity and risks identification.

2.2. Phase 2: Generative Optioneering

The second stage is the creation and assessment of possible courses of action with an excellent grasp of the situation created. The conventional strategic planning usually involves a small number of choices that are created by a group of leaders according to their experience and the study of historical data. Such a method is necessarily constrained through those limitations of human cognitive bias and the bounds of past precedent. With the introduction of AI augmentation, and the development of generative AI and more sophisticated simulation, this step is transformed into a process of "generative optioneering." In this case, leaders will be able to provide AI systems with not only the task to analyze existing options, but also to develop completely new ones. An example of this is a generative AI model that will suggest ten new market entry strategies on a new product, considering the existing competitive analysis, consumer trends and the core competencies of the company. These are not a repetition of strategies that have been used previously but new combinations and ideas that cannot always be seen at first sight by the human team (Korinek, 2023). The cloud is a testing ground, once a variety of options is created, not only by humans but also by AI. The digital twin technology allows leaders to run millions of Monte Carlo simulations by creating a virtual copy of their business operations or their market. They may ask: what are the chances that Option A will be successful in case of a recession in six months? What happens to option B in case a major supplier goes dead? What would happen were one of our competitors to introduce a disruptive product? Such simulations offer a more nuanced view of risks and rewards and the decisions made in them are less of a deterministic guess of a program's success (this will work, 100% certainty of success) and more of a probabilistic assessment of the potential results (this has a 70 percent chance of reaching our target, 10 percent chance of major loss). This stage radically increases the range of strategic palette the leaders can use, since they can traverse the adjacent possible and test their strategy in a risk-free digital space, before investing in the real world.

2.3. Phase 3: Data-Inherent Judgment

The last, and most important, step is the step where the purely human aspect of leadership is concerned the exercise of judgment. AI may present mind-blowing information and provide a myriad of ideas, and still it will not be able to make the decision. This stage is the integration of an algorithmic output and human wisdom. The leader in this case is supposed to use what is known as augmented gut feel, in this concept intuition is not substituted but instead informed and augmented by data. The leader has to approach the suggestions of the AI critically. This is implied by evaluating the quality and possible biases of the training data, being aware of the constraints of the model, and also taking into account outside the data-setting of the company-specifics-the subtleties of corporate culture, the silent rules of a boardroom, the moral values of the organization, and the welfare of the community and employees (Brynjolfsson and McAfee, 2017). As an example, an AI can suggest closing a poorly performing plant because it is a financially best choice. As a leader, one has to balance this with the social consequences to the local community, the possible loss of institutional knowledge, brand reputation impact, and morale of the workforce left behind. They are multi-faceted, complicated value judgments, which algorithms have limited capabilities to make. The leader is the unifier, one that performs the quantitative accuracy of the AI and applies it to the qualitative, human-based methods. Then they are required to convey the final decision in a clear and empathetic manner with creating a buy-in and trust among people, particularly when the data-driven decision is a challenging one. This step confirms that under the AI-enhanced paradigm, a decision and its outcomes are not on the algorithm, but on the human leader who has the ethos and the will to do so.

PART III: THE IMPLICATIONS OF AI-AUGMENTED LEADERSHIP

3.1. Managerial Implications

The use of AI to become a part of managerial work has far-reaching consequences to the organization itself, the way leaders are made, and the way performance is assessed. One of the major problems is the wide-spread skill gap. Most managers who were in their incumbent positions ascended to the top through experience and intuition rather than through literacy to the data. This has created a pressing urgent requirement by organizations to perform massive upskilling and reskilling programs to empower their leadership corps to be able to work with AI. It is not just a matter of simple data analysis, but it involves the development of the experimental attitude, the willingness to live with uncertainty, and the mental flexibility to turn around depending on new, information-driven conclusions (Besson, 2019). Besides, performance measures and evaluation systems need to be redesigned. What is your evaluation of an AI-enhanced leader? It is not on the business results only, but also the level of their ability to use the tools of the AI and the quality of the questions they ask AI systems, and how a hybrid human-AI team can be managed. The new behaviors that bring success might not be interested in traditional metrics. It is also a shift that drives the transformation of the organizational structure. Hierarchical bureaucracies are too slow to react to the quick knowledge created by AI. The most flexible organizations are moving to more flatter, networked, or team-based, which enable decentralized decision-making so that the team nearer to the information can make a decision without having to stop and go through management approval (Boudreau and Rice, 2015). This requires a considerable concern on change management. The greatest obstacle is resistance by managers who feel threatened or deskilled by AI. The top needs to have a clear vision, open communication regarding the additionary (not replacement) role of AI, and the managers involved in the design and implementation of these new systems to develop ownership and diminish anxiety.

3.2. Ethical Implications

The strength of AI-enhanced decision-making is also accompanied by the danger of critical ethical traps that should be prevented by leaders. The most popular risk is algorithmic bias and fairness. The AI models will be trained on past data, and as long as it suggests society or company biases (e.g. in hiring, promotions, or lending), it will be learned, and even exacerbated by the AI model (O'Neil, 2016). The leader who relied on an AI to manage talent might give in to solutions that unfavorably discriminate against some of the demographics. This will require strict bias auditing and the introduction of machine-learned methods that are fairness-conscious. Almost equally important is the issue of explainability (XAI). Most strong AI models are black boxes, or it is hard or impossible to realize the way they got to a particular output. A leader should be accountable and can do this by being able to justify his or her decisions. When they are not able to comprehend the rationale underlying AI, they cannot be in control. XAI is a scientific area concerned with the generation of models that give human-understanding explanations of model outputs, which is essential to the establishment of trust and ethical governance (Adabi and Berrada, 2018). This is what brings to the table the aspect of accountability. Even with a complex chain of decisions that should include data engineers, model developers, and the leader, who is legally and morally to bear responsibility when an AI-augmented decision is harmful? The human-in-command principle should be respected and the ultimate responsibility should be given to a human leader, rather than the algorithm (European Commission High-Level Expert Group on AI, 2019). Lastly, implementing AI to monitor employees and carry out performance analytics is a grave issue as it poses privacy and surveillance challenges. On the one hand, data may be applied to improve safety and productivity, but on the other hand, it may form a panopticon that will depreciate trust and autonomy. Managers should develop clear moral rules regarding the gathering and using data, as they are not supposed to invade the privacy of their employees, but rather empower them, not dominate.

3.3. Social Implications

The extensive implementation of AI-enhanced leadership will cause the effect that will reach much further than the corporate walls, transforming the social fabric of work and society. The future of work and that of the manager is one of the central issues of concern. Although this chapter proposes augmentation, the fear that the middle-management position that is characterized by routine supervision and relaying of information would be simplified or abolished is justified and may cause polarization of the workforce. The high-skill positions (AI specialists, data strategists) and the low-skill service ones might rise, and the traditional middle-tier jobs will be squeezed. This requires a social discussion on lifelong learning and social safety nets. At the global level, the AI augmentation may become a contributor to the digital divide. Very big and rich corporations and developed economies would be able to spend a lot of money on cloud and AI infrastructure, which would result in a substantial competitive edge over smaller businesses and emerging economies, and consequently would cause even more concentration in the market and economic inequality (Zwitter, 2021). Maybe the most significant social consequence is in the human contact sphere. Leadership is not just a functional aspect of a transaction but it is about inspiring, mentoring, culture building, and a sense of purpose. The excessive use of data-based, algorithmic directions may result in a sterile, hyper-performative, soulless and empathetic management. Unless leaders step outside the dashboards and leave their interpersonal roles to AI-analysed measurements, they will run the risk of producing disengaged, alienated workforces. The AI-enhanced leader has the challenge of using technology to process the computational complexity to be able to purchase additional time and cognitive bandwidth to engage in the highly human activity of connecting, developing, and inspiring their teams.

3.4. Practical Implications

In addition to the strategic and ethical factors, executives have to contend with a multitude of operational issues when deploying an AI-enhanced strategy. The former is the justification of implementation costs, and it is showing a clear return on investment (ROI). The expenses are not insignificant as they include the fees of cloud services, the license of AI software, the specific talent, which needs to be hired or trained, and the cost of change management. Leaders need to develop a business case connecting AI augmentation, with real-life achievements such as shorter decision-making time, less risk, greater innovation, or retention rates. Another complicated challenge is the vendor selection and tool integration environment. There are numerous AI platforms and point solutions out there, and it is a daunting challenge to select the appropriate, interoperable combination of tools that fit the existing tech stack of the company and its strategic objectives. The wrong decision may result in vendor lock-in, data silos and wasted resources. This is directly connected with the problem of data governance and security. The quality of the AI model depends on the data it is fed on. The data governance strategies such as defining data ownership, data quality, and data access mechanisms must be in place to ensure success. Furthermore, the subsequent risks to cybersecurity include the storage and processing of large volumes of sensitive corporate and personal data in the cloud. Data breach may prove disastrous that demands a lot of resource allocation in the security mechanisms and adherence to emerging data security laws such as GDPR. Lastly, leaders would need to strategize on how to keep the system resilient. To a high extent of dependency on AI systems is a weakness. What would occur in case of the cloud service failure, model error produces a catastrophic result, or system is hacked? To make sure that the organization will not lose its ability to make independent decisions in case of technological failure, it is necessary to implement contingency plans, strong model monitoring, and comprehensible human override policies.

CONCLUSION

The exploration of the ideas of the AI-enhanced leadership creates a terrain of both huge opportunities and considerable threat. The coming together of artificial intelligence and cloud computing has permanently transformed a toolkit of the contemporary leader, shifting decision-making as an art, based on intuition, to a hybrid science, combining science with human wisdom. The initial triad of AI, cloud and developed leadership, followed by a more practical model of increased awareness, generative optioneering and information guided judgment, serves as the map on this new land. But, as the implication debate has shown, the technological change is the simple bit. The actual difficulty is the human factor: in the process of managing the organizational change, maintaining ethical standards, addressing the social disruption and overcoming the practical hindrances of the implementation. The main argument of this chapter is: AI is not going to displace leaders, but leaders that use AI in an efficient manner will surely take the place of those that do not. Human leadership, as such, is not weakened but strengthened. When the artificially intelligent era arrives, the human ability to reason morally, empathize, form a creative vision of the world, and act in a peculiarly courageous way is the final sources of competitive advantage. The challenge to the current and future leaders is to actively participate in the use of these technologies not in the capacity of passive consumers, but in the capacity of active, critical and ethical creators. They need to understand its language, challenge its assumptions, advocate responsible use of it, and most importantly to keep in mind that the end is not to produce smarter machines, but to make decisions that are more intelligent in the good of their organizations, their workers and the society in general. The cloud age gives the strength; it is the wisdom of humans that has to give the meaning.

REFERENCES

Adadi, A., & Berrada, M. (2018). Peeking inside the black-box: A survey on explainable artificial intelligence (XAI). *IEEE Access : Practical Innovations, Open Solutions, 6*, 52138–52160.

Bersin, J. (2020). *The rise of the empathetic leader: How the pandemic is changing HR and leadership forever*. Josh Bersin Academy.

Besson, M. (2019). *The Upskilling Crisis: How to Build a Future-Ready Workforce*. Gartner.

Boudreau, J. W., & Rice, S. (2015). *The new organization: Different by design*. McGraw Hill Professional.

Brynjolfsson, E., & McAfee, A. (2017). *Machine, platform, crowd: Harnessing our digital future*. W. W. Norton & Company.

Davenport, T. H., & Ronanki, R. (2018). Artificial intelligence for the real world. *Harvard Business Review, 96*(1), 108–116.

European Commission High-Level Expert Group on Artificial Intelligence. (2019). *Ethics guidelines for trustworthy AI*. European Commission.

Iansiti, M., & Lakhani, K. R. (2020). *Competing in the age of AI: Strategy and leadership when algorithms and networks run the world*. Harvard Business Review Press.

Knight, R. (2023). How AI is transforming the role of the manager. *MIT Sloan Management Review, 64*(2), 1–5.

Korinek, A. (2023). *Generative AI and the future of work*. Brookings Institution.

Kumar, M. Yadav, M., & Sahoo, A. (2025). The Importance of Collaboration Skills in Workforce Preparation From Textbooks to Teamwork. In Revitalizing Student Skills for Workforce Preparation (pp. 333–352). https://doi.org/DOI: 10.4018/979-8-3693-3856-8.ch011

Kumar, M., Dagar, M., Yadav, M., Choudhury, T., & Luu, T. M. N. (2025). Navigating the future: Challenges and opportunities in intelligent RPA implementation. In Intelligent Robotic Process Automation: Development, Vulnerability and Applications (pp. 393–420). IGI Global. DOI: 10.4018/979-8-3693-4365-4.ch015

Kumar, M., Nain, N., Dagar, M., Luu, T. M. N., & Kaur, K. (2025). Redefining work and outsourcing in the AI era: Challenges and opportunities. In *Global Work Arrangements and Outsourcing in the Age of AI*. IGI Global. DOI: 10.4018/979-8-3373-1270-5.ch021

Kumar, M., & Yadav, R., Sushma, & Yadav, M. (2025). Investigating the influence of digital and social media marketing on university brand equity: An in-depth literature review and bibliometric study. *Global Knowledge, Memory and Communication*, 1–26. DOI: 10.1108/GKMC-07-2024-0426

Kumar, M., Yadav, R., Suresh, A. S., Singh, R., Yadav, M., Balodi, A., Vihari, N. S., & Srivastava, A. P. (2023). Mapping AI-Driven Marketing: Review of Field and Directions for Future Research. 2023 2nd International Conference on Futuristic Technologies (INCOFT), 1–5. DOI: 10.1109/INCOFT60753.2023.10425169

Kumar, V., Rajan, B., Venkatesan, R., & Lecinski, J. (2019). Understanding the role of artificial intelligence in personalized engagement marketing. *California Management Review*, *61*(4), 135–155.

Mell, P., & Grance, T. (2011). The NIST definition of cloud computing. *National Institute of Standards and Technology*, *53*(6), 50.

Nguyen, T. B., Kumar, M., Sahoo, D. K., Nain, N., Yadav, M., & Dadhich, V. (2025). Mapping the landscape of women entrepreneurs in micro and small enterprises: Trends, themes, and insights through systematic review and text mining. F1000Research, 14. DOI: 10.12688/f1000research.163030.1

O'Neil, C. (2016). *Weapons of math destruction: How big data increases inequality and threatens democracy*. Crown Publishing Group.

Raisch, S., & Krakowski, S. (2021). Artificial intelligence and management: The automation–augmentation paradox. *Academy of Management Review*, *46*(1), 192–210.

Shrestha, Y. R., Ben-Menahem, S. M., & von Krogh, G. (2019). Organizational decision-making structures in the age of artificial intelligence. *California Management Review*, *61*(4), 66–83.

Zwitter, A. (2021). The rise of the digital divide: AI and the global south. Journal of Cyber Policy, 6(2), 147-162.

Chapter 7
AI-Driven Sustainability: Automating ESG Reporting and Green Operations through the Cloud

Mohan Kumar
https://orcid.org/0000-0002-5171-0892
Faculty of Commerce and Management, SGT University, Gurugram, India

Sandeep Kumar Singh
https://orcid.org/0000-0002-1741-7254
JGBS, O.P. Jindal Global University, Haryana, India

Xuan-Hoa Nghiem
https://orcid.org/0000-0003-2292-0257
International School, Vietnam National University, Hanoi, Vietnam

Mananage Shanika Hansini Rathnasiri
https://orcid.org/0000-0002-6496-5106
Faculty of Management Studies, Sabaragamuwa University of Sri Lanka, Sri Lanka & Qasim Ibrahim School of Business, Villa College, Maldives

ABSTRACT

The escalating demand for robust Environmental, Social, and Governance (ESG) reporting, coupled with the urgent need for sustainable operations, has exposed the severe limitations of traditional manual methods. This chapter explores the transformative synergy of Artificial Intelligence (AI) and Cloud Computing in automating and enhancing corporate sustainability. It demonstrates how this convergence automates the entire ESG reporting lifecycle-from intelligent data ingestion to audit-ready assurance-transforming it from a compliance burden into a strategic

DOI: 10.4018/979-8-3373-7442-0.ch007

asset. Furthermore, it details how AI-driven analytics enable "green operations" by optimizing energy use, building smart supply chains, and facilitating proactive maintenance. Finally, the chapter addresses critical implementation considerations, including a strategic roadmap and the ethical imperatives required to navigate this technological shift responsibly and effectively.

INTRODUCTION

The New Imperative for Intelligent Sustainability

With the advent of the 21st century, the issue of corporate responsibility has entered a new level of scrutiny. A company no longer rests on its financial performance, but is being measured based on how it affects the environment, connects with stakeholders, and the strength of its governance- three aspects that are becoming known as Environmental, Social, and Governance (ESG) criteria (Tunio & Tunio, 2025). This has been driven by a combination of regulatory pressure (e.g. the Corporate Sustainability Reporting Directive, CSRD, by the European Union), investor need to be told about their risks, and consumer desire to buy an ethical brand, which has propelled sustainability in the boardroom to the centre of the boardroom agenda (Eccles, Ioannou, and Serafeim, 2014). Nonetheless, this increased attention has shown the sheer deficiency of conventional and paper-based methods of ESG management. Organizations are drowning in the flood of non-financial information languishing in silo systems-energy bill and supply chain logistics, employee survey and social media feeds (Kumar, Nain, et al., 2025; Pankaj et al., 2023). Gathering, certifying and summarizing this information is known to be laborious, costly and time-consuming, regularly leading to reports that are already outdated by the time of publication and subject to the charge of being inaccurate or greenwashing (Kumar et al., 2023). The chapter assumes that the symbiotic integration of Artificial Intelligence (AI) and Cloud Computing is a paradigm shift, and it is one of the ways to address these issues. We shall show how this technological convergence is not an incremental enhancement but it is a reinvention of corporate sustainability. The Cloud and AI are turning the compliance-costly, reputation-threatening, and seemingly tedious process of ESG reporting into a driver of strategic knowledge, operational effectiveness, and physical competitive edge.

THE ESG DATA CHALLENGE: WHY MANUAL PROCESSES ARE NO LONGER ENOUGH

The inherent complexity and volume of data needed is the root cause of the barrier to effective ESG management. As opposed to financial accounting, which functions in a rather standardized and closed system, ESG represents a huge and dynamic set of metrics obtained both internally and externally to the organization. The initial complex layer is due to the spread of ESG approaches and criteria. Companies have to struggle through the maze of reporting standards, such as Global Reporting Initiative (GRI), Sustainability Accounting Standards Board (SASB), Task Force on Climate-related Financial Disclosures (TCFD), and the recently influential CSRD in Europe. The various metrics and stakeholders of each framework result in creating a fragmented environment where data have to be gathered, computed and displayed in various formats at the same time. This inconsistency leads to high levels of redundancy and chances of discrepancy between various reports (Berg, Kolbel, and Pavlova, 2022).

Adding to this problem is the amounts of data which exemplify contemporary ESG. Environmental data may cover Scope 1 direct emissions of company-owned buildings or Scope 2 indirect emissions of purchased electricity and the perennial hard-to-detect Scope 3 emissions of the entire value chain, including those of suppliers and customers. To collect such data, it is necessary to communicate with thousands of external operators and internal software like Enterprise Resource Planning (ERP), Human Resources Management Systems (HRMS), supply chain management tools, and Internet of Things (IoT) sensors that control the energy and water consumption (Kumar, Dagar, et al., 2025; Mohan Kumar, Yadav & Sahoo, 2025). A lot of this valuable information is in an unstructured format, stored in PDF documents, utility billings, supplier contracts, and even satellite imagery, so it cannot be accessed by the normal database query (Kar et al., 2023). The time and resources invested in searching, mining and certifying this data is enormous and as such, which is the reason why manual reporting is not economical. Businesses spend months organizing a single sustainability report annually and as a result, waste valuable resources on strategic projects on data wrangling to satisfy administrative requirements (Wong, 2020).

It is an unwieldy exercise which is bound to provoke a credibility gap. The time gap between the collection of data and publication may take several months making reports only historical records not management instruments. The manual structure of the work leads to a very high risk of human error, both in transposition errors, to a calculation error of the emission factors. Moreover, a company is liable to a significant reputational loss and legal costs in case of allegations of greenwashing, i.e. making unproven or deceptive claims concerning environmental performance without a verifiable and transparent audit trail between the original data source and the published figure (Lyon and Montgomery, 2015). In this regard, the conventional model is not only inefficient but also becoming untenable in the sense of risk management.

THE POWERFUL SYNERGY: AI MEETS THE CLOUD

The constraints of the manual ESG process present an apparent need to find a technological solution. Neither AI nor Cloud Computing is new in isolation, but when they are put together, they form a strong, synergistic ecosystem, and the one that is uniquely applicable to the ESG data challenge. The Cloud is the essential facilitating tool and the backbone, giving the framework on which smart solutions are developed. In its simplest definition, the Cloud provides a Data Lake in its purest form that can be expanded to any size; a large unstructured repository which can store structured and unstructured data in their original form. This establishes a point of unified truth in all information related to ESG, which destroys the data silos inherent in manual operations (Hashem et al., 2015)(Mohit Yadav & Vihari, 2021). The data can be pushed into the cloud data lake, whether it comes via an on-premise sensor, a spreadsheet provided by a third-party supplier, or a global news feed, and provide a single view, which has never been the case previously.

In addition, unlimited, on-demand compute power is available on the Cloud. The large processing capacity needed to train advanced AI models and deep learning networks in particular would be prohibitively costly and slow to serve on-premise. Computational resources Cloud computing platforms such as Amazon Web Services (AWS), Microsoft Azure, and Google Cloud Platform (GCP) enable an organization to launch thousands of servers to execute a computational task and then terminate them, only pay on usage. This scalability which is elastic allows advanced data analytics to be available to a significantly broader group of organizations. Lastly, the nature of Cloud architecture enables teams worldwide to collaborate in sustainability, operational personnel, and external auditors to access, analyse and contribute to the same datasets in any part of the world with the strength of security and permission controls.

Whereas the Cloud is the stage, Artificial Intelligence plays the role of a star which derives value out of the data. AI is not a technology but a set of advanced capabilities. Natural Language Processing (NLP) enables machines to read, comprehend and read meaning of human language. NLP algorithms can be used in an ESG setting to scan thousands of news stories, social media posts, and corporate reports and understand how the people feel or how a particular ESG risk has emerged or find a data point such as carbon emissions in the sustainability report of a vendor

(Kumar et al., 2021). Machine Learning (ML) is a branch of AI that entails algorithms, which get better automatically with experience. ML models may be trained using the historical data to forecast the future energy consumption, detect some unusual patterns that can signal the existence of wasteful operations, or optimize more sophisticated logistics network to reduce the use of fuel and carbon footprint. In the case of environmental monitoring, Computer Vision can be used to process satellite or drone images to identify deforestation, water levels in dams, or leaks of methane in pipelines, and offer objective, verifiable information on a large scale.

It is, however, the virtuous cycle that is developed as a result of their union. The Cloud will offer the scaly data storage and computing capabilities that AI models need to learn and function. AI, in turn, offers the intelligence to automate the processes, generate predictive intelligence and optimize systems, thus unlocking the strategic value that hides in the cloud-hosted data. This synergy will turn the cloud into an active and smart sustainability nerve center rather than a passive storage component.

AUTOMATING ESG REPORTING: FROM BURDEN TO STRATEGIC INSIGHT

The use of the AI-Cloud synergy on ESG reporting transforms an inherently retroactive compliance operation into a proactive strategic task. It starts with smart data discovery and consumption. Rather than physically searching the files by employees, AI-driven agents can be used to crawl specific internal and external sources in search. These intelligent bots are capable of accessing utility portals to retrieve electricity and water consumption information, searching procurement databases to retrieve a spend-based emission factor, and scan internal servers to retrieve germane environmental and social compliance documents (Nguyen et al., 2025; M. Yadav et al., 2025). Such automation saves significant time and money of the preliminary data collection process and the human specialists can concentrate on the analysis and strategy.

Automated framework mapping is a huge challenge that follows after the ingestion of data. Since there are many standards of reporting, one data point, such as water withdrawal, might have to be reported differently by GRI, SASB, and TCFD. The machine learning models can be provided with the training to comprehend the context and demands of each framework. They are capable of automatically tagging consumed data and associating a figure of water usage with the appropriate metric of all applicable standards at once. This does not only remove the manual, error-prone cross-referencing but also future-proofs the reporting process; with the changes of standards, the ML models can be re-trained to keep up, and the process of compliance with minimum human input will be maintained.

In addition to structured data, sentiment and risk analysis NLP offers a critical view of the "Social" and the Governance" element of ESG. NLP algorithms have the ability to constantly scan news and social media sites and websites that review employees such as Glassdoor. These models can identify the change in the mood of people, the initial signs of the labor conflict, or a new scandal involving corporate governance by examining the speech, its tone, and frequency of conversations (Moniz and de Jong, 2021). This will give the management real time and data-driven information on the social license of their company to operate and they can mitigate their risks proactively instead of being reactive to the crisis. As an example, an outburst in the number of negative posts on social media regarding supply chain labor methods can lead to an inner audit, before the problem can become a global scandal.

Lastly, this data and automated approach allows audit ready assurance. All data in the cloud platform can be marked with metadata that specifies its origin, date when ingested, and any transformation done on it. This will generate an unalterable and transparent data thread age and both internal auditors and external assurance providers can trace any figure included in the final sustainability report to its source in just a few clicks. This openness is the most formidable remedy to the greenwashing accusations that fosters confidence among investors, regulators and society, as it shows that the ESG performance is rooted in verifiable information, rather than marketing talk.

ENABLING GREEN OPERATIONS: OPTIMIZING FOR EFFICIENCY AND IMPACT

Although streamlined reporting is a major advantage, the more important asset of the AI-based sustainability is to actively enhance the environmental performance by applying the core operations optimization. A good example is predictive energy management. Machine learning systems are able to process historical data on energy consumption, weather prediction, production programs and real-time IoT sensor measurements on a building or factory. These models are capable of predicting demand of energy with high precision and automatically controlling heating, ventilation, and air conditioning (HVAC) systems, lighting and machinery operation at the lowest possible consumption during peak tariffs or when the carbon intensity of the grid peaks. It has been demonstrated that these types of AI-powered systems can decrease the energy expenses and related emissions by 10-20 per cent without raising the output of the systems (Zhou and Yang, 2016).

The smart supply chain and circularity is perhaps the most complicated of the frontiers. The Scope 3 emissions also demand an insight into the activities of hundreds or thousands of suppliers. This can be automated by AI, which can analyze the procurement data, supplier self-reported information, and external ESG ratings to generate a holistic carbon footprint of the supply chain. Also, ML algorithms can streamline the logistics networks by computing the variables such as traffic, weather, and fuel prices in order to identify the most efficient shipping routes, thus lowering Scope 1 transportation emissions. In addition to efficiency, AI can be used to achieve a switch to a circular economy. Recycling facilities can be enhanced with the help of computer vision systems, which will help to sort products better, whereas ML models can be utilized to examine product design and sales statistics and transform waste into value by offering remanufacturing, refurbishment, and waste reduction opportunities.

Proactive asset maintenance is another highly important use. In industries such as water management, energy distribution and manufacturing, the failure of important assets such as pumps, transformers or turbines is not only operationally disruptive but also harmful to the environment. ML models may be trained on sensor data (vibration, temperature, pressure) to identify minor abnormalities that lead to a failure. This enables maintenance to be planned at the appropriate time thus avoiding a disaster that causes a lot of wastage of resources, pollution and wastage of energy (Pankaj et al., 2023). This predictive method is much more superior than the traditional run to failure or set frequency maintenance schedules.

Lastly, AI is transforming green product design. As a type of AI, generative design algorithms allow engineers to produce best product designs according to the desired objectives, e.g. reducing material consumption or improving energy efficiency. In the meantime, Life Cycle Assessment (LCA) models that are based on ML can quickly simulate the environmental impact of various design options, material choices, and production methods at the concept stage. This enables the companies to design out carbon and waste even before a product is even produced and thus the concept of sustainability is entrenched at the bottom of the value chain.

IMPLEMENTATION ROADMAP: FROM STRATEGY TO DEPLOYMENT

Implementation of AI-based sustainability model is an important endeavor that must be planned and executed. The risk can be reduced with the help of a staged implementation plan, which will guarantee the consistency with business goals. Phase 1: Assess and Align is background. It starts with a materiality analysis to determine the most important ESG issues to the business and its stakeholders. This is to ensure that technological activities are concentrated in areas of maximum impact. At the same time, the executive buy-in is the most crucial thing, as such transformation needs investment, and even collaboration on an inter-organizational level. It is important to consider creating a cross-functional team that includes the representatives of the sustainability, IT, finance, and operations to address the knowledge gaps and be sure that the solution will address the needs of various users.

Phase 2: Develop Data Foundation entails making major technological decisions. The organization will have to choose the cloud platform (ex: AWS, Azure, GCP) depending on its unique abilities in AI/ML services, sustainability-focused tools (such as the Google Cloud Carbon Footprint tool or Microsoft Cloud for Sustainability) and already existing enterprise agreements. The essence of this stage is to design the ESG data architecture - the design of the data lake structure, data governance policies to guarantee quality and security, and connectors and pipelines to ingest data to the most important source systems. An effective groundwork is paramount; failing to get reliable information in a clean format, the best AI models cannot take too far.

Phase 3: Prioritize and Pilot: This one advocate a practical, use-case-based approach. Organizations should not go all out and roll out a pilot project but to choose a pilot project that is high impact, manageable. This may be the automation of the carbon accounting of the one manufacturing plant or the computer vision of waste separation at a given plant. This pilot is a pilot that can be used as a proof-of-concept, it will provide a quick win that will accelerate and it will give some valuable lessons. The first AI models can then be chosen and trained on this targeted data by the team and the strategy can be refined and improved in a controlled setting.

Phase 4: Scale and Integrate is about making the successful pilot go throughout the company. This implies linking additional sources of data, extending the models to new business units or geographical locations, and addressing even more complex uses cases such as Scope 3 emissions or predictive analytics. The eventual aim of this stage is not to remain in an isolated so-called sustainability system but to incorporate entirely AI-based wisdom into fundamental business activities and decision-making platforms. It is at this point that sustainability indeed becomes a strategic benefit, and makes capital allocation, product development, and risk management on the highest level of information.

NAVIGATING THE PITFALLS: ETHICS, GOVERNANCE, AND GREENWASHING

There is no AI-driven sustainability endeavor that can be undertaken without its associated risks and ethical concerns, which should be managed ahead of time. One of the major issues is the cost of carbon of compute. It takes a lot of energy to train large AI models, which can also be a contributing factor to the issue that the technology is intended to resolve. A conscientious practice will entail considering this footprint and making efforts to reduce it. These involve choosing cloud providers who are determined to run their data centers with 100 per cent renewable energy, applying more energy-efficient AI algorithms, and timely scheduling intensive training tasks when the carbon intensity of the grid is low (Strubell, Ganesh, and McCallum, 2020).

The next problem is the algorithmic bias in ESG. Artificial intelligence is trained on the information, and in case the information contains historical biases, it will be repeated by the models and may even be exaggerated. As an illustration, an AI reporter on the screening of socially responsible suppliers may be biased against companies in the developing world in case the training examples skew towards Western norms and behaviour. Equally, an NLP model that looks at employee sentiment may be predisposed to reject some dialects or forms of communication. To counter this, a variety of training data is required and ongoing observation of model output patterns of discrimination and the human component in important decision loops.

Privacy and security of data are also the most important. The centralized cloud data lake may be a high-value organization, but it also makes it an easy target of cyberattacks. This storage might be storing confidential operational information, staff details, and confidential supply chain information. The use of powerful cybersecurity enhancers, such as encryption, multi-factor authentication, and rigorous access control, is non-negotiable. Besides, gathering and utilizing data on employees or communities used in social measurements should adhere to privacy laws such as the GDPR and must be done in a transparent and ethical manner.

Lastly, organizations should keep an eye on preventing a new kind of the so-called green-hushing or techno-greenwashing. This is so when firms operate AI and sophisticated models as a black box to announce fantastic sustainability claims without delivering transparency to the data and methodology. In order to establish true trust, the companies should be transparent about the capabilities as well as the limitations of their AI systems. It includes open disclosure of the data sources, model assumptions, and uncertainty bounds of their AI-based sustainability measures, and through which technological development would always be accompanied by accountability.

CONCLUSION

The Future is Automated, Intelligent, and Sustainable

To sum up, the intersection of Artificial Intelligence and Cloud computing is a watershed point in terms of the sustainability of the corporation. The chapter has also claimed that this technological synergy provides the sole scalable avenue to address the paralyzing inefficiencies and risks of manual ESG processes. The implementation of AI and the Cloud will turn ESG into a costly and backward-thinking compliance process into an efficient and transparent strategic business ability. More fundamentally, because they facilitate predictive energy management, intelligent supply chains, and proactive maintenance of assets, they help organizations to take a proactive role in the reduction of their footprint, making sustainability targets a reality in the operational performance.

The scenery will keep changing at a very high rate. With the introduction of Generative AI, we will have a further automatized process of drafting sustainability stories and an unfolding of insights out of multifaceted data. The rise of the Internet of Things (IoT) will serve as an enormous and perpetually updated stream of data to drive AI models, offering increasingly more and more details on the use of resources and environmental impact. Moreover, the incorporation of blockchain technology may be an immutable registry of ESG information that offers an alternative degree of verifiability and trust.

The call to action is clear. Companies could not afford to consider sustainability and technology as two differing spheres. To make the shift towards AI-powered sustainability, one must begin now, have a plan, a roadmap, and invest in ethical execution. Those organizations that manage this power the most effectively will not only reduce risk and guarantee compliance but they will also create resiliency never seen before, unlock new efficiencies, and gain a potent competitive edge in an ever-increasingly demanding world that will place demands on demonstrating purpose as well as profit.

REFERENCES

Berg, F., Kölbel, J. F., & Pavlova, A. (2022). The ESG-returns relationship: A meta-analysis. *Journal of Financial Economics*, *144*(1), 1–32.

Eccles, R. G., Ioannou, I., & Serafeim, G. (2014). The impact of corporate sustainability on organizational processes and performance. *Management Science*, *60*(11), 2835–2857.

Hashem, I. A. T., Yaqoob, I., Anuar, N. B., Mokhtar, S., Gani, A., & Khan, S. U. (2015). The rise of "big data" on cloud computing: Review and open research issues. *Information Systems*, *47*, 98–115.

Kar, S., Yadav, M., & Panda, T. (2023). Inclusive organizational behaviour – the dynamic rules of building new workplaces. *VINE Journal of Information and Knowledge Management Systems*, *55*. Advance online publication. DOI: 10.1108/VJIKMS-05-2022-0155

Kumar, M. Yadav, M., & Sahoo, A. (2025). The Importance of Collaboration Skills in Workforce Preparation From Textbooks to Teamwork. In Revitalizing Student Skills for Workforce Preparation (pp. 333–352). DOI: 10.4018/979-8-3693-3856-8.ch011

Kumar, M., Dagar, M., Yadav, M., Choudhury, T., & Luu, T. M. N. (2025). Navigating the future: Challenges and opportunities in intelligent RPA implementation. In Intelligent Robotic Process Automation: Development, Vulnerability and Applications (pp. 393–420). IGI Global. DOI: 10.4018/979-8-3693-4365-4.ch015

Kumar, M., Nain, N., Dagar, M., Luu, T. M. N., & Kaur, K. (2025). Redefining work and outsourcing in the AI era: Challenges and opportunities. In *Global Work Arrangements and Outsourcing in the Age of AI*. IGI Global., DOI: 10.4018/979-8-3373-1270-5.ch021

Kumar, M., Yadav, R., Suresh, A. S., Singh, R., Yadav, M., Balodi, A., Vihari, N. S., & Srivastava, A. P. (2023). Mapping AI-Driven Marketing: Review of Field and Directions for Future Research. 2023 2nd International Conference on Futuristic Technologies, INCOFT 2023, March 2024, 1–5. DOI: 10.1109/INCOFT60753.2023.10425169

Kumar, S., Singh, S. K., & Sharma, R. (2021). A review of natural language processing in ESG. *Journal of Sustainable Finance & Investment*, •••, 1–18.

Lyon, T. P., & Montgomery, A. W. (2015). The means and end of greenwash. *Organization & Environment*, *28*(2), 223–249.

Moniz, A., & de Jong, M. (2021). The use of NLP for corporate social responsibility monitoring. In *Proceedings of the International Conference on Web Intelligence* (pp. 450-457).

Nguyen, T. B., Kumar, M., Sahoo, D. K., Nain, N., Yadav, M., & Dadhich, V. (2025). Mapping the landscape of women entrepreneurs in micro and small enterprises: Trends, themes, and insights through systematic review and text mining. F1000Research, 14. DOI: 10.12688/f1000research.163030.1

Pankaj, Yadav, R., Naim, I., Kumar, M., Misra, S., Dewasiri, N. J., Rathnasiri, M. S. H., Yadav, M., Balodi, A., Kar, S., & Vihari, N. S. (2023). An Exploratory Study to Identify Effects of Blockchain Technology on Organizational Change and Practices. 2023 IEEE Technology & Engineering Management Conference - Asia Pacific (TEMSCON-ASPAC), 1–8. DOI: 10.1109/TEMSCON-ASPAC59527.2023.10531372

Strubell, E., Ganesh, A., & McCallum, A. (2020). Energy and policy considerations for deep learning in NLP. *Proceedings of the 58th Annual Meeting of the Association for Computational Linguistics*, 3645–3653.

Tunio, M. N., & Tunio, S. (2025). *Navigating Digital Transformation: A Systematic Review of SMEs' Transitions and Growth in the Digital Age*. Sukkur IBA Journal of Management and Business., DOI: 10.30537/sijmb.v11i2.1509

Wong, J. (2020). The cost of ESG reporting. *Journal of Applied Corporate Finance, 32*(2), 1–8.

Yadav, M., Kumar, M., Alyousef, M. I., Dagar, M., Dewasiri, N. J., & Amoozegar, A. (2025). Mapping Research Trends on Gender Diversity, Work Stress, and Job Satisfaction in Higher Education: A Bibliometric Analysis. *International Journal of Learning. Teaching and Educational Research, 24*(8), 444–472. DOI: 10.26803/ijlter.24.8.19

Yadav, M., & Vihari, N. S. (2021). Employee Experience: Construct Clarification, Conceptualization and Validation of a New Scale. *FIIB Business Review, 12*(3), 328–342. DOI: 10.1177/23197145211012501

Zhou, K., & Yang, S. (2016). Understanding the power of smart energy management. *Renewable & Sustainable Energy Reviews, 57*, 1–10.

Chapter 8
AI-Powered Supply Chains:
Cloud-Native Intelligence for Resilient and Adaptive Operations

Mohan Kumar
https://orcid.org/0000-0002-5171-0892
Faculty of Commerce and Management, SGT University, Gurugram, India

Xuan-Hoa Nghiem
https://orcid.org/0000-0003-2292-0257
International School, Vietnam National University, Hanoi, Vietnam

Sandeep Kumar Singh
https://orcid.org/0000-0002-1741-7254
JGBS, O.P. Jindal Global University, Haryana, India

Mananage Shanika Hansini Rathnasiri
https://orcid.org/0000-0002-6496-5106
Faculty of Management Studies, Sabaragamuwa University of Sri Lanka, Sri Lanka & Qasim Ibrahim School of Business, Villa College, Maldives

ABSTRACT

Global supply chains face unprecedented volatility, rendering traditional, linear models obsolete. This chapter argues that resilience and adaptability are now critical competitive advantages, achievable through the integration of artificial intelligence (AI) and cloud-native architectures. It explores the evolution from legacy systems to cognitive networks, detailing how AI enables predictive demand sensing, autonomous planning, and proactive risk management. The cloud-native paradigm—built on microservices and containers—provides the essential scalable infrastructure

DOI: 10.4018/979-8-3373-7442-0.ch008

for these AI capabilities. The chapter further outlines the core technology stack, implementation challenges, and key performance indicators for success. Finally, it envisions the future of hyper-automation and the autonomous supply chain, concluding that this technological fusion is indispensable for building the intelligent, adaptive enterprise capable of thriving in a complex global landscape.

INTRODUCTION

The early 21st century has gone through a crucible of global supply chains, putting them to the test and revealing deep structural weaknesses. The COVID-19 crisis was not a one-off event but the eye-opener in an ever-growing list of disruptive events, such as geopolitical tensions, weather-driven by climate change, and chronic port congestions. These experiences have shown that the classical, linear, and sequential supply chain model, which is optimal at cost-effectiveness and lean inventories in a setting of stability, is inherently unfit to be applied in the world that is volatile, uncertain, complex, and ambiguous (VUCA). The pandemic as Ivanov and Dolgui (2020) suggest necessitated the shift in paradigm away from efficiency to supply chain viability, which is the capacity to sustain itself and endure extreme, prolonged disruptions. The legacy systems which are frequently monolithic Enterprise Resource Planning (ERP) and which are dependent on inactive, historic data sets establish a reactive stance. Moving forward, decision-makers find themselves examining the past to see what did go wrong, instead of looking ahead to what could go wrong (Ahuja et al., 2022; Mohit Yadav & Rangnekar, 2016). Such response delay is a very serious liability in a world where competitive advantage is decided in seconds and adaptability.

This chapter hypothesizes that resilience and adaptability have ceased to be a minor aspect of consideration but the key competitive points of distinction in contemporary commerce. The solution to these features is in the harmonious combination of the two transformational technological forces, which are Artificial Intelligence (AI) and cloud-native computing. With the ability to shift descriptive analytics to predictive foresight and prescriptive decision-making, AI offers the cognitive ability to do so. Nevertheless, artificial intelligence models are data and power hungry. It is at this point when the cloud-native paradigm will be critical. It offers the adaptable, scalable, and resilient infrastructure that is required to power and launch AI on an enterprise scale (Kumar, Rani, et al., 2025; Trivedi, Singh, et al., 2023; M. Yadav et al., 2024) Collectively, they facilitate a shift in a brittle, chain-like approach to a dynamic, networked one that has the ability to sense, interpret and respond to change in near real-time. The given transformation will be thoroughly explored in this chapter. We will start by following the development of the supply chain management and establishing the essential ideas of the AI-driven, cloud-based architecture. And then we will explore the actual uses of this technology in each core supply chain operation, look at the technology stack behind it, give a realistic roadmap of its implementation and finally look into the future to see what will become of the autonomous, self-healing supply chains of tomorrow.

THE FOUNDATIONS: FROM LINEAR CHAINS TO COGNITIVE NETWORKS

One has to be able to comprehend the development of AI-powered supply chains before fully realizing the revolutionary capacity it holds. It started with manual systems which were paper based with slowness in communication, high error rates and low visibility. Enterprise Resource Planning (ERP) systems which began to emerge in 1990s was a massive step in the right direction, combining the key business processes such as finance, human resources and inventory management into one unit. On the one hand, ERPs have digitalized the processes and minimized the data redundancy, but, on the other hand, new issues have emerged. Their monolithic architectures were also hard, costly to change, and often resulted in data being siloed within the ERP itself which could not be used in cross-functional analytics (Kumar, Dagar, et al., 2025; Pankaj et al., 2023). The following innovation had the Internet of Things (IoT) and the notion of the digital twin. IoT sensors in containers, cars, and shelves in the warehouse started producing the constant flow of real-time information about the place, temperature, vibration, and other essential characteristics. This enabled the establishment of a so-called digital thread, which is a comprehensive, data-driven perspective on the lifecycle of a product, which involves the raw material to the final client (Tao et al., 2018). This data could be used by a digital twin, a virtual dynamic representation of a physical supply chain, to simulate operations.

With these improvements, there was still a severe gap. Most organizations were in the data-rich and insights poor state. They received loads of data on ERPs, IoT, Transportation Management systems (TMS) and other sources, yet the tools to analyze it using sophisticated analytical tools were lacking. It was mostly a descriptive kind of reporting responding to the question What happened? but not answering the questions of Why did it happen? and What will happen next? or (this is the most important) What should I do about it? It is this loophole that AI-based, cloud-native supply chain is created to address.

The cloud-native is a term that has been misunderstood and taken to imply only moving the already existing applications to a cloud provider such as Amazon Web Services (AWS), Microsoft Azure or Google Cloud Platform (GCP). As a matter of fact, it is a basic architectural and cultural transformation (Tayal et al., 2019; Trivedi,

Vishnu, et al., 2023). Cloud-native development is a development and operational method that uses the distributed and scalable characteristics of the cloud computing model to their fullest. It is defined by a number of major principles. Originally, the applications are broken down into small, independent, and loosely coupled services called microservices. A microservice, such as a service in charge of computing carbon footprint or a service in charge of truck loading, is created, deployed, and scaled separately. The combination of containerization (such as Docker (a system that packages an application and its dependencies into a standardized unit) and orchestration systems (such as Kubernetes (a system that automates the deployment, scaling, and management of containers)) makes this possible (Villamizar et al., 2015). DevOps practices underpin this architecture by removing the walls between the development and operations teams so that they can deliver software faster, continuously and reliably.

And at the same time, an AI-driven system makes this agile infrastructure be cognitive. It uses a subdivision of AI called machine learning (ML) to discover intricate patterns in large datasets that are not perceived by human analysts or conventional statistical analysis. This makes it possible to move up the analytics maturity curve descriptive (what happened) to predictive (what is likely to happen) and finally to prescriptive (what is the best action to take). There is a very deep synergy between these two paradigms. The cloud-native infrastructure has the fuel necessary to run sophisticated AI models, namely, huge, consolidated sources of data, and the engine necessary and powerful namely, scalable compute resources. The cost and time of the large-scale AI would be prohibitive in many organizations without the cloud. On the contrary, in the absence of AI, the cloud is a data warehouse that cannot provide the insightfulness required to support adaptive operations. It is this strong marriage that makes a dead digital thread a living, cognitive network in a position to engage in autonomous learning and intelligent reaction.

THE CORE PILLARS OF INTELLIGENT OPERATIONS

The hypothetical advantages of AI and cloud-native architecture become actual with the help of the transformative influence of these technologies on the fundamental supply chain processes. These applications are the building block of an intelligent, resistant, and adaptive operational model, which has altered the way planners, logisticians, and strategists operate.

The former pillar, Predictive Demand and Inventory Intelligence, transforms the organization out of the shortcomings of time-series-based forecasting that relies only on the past experiences in sales. Conventional models are not used in responding to a once in a lifetime occurrence, and in the fine changes in market feeling (Kumar et al., 2023; M. Yadav et al., 2025). AI-based demand sensing, in its turn, takes into consideration a multitude of external and unstructured data signals to form a much more detailed and precise forecast. Social media patterns, weather predictions, localized events, scraped web competitor price adjustments, or even macroeconomic data can be analyzed by machine learning models to identify their initial demand changes (Fildes et al., 2019). An example is the fact that an unseasonally warm prediction in the area will cause automatic correction on the forecast of drinks and seasonal clothes. This probabilistic forecasting can be used to have a radical new approach to the management of inventory. The AI can calculate and suggest optimal inventory positions of each Stock-Keeping Unit (SKU) in every node of the network instead of fixed stock levels determined by historical variability. This system incessantly trades the cost of inventory as holding it with the likely risk of a stock-out, and in this way, it is able to maintain service level goals with reduced overall capital commitment and significantly low excess and obsolete inventory.

The second pillar, Autonomous Planning and Sourcing, transforms strategic decision making, a periodic, manual process, into an automated process. In the field of supplier management, AI allows advancing with risk assessment on a multi-level level. AI systems have the unlimited capacity to scan volumes of supplier data and sub-supplier data on suppliers instead of taking an annual audit. It involves tracking news feeds and satellite data of geopolitical instabilities or natural disasters, financial data of distress, and ESG (Environmental, Social, and Governance) performance against regulatory demands and corporate principles. Upon detecting a possible risk, the system does not simply put a flag on it but passes to prescriptive sourcing. It has the ability to automatically model the effects of supplier disruption and prescribe the most effective alternative sourcing plans, which presents a balanced scorecard to evaluate alternative sourcing methods by cost, lead time, quality history, and carbon footprint. This makes cognitive procurement, where AI-assisted so-called smart agents are able to automate the request-for-quotation (RFQ) process, that is, screening potential suppliers, negotiating standard terms, and even creating and evaluating contracts, and leaving human procurement professionals to deal with strategic relationships and high-level negotiation.

Smart Logistics and Fulfillment is the third pillar and it bring intelligence and automation to the actual flow of goods. The classical example of AI in real-time applications is dynamic route and mode optimization. The present-day AI-driven transportation management systems (TMS) are not based on fixed routes. They keep consuming real-time information about traffic, weather, port congestion and even carrier-specific performance indicators in order to dynamically reroute shipments away. This does not only save fuel and transportation expenses but also enhances delivery reliability and ETA accuracy which is a major force in terms of customer satisfaction. The integration of AI, IoT and robotics within the four walls of the warehouse is forming Lights-out distribution centers. The autonomous mobile robots (AMRs) will cooperate with human workers and carry shelves to picking stations, whereas computer vision detects the correctness of picks and reads barcodes. Automated quality inspection also applies to computer vision where damaged goods are identified without the need of human labor (Bogue, 2016). Last-mile delivery, which stands as the last frontier being transformed by AI, is optimising, complex, delivery routes of vehicles fleets and controllable integration of the emerging autonomous delivery technologies, including drones and sidewalk robots, into the fulfilment network.

The fourth, and, possibly, the most important pillar is Proactive Risk Management and Resilience. This is a paradigm change, whereby they respond to the disruptions rather than prevent them. Here the idea of the digital twin is involved. A supply chain digital twin is not a map, but a breathing-living simulation model of the whole end-to-end network. It uses AI and real-time data to enable planners to operate within a risk-free virtual environment and execute what-if scenarios (Ivanov et al., 2019). They are able to predict the effect of a hurricane striking a major port, a strike at a factory, or an emergency increase in tariffs, and monitor the rapidly spreading effects across the network. In tandem with this, there is AI-based anomaly detection. The unsupervised learning models are trained on the historical data of operations in order to learn about normal performance. Then they stream live data incessantly, shipping times, volumes of orders, sensors of machines and start automatically indicating the aberrations that might suggest that some issue is starting to emerge, e.g., a custom clearance delay, or a supplier beginning to be behind on their promises. This type of system then becomes prescriptive, in that it does not just report the anomaly but also provides a pre-vetted contingency plan, or even automatically executes a predefined response, like rerouting a shipment or automatically initiating an alternative sourcing request. This results in a supply chain that is not merely resilient, but indeed robust enough to take a shock and reconfigure itself to continue its operation.

THE TECHNOLOGY STACK: BUILDING BLOCKS OF INTELLIGENCE

The smart processes outlined in the preceding section are made possible by an intelligent, multi-layered technology stack. This stack is a radical departure, ground-up agile, scalable, and intelligent stack of the monolithic ERP systems of the past.

The bottom of this stack is The Data Foundation. The quality of data that an AI model is trained on is the only thing that can make the model good, and thus, a unified and high-quality data source is the non-negotiable precondition of success.

The new answer is a cloud-based data lake. A data lake can store huge amounts of raw, structured and unstructured data in their natural form as compared to traditional data warehouses where data is required to be structured, pre-processed and organized. This is necessary in order to consolidate the various streams of data of a contemporary supply chain: structured transactional information of ERPs and TMS; semi-structured data of IoT sensors and GPS trackers; and unstructured data of news articles, supplier contracts, and customer emails. The data lake eliminates the so-called organizational silos, consolidating this information into one accessible repository and forms a single source of truth, which can drive the entire analytics ecosystem. The data ingestion occurs continuously and automatically, through cloud-native tools to deliver data, as it is generated by source systems, to the lake, such that AI models can always be trained on the most recent information at hand.

On top of this data base is The AI/ML Toolkit which provides the analytical brain of the operation. This toolkit does not represent one technology but a set of specialised techniques. The workhorse is machine Learning, and the supervised learning algorithms are applied in the tasks of demand forecasting and demand classification (e.g., classifying a supplier as high-risk) and unsupervised learning in the tasks of clustering and anomaly detection (e.g., identifying unusual patterns in shipping lanes). Natural Language Processing (NLP) enables the comprehension and processing of human language by the systems. Within a supply chain setting, NLP is able to search through geopolitical or reputational risks of a supplier by scanning through thousands of news articles and social media postings in real-time. It is also capable of processing unstructured and complicated shipping documents, such as bills of lading and certificates of origin, and automatically harvesting major data points and removing manual data entry. Computer Vision allows machines to extract meaning on visual images. It is used to do automated quality inspection of production lines and it can detect defects faster and more accurately than a human being. Computer vision algorithms used in yard management have the potential to monitor video feeds to track truck and container arrivals and departures, automating the gate checks to optimally schedule the docks. Reinforcement Learning is also being applied on the frontier to train autonomous systems to make complex and dynamic decisions, like optimizing the transport of containers in a busy port terminal where the conditions continuously vary.

Planning and executing this intelligence is The Cloud-Native Platform. Organizational agility is in the microservices architecture. The supply chain platform is made up of dozens of autonomous microservices instead of a single application that is massive and where any change in one feature poses a risk of disrupting the whole system. It is possible to create, maintain, and scale a demand forecasting microservice, a carrier selection microservice, and a sustainability calculation microservice separately by teams of specialists. This architecture is one which develops an API-economy. Clear Application Programming Interfaces (APIs) serve as a contract between these microservices, and, most importantly, external partners. This makes it easy to integrate seamlessly and real-time with carrier systems to track, supplier portals to check the status of orders, and customer system to place orders, bringing about a real ecosystem. Lastly, in order to be able to control the end-to-end lifecycle of machine learning models, the field of MLOps is needed. MLOps is DevOps applied to ML, which provides a data preparation, model training, validation, deployment, monitoring, and retraining pipeline that is automated (Sato et al., 2019). This guarantees that deployed models do not experience model drift - a reduction in their performance with changing real-world data over time - and new and better models can be deployed quickly and dependably to complete the circle of continuous intelligence.

THE IMPLEMENTATION JOURNEY: FROM STRATEGY TO VALUE

AI-powered supply chains have enormous potential in terms of technology, yet achieving this potential must be a managed, strategic, and conscious implementation process. It is a plan to fail when one goes into a wholesale change without a road map.

The first stage is Developing a Roadmap. The initial one is an open evaluation of the existing supply chain maturity of the organization on such dimensions as data governance, technology infrastructure, process standardization, and workforce capabilities. This analysis points out the biggest gaps and opportunities. Instead of seeking a big bang transformation, the most effective strategies would find and implement high ROI and low complexity pilot projects. These short-term wins, e.g. the introduction of a predictive ETA system of inbound deliveries, or an AI-based tool to auto-audit and auto-pay freight invoices, are important in that they provide an actual value, a sense of trust in the innovation, as well as the impetus and the budget required to roll out bigger projects. Reframing of the business case is also critical. Although the cost cutting is one of the main results, the resilience (the decreased cost and effect of disruptions), adaptability (the possibility to respond quickly to changes in the market), and the improved level of customer service should become the primary reasons to invest in it, and all these are directly connected with the revenue protection and growth.

This process comes with a lot of challenges that have to be actively handled. The concept of garbage in, garbage out is especially applicable, as the main contributor to the failure of AI projects is poor quality of data. Thus an effective Data Quality and Governance program is not an IT project but one of the business projects where the data is correct, complete and uniform. The second and, on many occasions, a more challenging challenge is Talent and Culture. The entity will need to close the divide between the supply chain domain specialists and data scientists. This involves training the available employees to be data literate and adopting a culture of trusting and taking data-related insights into action, even when they do not align with the underlying wisdom or the gut feeling (Ransbotham et al., 2017). This is a direct connection with the Change Management. The workers will be afraid that AI will replace them, and therefore, it is necessary that the importance of AI being an augmentative technology, which performs repetitive analysis and liberates people to do more high-value strategic work, is explained. To achieve buy-in of experienced professionals who are used to legacy processes, it is necessary to demonstrate apparent value and engage them in the design process. Lastly, the problem of Ethical AI and Trust should be mentioned. It is possible that the existing biases may be reproduced or even increased by models (e.g. favoring particular suppliers due to non-performance factors). It is essential that organizations invest in Explainable AI (XAI) methods in order to create transparent and understandable model decisions to human users and establish the trust to be able to adopt it on a large scale.

Organisations need to transform their Key Performance Indicators (KPIs) to measure the effectiveness of the change. Classical indicators such as Perfect Order Rate and Inventory Turnover, Cash-to-Cash Cycle Time will always be significant to gauge operation efficiency. They should however be complemented by the next generation KPIs which are aimed at measuring resilience and adaptability. The new resiliency measures are Time to Recovery (TTR) the time taken by the supply chain to get back to normal operational levels following a disruption, and a Risk Exposure Index, which is the overall vulnerability of the network. The measure of adaptability is determined by Decision Velocity, or the speed of an AI-generated insight to a reality action, and Plan Adherence, or the capacity of an organization to stick to the action plans that are changing under volatility. Monitoring this balanced scorecard, leaders would be able to make sure that their transformation does not only provide efficiency, but also such strategic abilities as resilience and adaptability.

THE FUTURE HORIZON: WHAT'S NEXT?

The development of AI-driven supply chains is a continuous process, and various new trends are on the verge of transforming the boundaries of the possibilities in the next few years. The trend is towards more and more automation, intelligence, and alignment with larger societal objectives.

One obvious direction is the shift toward Hyper-Automation, i.e. the integration of several technologies - AI, Robotic Process Automation (RPA), and IoT - to automate processes on an end-to-end basis. Although nowadays discrete tasks are frequently automated, hyper-automation is aimed at building complete operational processes that are touchless. To give an example, a low inventory sensor in the IoT may send a purchase order to an AI, which, in turn, will generate a purchase order in the supplier portal, an RPA bot will take it and automatically make a purchase, and a smart contract will automatically run payment after delivery is verified, all without any human intervention. One such revolutionary technology, which has swept the globe, is that of Generative AI. In addition to the conversational skills of such models as ChatGPT, Generative AI can be used in supply chain management in the most significant way. It may drive natural language interfaces, enabling managers to just ask their supply chain system complicated questions in plain English, like, What are the top three things that affect my on-time delivery to Europe this quarter, and what mitigation measures do you recommend?

Synthetic data is another type of data produced with generative AI as an artificially generated data that mimics real-world statistics. It would be invaluable in the process of training AI models to deal with the rare but high-impact events known as black swans that have little to no real historical data (van Dis et al., 2023). Moreover, it can be used to automatize the production of complex reports, scenario analyses and even draft communications to suppliers and customers and can dramatically improve the administrative process. With the increasing severity of the climate crisis, the importance of AI in the creation of the Sustainable and Ethical Supply Chain will gain the first order. The AI-based optimization models will be developed to be multi-objective by balancing cost, service, time, and carbon emission to find the most sustainable course of action. This will prove key towards achieving corporate decarbonization objectives and adhering to changing regulations. AI will also play the key role of facilitating the circular economy of maximizing reverse logistics of returns and recycling, as well as through computer vision to automatically sort materials and remanufacture.

The final destination state in this evolution is The Autonomous Supply Chain. It is a self-organizing network which is capable of sensing, interpreting, making decisions and taking action with a minimum human intervention. It is a system which stands out to be continuously learning with reference to the surrounding environment and its own choices, which corrects and improves with time. Within this future condition, the supply chain is a resilient, adaptive, and autonomous service to the enterprise, able to not only anticipate the direction of the hurricane, but reroute inventory, reschedule production, and send alerts to the customers about the delay, even before the storm hits the shore. This is the ultimate level of maturity in supply chains where the network is itself sustainable as a source of competitive advantage.

CONCLUSION

The Adaptive Enterprise is an Intelligent Enterprise

To sum it up, supply chain management has turned the strategic focus irrevocably due to the volatile and interdependent world economy. The decades of the quest of leanness and cost efficiency, though are still significant, are no longer enough. The ability to recover and endure upsetting changes is called resilience and the ability to change direction quickly in reaction to opportunity and threat is called adaptability, which has become the keystone capabilities to business continuity and growth. These attributes, as explained in this chapter in detail, cannot be glued onto the old, linear, and siloed supply chain models. They need a complete re-architecturalization.

The combination of the strong integration of artificial intelligence and cloud-native computing becomes the catalyst of the paradigm shift. AI offers the mental abilities to shift the hindsight to the foresight to convert huge volumes of data into predictive and prescriptive intelligence. The cloud-native platform offers the agile, resilient and scalable infrastructure that is needed to roll out this intelligence immediately and at scale to a worldwide operation. This is much more than an IT upgrade; it is an essential strategic enabler that reinvents the nature of the way an enterprise function. It enables the supply chain to become dynamic and cognitive and value-creating as opposed to a cost centre and a source of risk. The call to business leaders is thus very emphatic and immediate. The decision to create the adaptive enterprise, which will be able to exist under the conditions of constant uncertainty, should be started. And that path is tightly coupled with the creation of an intelligent, cloud-native, supply chain in the center of it.

REFERENCES

Ahuja, J., Yadav, M., & Sergio, R. (2022). Green leadership and pro-environmental behaviour: A moderated mediation model with rewards, self-efficacy and training. *International Journal of Ethics and Systems*, *39*. Advance online publication. DOI: 10.1108/IJOES-02-2022-0041

Bogue, R. (2016). Growth in warehouse robotics: The success stories continue. Industrial Robot. *International Journal (Toronto, Ont.)*, *43*(6), 583–587. DOI: 10.1108/IR-07-2016-0184

Fildes, R., Ma, S., & Kolassa, S. (2019). Retail forecasting: Research and practice. *International Journal of Forecasting*, *35*(1), 1–22. DOI: 10.1016/j.ijforecast.2018.08.004 PMID: 36217499

Ivanov, D., & Dolgui, A. (2020). Viability of intertwined supply networks: Extending the supply chain resilience angles towards survivability. A position paper motivated by COVID-19 outbreak. *International Journal of Production Research*, *58*(10), 2904–2915. DOI: 10.1080/00207543.2020.1750727

Ivanov, D., Dolgui, A., & Sokolov, B. (2019). The impact of digital technology and Industry 4.0 on the ripple effect and supply chain risk analytics. *International Journal of Production Research*, *57*(3), 829–846. DOI: 10.1080/00207543.2018.1488086

Kumar, M., Dagar, M., Yadav, M., Choudhury, T., & Luu, T. M. N. (2025). Navigating the future: Challenges and opportunities in intelligent RPA implementation. In Intelligent Robotic Process Automation: Development, Vulnerability and Applications (pp. 393–420). IGI Global. https://doi.org/DOI: 10.4018/979-8-3693-4365-4.ch015

Kumar, M., Rani, P., Mangla, N., Nguyen, N. T., & Kaur, K. (2025). Sustainable outsourcing in an AI-driven world: Ethical, economic, and environmental perspectives. In *Global Work Arrangements and Outsourcing in the Age of AI*. IGI Global., DOI: 10.4018/979-8-3373-1270-5.ch001

Kumar, M., Yadav, R., Suresh, A. S., Singh, R., Yadav, M., Balodi, A., Vihari, N. S., & Srivastava, A. P. (2023). Mapping AI-Driven Marketing: Review of Field and Directions for Future Research. 2023 2nd International Conference on Futuristic Technologies (INCOFT), 1–5. https://doi.org/DOI: 10.1109/INCOFT60753.2023.10425169

Pankaj, Yadav, R., Naim, I., Kumar, M., Misra, S., Dewasiri, N. J., Rathnasiri, M. S. H., Yadav, M., Balodi, A., Kar, S., & Vihari, N. S. (2023). An Exploratory Study to Identify Effects of Blockchain Technology on Organizational Change and Practices. 2023 IEEE Technology & Engineering Management Conference - Asia Pacific (TEMSCON-ASPAC), 1–8. https://doi.org/DOI: 10.1109/TEMSCON-ASPAC59527.2023.10531372

Ransbotham, S., Kiron, D., Gerbert, P., & Reeves, M. (2017). Reshaping business with artificial intelligence: Closing the gap between ambition and action. *MIT Sloan Management Review*, *59*(1), 1–17.

Sato, D., Wider, A., & Windheuser, C. (2019). Continuous delivery for machine learning. Martin Fowler. https://martinfowler.com/articles/cd4ml.html

Tao, F., Zhang, M., & Nee, A. Y. C. (2018). *Digital twin driven smart manufacturing*. Academic Press.

Tayal, R., Upadhyay, R., Yadav, M., & Singh, R. (2019). The Development and Validation of the Organizational Effectiveness Scale Using Confirmatory Factor Analysis. *Global Business Review*, *22*, 097215091881172. DOI: 10.1177/0972150918811720

Trivedi, S., Singh, A., Yadav, M., Kumar, A., & Srivastava, P. (2023). Intention to Use Social Media in Online and Blended Learning: Potential Predictors and Mediation of Collectivism. IEEE Transactions on Engineering Management, PP, 1–14. https://doi.org/DOI: 10.1109/TEM.2023.3257422

Trivedi, S., Vishnu, S., Singh, A., & Yadav, M. (2023). Research Trends in Sustainable E-Payment Systems: A Study Using Topic Modeling Approach. IEEE Transactions on Engineering Management, PP, 1–15. https://doi.org/DOI: 10.1109/TEM.2023.3280216

Van Dis, E. A., Bollen, J., Zuidema, W., van Rooij, R., & Bockting, C. L. (2023). ChatGPT: Five priorities for research. *Nature*, *614*(7947), 224–226. DOI: 10.1038/d41586-023-00288-7 PMID: 36737653

Villamizar, M., Garces, O., Castro, H., Verano, M., Casallas, R., Gil, S., & Salamanca, L. (2015). Evaluating the monolithic and the microservice architecture pattern to deploy web applications in the cloud. In 2015 10th Computing Colombian Conference (10CCC) (pp. 583–590). IEEE. https://doi.org/DOI: 10.1109/ColumbianCC.2015.7333476

Yadav, M., Kumar, M., Alyousef, M. I., Dagar, M., Dewasiri, N. J., & Amoozegar, A. (2025). Mapping Research Trends on Gender Diversity, Work Stress, and Job Satisfaction in Higher Education: A Bibliometric Analysis. International Journal of Learning. *Teaching and Educational Research*, *24*(8), 444–472. DOI: 10.26803/ijlter.24.8.19

Yadav, M., Mittal, S., Kumar, M., Sahoo, A., Jayarathne, P. G. S. A., Yadav, M., Mittal, S., Kumar, M., Sahoo, A., & Jayarathne, P. G. S. A. (2024). From Textbooks to Teamwork: The Importance of Collaboration Skills in Workforce Preparation. In *Revitalizing Student Skills for Workforce Preparation* (pp. 333–352). IGI Global Scientific Publishing. https://doi.org/DOI: 10.4018/979-8-3693-3856-8.ch011

Yadav, M., & Rangnekar, S. (2016). Role Clarity and Organizational Citizenship Behaviour: Does Tenure Matter? A Study on Indian Power Sector. *Global Business Review*, *17*(3, suppl), 207S–224S. DOI: 10.1177/0972150916631216

Chapter 9
Cross-Cultural Management in AI-Augmented Workplaces: Opportunities and Challenges

Himanshi Rajora
https://orcid.org/0000-0001-5686-3030
JGBS, O.P. Jindal Global University, India

Dilshad Jeffery
Qasim Ibrahim School of Business, Villa College, Maldives

Azadeh Amoozegar
https://orcid.org/0000-0003-4988-5113
INTI International University, Malaysia

ABSTRACT

The contemporary global business environment is being simultaneously reshaped by two powerful forces: the deepening integration of cross-cultural teams and the rapid proliferation of artificial intelligence (AI). This chapter examines the intersection of these trends, exploring the opportunities and challenges of managing culturally diverse, AI-augmented workforces. We argue that AI is not culturally neutral but, when guided by cultural competence, can strengthen global collaboration, decision-making, and inclusion. Conversely, if implemented without regard for cultural nuances, AI risks amplifying biases, eroding trust, and creating new forms of digital exclusion. This chapter begins by establishing the theoretical foundations of cross-cultural management and AI augmentation. A framework for culturally-competent AI integration is proposed, focusing on managerial strategies and tool

DOI: 10.4018/979-8-3373-7442-0.ch009

selection. The chapter concludes with a discussion of the social, ethical, and managerial implications, calling for a more inclusive and critically aware approach to AI-augmented workplaces.

1. INTRODUCTION: THE NEW FRONTIER OF GLOBAL WORK

The global workplace is the essence of the 21st century workplace. With the spur of digital connectivity and economic interdependence, organizations have come to regularly consist of teams spread across the continents, time zones and cultures (Adler and Gundersen, 2008). This changing pattern in demographics has transformed cross-cultural management the knowledge and handling of cultural distinctions into a competitive ability of the contemporary leaders. At the same time, there is a second revolution: the extensive application of the concept of artificial intelligence (AI) to business operations. AI is replacing the human touch in the modern enterprise, whether as a resume screening tool, an automation of workflows, or a natural language processing tool to translate communications and understand their tone (Wilson and Daugherty, 2018). In this chapter, the view is that these two revolutions converge to form a new frontier in management that is both promising as never before and at times multifaceted with a great deal of complexity.

The notion of AI augmented workplace needs to be defined. Here we differentiate between automation, which entails machines taking the place of human work, and augmentation, which entails human and machine companies to produce better results (Raisch and Krakowski, 2021). We are interested in augmentation, the formation of human cultural intelligence and machine computational teams, hybrid intelligence. Such synergy is bound to break the conventional boundaries in the international management, but also it brings the new challenges, which are based on the cultural embeddedness of the technology itself. Technology is always created and implemented in a cultural context, and it is indicative of values, assumptions, and prejudices of the people who make it (Borenstein and Howard, 2021). Consequently, the key argument of this chapter consists of the idea that the successful management of cross-cultural teams in the AI era requires a twofold emphasis: taking advantage of the potential of AI to reduce the cultural differences and paying extra attention to avoiding the further enhancement of the latter.

The chapter is divided into sections that attempt to give a detailed discussion of this dynamic. We start by examining the fundamental theoretical underpinnings of cross-cultural management and AI at the workplace. Thereafter, we attribute considerable space to a balanced discussion of the opportunities the AI offers in developing cross-cultural collaboration and subsequently undertake the critical analysis of the natural challenges and threats. Based on this discussion, we can suggest a viable model that managers can use to adopt AI tools in a culturally competent way. The chapter proceeds to elaborate further the extended implications of this convergence, both social and practical, ethical and managerial which these convergences bring to organizations and society. We conclude the paper finally by summarizing the main arguments and locating the critical directions of future research and practice.

2. THEORETICAL FOUNDATIONS: CULTURE AND TECHNOLOGY

The dynamics of AI and cross-cultural management can only be comprehended following the existing theories used to explain cultural variation and the emerging conceptualization of AI as a social actor.

2.1. Key Frameworks of Cross-Cultural Management:

The scholarship in cross-cultural management offers powerful theory and schemes of classifying and comprehending systematic variation of work-related values and behaviors. The most popular is a cultural dimensions theory by Geert Hofstede, which was created as a result of extensive research on IBM workers in more than 50 countries. Hofstede model determines various major spectrums according to which cultures differ such as Power Distance (the acceptance of the hierarchical inequality), Individualism versus Collectivism (degree of integration into groups), Uncertainty Avoidance (tolerance to the ambiguous situation), and Masculinity versus Femininity (the focus on achievement versus nurturing) (Hofstede, 2001). Thcse dimensions have an enormous implication to the management practice; then a leader in a low power distance culture such as Sweden may invite his team to use a collaborative AI tool with the expectation of a flat involvement, team members in a high power distance culture such as Malaysia may be reluctant to actively participate, waiting to be instructed by the manager.

Following Hofstede, GLOBE (Global Leadership and Organizational Behavior Effectiveness) study further elaborated on these dimensions and interrelated them to the prototypes of leaders, which attributes of a leader are universally accepted (e.g. integrity) and which dimensions depend on the culture (e.g. ambition) (House et al., 2004). The next important theory is that of high and low context cultures developed by Edward T. Hall (Hall, 1976). In high-context cultures (e.g. Japan, Arab countries) communication is subtle, unspoken and largely based on non-verbal communication and common ground. Communication in low-context culture (e.g. in Germany or in the United States) is explicit, direct, and literal. This aspect is highly dependent on the performance of AI tools, especially natural language processing-based ones, which are commonly trained on low-context and explicit data and do not have the ability to contact the intricacies of high-context interaction.

2.2. An Introduction to AI in the Workplace:

Artificial Intelligence In the workplace, artificial intelligence is not a single entity but a collection of technologies. The first and most important one is Machine Learning (ML) where algorithms are trained on data to make a prediction; Natural Language Processing (NLP), which teaches machines to understand, interpret, and generate human language; and Robotic Process Automation (RPA), which automates repetitive, rule-based digital work (Davenport and Ronanki, 2018). The concept involved in management is the shift to the hybrid intelligence systems which are specifically created to capitalize on the synergistic capabilities of machines and humans. According to Dellermann et al. (2019), AI is faster, more scalable, and able to be analytic, and human beings are more creative and can reason ethically, as well as possess social and cultural insights. The management problem, thus, is no longer the application of technology but the coordination of successful interaction of human and artificial team members and is more than exponentially complicated when it comes to culturally diverse human team members.

3. THE OPPORTUNITIES: AI AS A CROSS-CULTURAL BRIDGE

After implementation with cultural awareness, AI can become a potent facilitator and assist in addressing some of the most enduring problems in managing multinational teams.

3.1. Enhancing Communication and Overcoming Language Barriers:

Barrier One of the most direct ways to use AI in cross-cultural environment would be to break the language barrier. On-the-fly translators, which are built on advanced NLP models, are getting more and more precise and fluent, and they are integrated into email and video conferencing services, as well as into collaborative documents (Brynjolfsson and McAfee, 2014). This will enable communication to be smoother between team members who may not have a common first language and the number of misunderstandings can be reduced as well as making the environment more inclusive in which the non-native speakers may feel more confident to participate. In addition to direct translation, AI-based writing assistants also give people more or less direct, formal or persuasive in their communication style, as per the cultural norms of their intended audience. As an example, a tool might recommend that a high-context Korean employee (managed by an American direct and low-context manager) be softened to withhold harm and respect in the case of a critical feedback message.

3.2. Mitigating Unconscious Bias in Hiring and Promotions:

One such area in which AI has a lot of potential in ensuring the promotion of fairness within culture is in mitigating Unconscious Bias in Hiring and Promotions Human resources. Human recruiters and managers are prone to unconscious prejudice that may discriminate against any individual who is of another cultural, ethnic, or educational background. It is possible to develop AI-driven systems that will allow conducting blind recruitment, deleting names, photos, and university names on the applications and viewing only the skills and qualifications (Cowgill, 2018). Moreover, interviews can be structured and analyzed by AI following pre-defined culturally-intelligent sets of competencies so that all candidates would be assessed based on the same objective requirements. Performance management It can be used to aggregate data on project work, commits, or sales data to enable more data-driven promotions, potentially overcoming the similar-to-me bias where managers prefer employees with whom they have a cultural background.

3.3. Data-Driven Insights into Cultural Dynamics:

Data-Driven Insights into Cultural Dynamics ranks among the most potent analytical tools the managers have ever encountered as they allow them to gain an unprecedented insight into the cultural dynamics of the teams they are leading. Using metadata of communication tools (e.g. response rates in emails, the pattern of meeting attendance, maps of collaboration networks), AI will be able to identify possible remote team member isolation or communication silos that emerge based on a cultural or geographical boundary (Choudhury et al., 2020). As an example, an algorithm can notice that the members of the Brazilian office contribute to the asynchronous discussion less frequently than the ones in the German office, not because they are not interested in the discussion, but because of the cultural differences between the two cultures (synchronous and asynchronous communication). With such information, a manager can take the initiative of redressing the situation and come up with fairer communication procedures.

3.4. Standardizing Processes while Allowing for Local Adaptation:

Global organizations are always faced with the dilemma of standardization of the best practice as well as providing a local adaptation. The AI-powered enterprise systems have the potential to offer a single, data-supported framework of core business processes, such as project management, budgeting, and compliance, which will be consistent and efficient throughout the organization. Nevertheless, unlike hard, pre-AI based systems, new, intelligent, platforms can be set up to fit local differences. One example is an AI project management tool that, based on its learning, would know that teams in Spain tend to have a later start to the workday, and automated prompts can be arranged, or it may provide flexibility in the style of deadline negotiation between high and low uncertainty avoidance cultures. This enables a glocalized strategy, with global standards and adherence to local cultural processes.

4. THE CHALLENGES: WHEN AI MEETS CULTURAL COMPLEXITY

Despite all of the potential, AI poses a plethora of new challenges and threats that are bound to destroy cross-cultural management, unless addressed. These issues are typically based on the fact that AI models are culturally biased.

4.1. The Myth of Cultural Neutrality:

Algorithmic Bias Amplified The issue of algorithmic bias is the most important. The AI models will be trained on historical data, and in case that old data is not representative or is based on the current biases of the society, the AI will reproduce and tend to extend those biases (Buolamwini and Gebru, 2018). This is especially dangerous when it comes to cross-cultural situation. A recruitment algorithm that has been trained on successful resumes in North America and Western Europe may bias against applicants with educational histories in an elite university in Asia or Latin America, or whose career history is non-Western (e.g. more job changes, because of family or cultural reasons). Equally, even emotion recognition software, which is frequently trained on stereotypical faces of Western individuals, can often be misreading the expressions of people of East Asian cultures, in which suppression of emotion is more highly regarded (Barrett et al., 2019). This is a vicious circle of AI systems in which history of Western-centric bias is replicated into the future of global work.

4.2. Variations in Technology Acceptance and Trust:

Cultural dimensions have a direct impact on employee perception of AI systems and trust. Uncertainty Avoidance is the dimension that is particularly relevant by Hofstede. the black box of certain AI-based decision-making can make employees of cultures with high uncertainty avoidance (e.g., Japan, France) incredibly uncomfortable and unwilling to give up authority to an algorithm whose logic fails to be understood by them (Kaplan and Haenlein, 2020). In contrast, higher pragmatism and innovation acceptance will result in cultures being more trusting. In addition, cross cultural attitudes to surveillance are quite different. In some low-context and individualistic societies, Keystroke logging, or tracking people's movements on websites with AI might be accepted but perceived as a grave trust and privacy breach in others with higher collectivist or high-context, therefore negatively affecting morale and psychological security.

4.3. Human-AI Communication Divide:

The existing abilities of AI pose a large communication gap especially among high-context cultures. Even heavily developed NLP models cannot cope with the nuances that constitute a high-context communication sarcasm, irony, indirectness or the entire world of non-verbal communication (tone of voice, facial expressions, body language), which conveys the actual meaning of a message (Glikson and Woolley, 2020). When a group is over-dependent on an AI to condense meeting documentation or gain an understanding of sentiment in a chat room, it runs the danger of entirely failing to uncover the subtle conflict or unspoken issue conveyed by a member of a high-context society. This has potential to create a systematic marginalization of voices with these cultures, since their input is not properly heard or received by the AI systems that managers are increasingly relying on to monitor the team.

4.4. Erosion of Cultural Nuance and Homogenization:

Excessive trust in AI in creating standardization runs the risk of cultural homogenization: unique local ways of solving problems, communication patterns, and creative practices are flattened out, replaced by one, algorithmically-defined best practice. This does not only suppress innovation, but can also result in declining employee engagement and a feeling of cultural imperialism, the practices of the home country of a corporation are imposed on the rest of the world through technology (Stahl et al., 2023). The diversity that makes global teams competitive, i.e., it is possible to think in a variety of cultural ways, can be destroyed in case AI systems are applied to impose a monolithic method of working.

4.5. The Black Box Problem and Responsibility:

The lack of transparency of complex AI models (particularly deep learning networks) introduces a significant accountability problem. In the event that the AI system takes a decision with adverse cross-cultural implications, such as automatically classifying a culturally-specific presentation as inappropriate or unfairly distributing resources outside of a particular team in the region, it may be hard or even impossible to explain why. This issue is a complex black box problem that makes it difficult to hold people responsible (Pasquale, 2015). Does the manager have a duty of taking action on the defective product? Is the biased system being purchased by the HR department? Or does the software developer have any case to answer? In a cross-cultural environment, where the very concept of fairness and accountability differs, it will be an urgent management and legal issue to define how decisions made on the basis of AI responsibility should be accounted.

5. A FRAMEWORK FOR CULTURALLY-COMPETENT AI INTEGRATION

To take the opportunities and suppress the threats, organizations should make a conscious and critical approach to the application of AI to their cross-cultural management practices. The table below illustrates major principles and strategies.

5.1. Principles for Developing and Selecting Cross-Cultural AI Tools:

The first defence is at the point of procurement and development. Organizations should no longer rely on technical specifications in their regard but require cultural transparency on the part of the vendors. This involves questions concerning the diversity of the training data, the cultural background of the development team and the measures that were taken to detect and remove bias in the development of the model (Rakova et al., 2021). The philosophy should be augmentation as opposed to automation; the tools that are chosen should be those that augment human cultural intelligence and not the ones that substitute it. A tool that indicates the possible cross-cultural misconceptions that a human manager should be aware of is better than that which fixes communication errors without providing the reasons.

5.2. Managerial Strategies

Managerial strategies are the vital connection between the team dynamics, policy and technology. Their position transforms to that of a team conductor that is a hybrid.

- Training & Upskilling: Employees should be trained not only on how to use new AI tools, but also on their cultural weaknesses. This develops a state of cultural AI literacy, which allows the team members to critically analyze AI outputs and know when a machine recommendation may be culturally tone-deaf (Tarafdar et al., 2019).
- Hybrid Decision-Making: Managers should come up with explicit procedures that stipulate how labor is shared between human and AI. As an example, a shortlisting of candidates can be performed using an AI, but final interviews should be held by a culturally diverse human panel. An AI will recommend a project schedule, but a human manager will have to complete the schedule by consulting the global team taking into account local holidays and working culture.
- Developing a Culture of Feedback: A psychologically safe system of employees to report cases of perceived AI bias or cultural insensitivity should be established. It is an essential source of feedback that will be used to constantly refine the application of the technology as well as the managerial strategy, making the very team a sensor of the cultural misfires.

6. IMPLICATIONS

The intersection of AI and cross-cultural management has much more to it than just the direct team dynamics, as it has extensive implications both on society, management practices, and ethics and on the nature of management itself.

6.1. Social Implications:

On a societal level, AI-enhanced work environments would increase the global interconnection, but also introduce novel social segregations. On the one hand, they are able to enhance inclusion to a greater extent by providing a voice to employees that are geographically and culturally remote. Conversely, there is a danger that they will result in a digital caste system, with all individuals who are technically capable of benefiting successfully through the application of AI (who are often not of the dominant cultures) getting ahead, and those who are not getting left behind (Frey & Osborne, 2017). This would further increase inequalities across the world.

Moreover, the professional identities would be mixed with AI tools, which would result in the potential risk of the disappearance of specific cultural approaches to professions (e.g., consensus-based Japanese management vs. directive American management) in the favor of an even more homogenized and globally standardized professional culture.

6.2. Practical Implications:

To organizations, the practical implications are far between. The approach to Talent Management strategies needs to be radically changed, making cultural AI literacy, the capacity to operate in the overlap between culture and technology, a key competency to be either hired or developed. The functions of System Design and Procurement are no longer the preserve of IT but it must have cross-cultural experts, ethicists and cross-cultural end-users to check the cultural suitability of tools. The Workflow Re-engineering is made an ongoing process since the organization has to continuously modify its processes to take advantage of new AI features without unintentionally disadvantaging a particular cultural group.

6.3. Ethical Implications:

The ethical environment is especially perilous. The concept of Transparency & Explainability (XAI) turns into the question of justice. The rights of the employees who are the targets of AI-made decisions would be to an explanation that is technically feasible but also culturally reasonable. The issues of Privacy and Data Sovereignty are exaggerated, since the AI systems tend to demand a significant number of employee data. Companies have to work around the intersected landscape of international laws (such as the GDPR in Europe compared to weaker regulations in other places) and, more to the point, cultural understanding of the privacy. The general issue of ethics is Fairness & Justice. It is imperative to make sure that the implementation of AI in international workplaces does not systematically deprive certain cultural, ethnic or national groups of opportunities hence continuing the colonial-era power relations on a digital front.

6.4. Managerial Implications:

To the individual manager, this new reality dictates that he or she has to change his or her role and competence considerably. Emerging Leadership Skills are needed; managers are to become translators and integrators, able to comprehend AI-generated data through the cultural prism and clarify human cultural backgrounds to those who write and program AI programs. The measures of Performance Evaluation should be redefined to perform better in rewarding soft skills such as cultural mediation and the capacity to achieve effective human-AI collaboration, and stop at the quantitative outputs. Lastly, AI implementation should be carried out on a profoundly culturally conscious level under Change Management. A standardized roll-out package by the head office is the formula of opposition. Rather, the managers should have a discussion with their local teams, and they should jointly develop implementation plans that address local norms and issues.

7. CONCLUSION AND FUTURE DIRECTIONS

This chapter has maintained that AI as a part of working in a cross-cultural environment is a two-sided sword and has offered a paradigm of both strong opportunities and serious challenges. The potential of AI as a powerful bridge, where the communication is improved, prejudice is minimized, and the insights about the dynamics of the team are offered, is high. However, it might also be an insurmountable obstacle, increasing the historical bias, ruining trust, misunderstanding the cultural subtlety, and developing an intricate zone of accountability. Technology is not the key determinant of which direction to follow, but rather the cultural competency in the choice, application, and administration of technology.

The message of the future is clear: we need to shift towards more inclusive AI creation as well as more critically-aware deployment. This involves interdisciplinary work of computer scientists, organizational psychologists, anthropologists and ethicists. Some of the new frontiers that need to be explored in future studies are the effect of generative AI (such as ChatGPT) on creative and knowledge work in global teams; the interface between AI and culture; and the sociological long-term effects of AI on the future of organizational cultures.

Finally, it is not to make the workplace a sterile place where AI does all the tasks, culturally different people are erased. Instead, it is to embrace the power of AI to create more dynamic, diverse, and functional global teams in which cultural diversity is not merely managed, but actually used as a source of innovation and power. Those managers and organizations that excel in this new frontier will be those who see AI not as a tool, but as a sophisticated cultural actor that will have to be incorporated with wisdom, empathy, and a firm adherence to human dignity.

REFERENCES

Adler, N. J., & Gundersen, A. (2008). *International dimensions of organizational behavior* (5th ed.). Thomson South-Western.

Barrett, L. F., Adolphs, R., Marsella, S., Martinez, A. M., & Pollak, S. D. (2019). Emotional expressions reconsidered: Challenges to inferring emotion from human facial movements. *Psychological Science in the Public Interest, 20*(1), 1–68. DOI: 10.1177/1529100619832930 PMID: 31313636

Borenstein, J., & Howard, A. (2021). Emerging challenges in AI and the need for AI ethics education. *AI and Ethics, 1*(1), 61–65. DOI: 10.1007/s43681-020-00002-7 PMID: 38624388

Brynjolfsson, E., & McAfee, A. (2014). *The second machine age: Work, progress, and prosperity in a time of brilliant technologies.* W. W. Norton & Company.

Buolamwini, J., & Gebru, T. (2018). Gender shades: Intersectional accuracy disparities in commercial gender classification. *Proceedings of the 1st Conference on Fairness, Accountability and Transparency, 81*, 77-91. http://proceedings.mlr.press/v81/buolamwini18a.html

Choudhury, P., Foroughi, C., & Larson, B. (2020). Work-from-anywhere: The productivity effects of geographic flexibility. *Strategic Management Journal, 42*(4), 655–683. DOI: 10.1002/smj.3251

Cowgill, B. (2018). *Bias and productivity in humans and algorithms: Theory and evidence from resume screening.* Columbia Business School Research Paper No. 18-19. https://ssrn.com/abstract=3181120

Davenport, T. H., & Ronanki, R. (2018). Artificial intelligence for the real world. *Harvard Business Review, 96*(1), 108–116.

Dellermann, D., Ebel, P., Söllner, M., & Leimeister, J. M. (2019). Hybrid intelligence. *Business & Information Systems Engineering, 61*(5), 637–643. DOI: 10.1007/s12599-019-00595-2

Frey, C. B., & Osborne, M. A. (2017). The future of employment: How susceptible are jobs to computerisation? *Technological Forecasting and Social Change, 114*, 254–280. DOI: 10.1016/j.techfore.2016.08.019

Glikson, E., & Woolley, A. W. (2020). Human trust in artificial intelligence: Review of empirical research. *The Academy of Management Annals, 14*(2), 627–660. DOI: 10.5465/annals.2018.0057

Hall, E. T. (1976). *Beyond culture*. Anchor Books.

Hofstede, G. (2001). *Culture's consequences: Comparing values, behaviors, institutions, and organizations across nations* (2nd ed.). Sage Publications.

House, R. J., Hanges, P. J., Javidan, M., Dorfman, P. W., & Gupta, V. (Eds.). (2004). *Culture, leadership, and organizations: The GLOBE study of 62 societies*. Sage Publications.

Kaplan, A., & Haenlein, M. (2020). Rulers of the world, unite! The challenges and opportunities of artificial intelligence. *Business Horizons, 63*(1), 37–50. DOI: 10.1016/j.bushor.2019.09.003

Pasquale, F. (2015). *The black box society: The secret algorithms that control money and information*. Harvard University Press.

Raisch, S., & Krakowski, S. (2021). Artificial intelligence and management: The automation–augmentation paradox. *Academy of Management Review, 46*(1), 192–210. DOI: 10.5465/amr.2018.0072

Rakova, B., Yang, J., Cramer, H., & Chowdhury, R. (2021). Where responsible AI meets reality: Practitioner perspectives on enablers for shifting organizational practices. *Proceedings of the ACM on Human-Computer Interaction*, 5(CSCW1), 1-23. https://doi.org/DOI: 10.1145/3449081

Stahl, B. C., Brooks, L., Hatzakis, T., & Santiago, N. (2023). The ethics of artificial intelligence in international business: A review of the literature and a research agenda. *Journal of Business Ethics, 182*, 1–22. DOI: 10.1007/s10551-022-05239-2

Tarafdar, M., Page, X., & Marabelli, M. (2019). Algorithms as co-workers: Human-algorithm role interactions in professional work. *Journal of the Association for Information Systems, 20*(9), 1–38. DOI: 10.17705/1jais.00565

Wilson, H. J., & Daugherty, P. R. (2018). Collaborative intelligence: Humans and AI are joining forces. *Harvard Business Review, 96*(4), 114–123.

Chapter 10
Shaping Entrepreneurial Creativity and Innovation Through Human-AI Cognition

Khushboo Jain
Shri Vaishnav Institute of Management and Science, Indore, India

Deepa Katiyal
Shri Vaishnav Institute of Management and Science, Indore, India

Shashank Mittal
https://orcid.org/0009-0007-8643-4350
JGBS, O.P. Jindal Global University, Sonipat, India

Mukhtaruddin Mukhtaruddin
Faculty of Economics, Universitas Sriwijaya, Palembang, Indonesia

ABSTRACT

Artificial Intelligence (AI) is reshaping entrepreneurial dynamics in modern times by fostering creativity, innovation and strategic decision-making. The current study examines the contribution of AI tools and technologies in the enhancement of entrepreneurial creativity and innovation. The data-driven insights provided by AI helps in automating cognitive processes and supporting idea generation. The qualitative and conceptual analysis of literature done using NVivo 14 software coding generated five themes for the study. Idea generative support of AI, cognitive intelligence, fostering innovation, human-AI's ethical alignment and entrepreneurial ecosystem readiness

DOI: 10.4018/979-8-3373-7442-0.ch010

for AI. These themes helped identify the ways AI tools such as Machine Learning, natural language processing, generative AI models etc., support entrepreneurs in ideation, development, decision-making and implementation.

INTRODUCTION

Entrepreneurial creativity and innovation are vital drivers of economic growth and competitiveness in today's dynamic business landscape. With technological advancements, Artificial Intelligence (AI) has emerged as a supportive tool for re-shaping entrepreneurs' imagination. AI tools and techniques provide better design that helps in visualizing human concepts and deliver innovative solutions based on large data sets. With its ability to analyze vast amounts of data, uncover hidden patterns and generate meaningful insights, AI empowers entrepreneurs to think more creatively and make smarter decisions in uncertain markets.

AI Technologies such as generative AI, predictive analytics, natural language processing etc. are revolutionizing every stage of the entrepreneurial journey. From ideation and design thinking to market testing and scaling AI is assisting entrepreneurs in providing better and more realistic solutions backed by facts. Bringing innovation amidst the challenges of modern times is crucial. However, entrepreneurs are increasingly turning to AI tools and techniques for creative brainstorming, opportunity identification and customer engagement. Though, this growing reliance on AI brings challenges concerning to diminishing human creativity, ethical biases and excessive dependence on automation. Therefore, understanding how AI enhances entrepreneurial creativity and innovation is both timely and crucial for shaping the future of entrepreneurship.

LITERATURE REVIEW

Entrepreneurial creativity involves the generation of ideas that stand different and address consumer needs and/or create new market opportunities. Entrepreneurial innovation involves successful implementation of these ideas to provide practical and marketable solutions. In modern entrepreneurship ecosystem entrepreneurial creativity and innovation are increasingly shaped by AI technologies. AI has emerged as a transformative driver of entrepreneurial creativity and innovation over the past decade. Research consistently highlights AI's potential to augment entrepreneurial initiatives by expanding operational capacities, accelerating decision-making and enabling novel product and service development (Sayah, 2025). For instance, AI facilitates data-driven decision-making, predictive analytics and automation of routine tasks, which frees entrepreneurs to focus on creative and strategic ventures, thereby enhancing innovation (Abu Hashima & Al-Azab, 2024).

However, literature also stresses the complementary role of human creativity in AIenabled entrepreneurship. AI enhances cognitive processes but cannot fully replicate the nuanced originality intrinsic to human innovation (Hadhli & Sheikh, 2022). This duality positions AI as a catalyst that amplifies human creative potential rather than a substitute. Prompt engineering within AI platforms has emerged as a critical skill, allowing entrepreneurs to harness AI's capabilities effectively by tailoring interactions and extracting precise, innovative outputs (White et al., 2023; Obschonka & Audretsch, 2020).

Studies show that AI driven tools and techniques including Machine Learning, LLMs and multimodal generators augment human creativity by offering novel prompts, synthesizing cross-domain insights, thus supporting idea generation. Hence, AI acts as a creative assistant expanding cognitive boundaries by synthesizing information beyond human capacity (Huang & Rust, 2021). AI enhances innovation management through predictive analytics and trend forecasting. Studies show that AI assists entrepreneurs in identifying market gaps, optimizing R&D and enhancing product differentiation (Zeng et al., 2022). By analyzing customer sentiment, competitors and macroeconomic indicators, AI tools enable more strategic innovation choices. The empirical findings of AI tools assist conceptual understanding of humans by speeding iteration of ideation cycles and enhancing product-market fit through predictive analytics and automation of routine R&D tasks (Mariani et al., 2023).

Generative AI use expands the ideation stage by surfacing more diverse concepts and increase ideational throughput along with considering authorship and IPR (Mariani & Dwivedi, 2024). Recent studies emphasize generative AI (GenAI) as a powerful tool that supports entrepreneurial exploration by synthesizing vast, diverse datasets into actionable insights, thus reducing uncertainty and enabling rapid opportunity identification (Ferreira et al., 2025). GenAI facilitates brainstorming, prototype development, market research and content creation, stimulating creativity and collaborative innovation (Wang et al., 2022).

The entrepreneurial lifecycle benefits variably from AI applications. AI markedly enhances the exploration and development phases through improved information foraging, problem-solving and business model innovation (Cantner et al., 2021). Yet, the exploitation phase, which demands complex strategic decision-making, remains a challenge for current AI technologies due to their limited contextual understanding, indicating a need for advanced AI improvements (Chalmers et al., 2021).

Global investment trends showcase significant growth in AI technologies, led by major companies like Google, Microsoft and Baidu, underpinning AI's central role in driving economic expansion and entrepreneurship worldwide (McKinsey, 2020). However, challenges include the digital divide exacerbated by unequal AI adoption between developed and developing countries, as well as ethical concerns surrounding employment and privacy (Hiba & Mohamed, 2021).

There are experimental evidences suggesting that teams those who have adopted AI assistance produce higher-quality ideas in less time. Although outcomes depend on human–AI workflow design and team processes. But it has been noteworthy that the success of AI–human collaboration depends on ethical integration, trust and human oversight to preserve originality and authenticity. AI is a collaborator and not a replacement of human intelligence. (Gindert & Müller, 2024). Thus, AI functions as a cognitive collaborator by enhancing creative possibilities and accelerating innovation, contingent on governance, skill development and human–machine collaboration design.

In conclusion, the literature underlines AI as an indispensable tool in enhancing entrepreneurial creativity and innovation by improving efficiency, fostering knowledge creation and enabling sustainable, scalable ventures. Future research directions include deepening theoretical frameworks for AI's impact on entrepreneurship, advancing AI technologies for strategic exploitation phases and exploring optimal human-AI collaborations that balance algorithmic capabilities with human ingenuity (Kusetogullari et al., 2025). It augments the idea generation, accelerates development, facilitates decision-making and soothes implementation. Economic and innovation scholars argue that AI can act as a collaborator to human conceptualization. It strengthens R&D by using patterns from large datasets. The integration of AI in entrepreneurship contributes significantly to overcoming traditional barriers such as information asymmetry and resource constraints.

RESEARCH OBJECTIVES

- To study the role of AI in enhancing entrepreneurial creativity and innovation.
- To analyze the challenges in integrating AI into entrepreneurship.
- To propose a framework for the role of AI in fostering creativity and innovation for decision-making.

METHODOLOGY

The research design adopted for the present study is qualitative and conceptual. For secondary data collection a systematic review of journal articles, working papers, case studies and industry reports was done available on Scopus, Web of Science, Google Scholar etc.. To analyze the AI's role in enhancing entrepreneurial creativity and innovation, Thematic Analysis of the collected secondary sources was done using NVivo 14 software. Publications between 2015 and 2025 were referred for the study.

Documents were open-coded in NVivo by assigning descriptive nodes (codes) to short meaningful text segments. Frequently used words like AI-enabled creativity, entrepreneurial innovation, cognitive augmentation, human–AI collaboration were then categorized. Further NVivo matrix coding and cluster analyses were used to explore cooccurrence of codes across documents and to visualize relationships between concepts. After establishing this relationship Braun & Clarke's (2006) six-phase approach was used. Around 30 initial nodes were generated such as *idea generation, generative models, predictive insight, human–AI collaboration, skill gaps, ethical risk, speed to market, R&D automation* and *creative constraints*. These were further clustered into five robust themes which are mentioned below.

FINDINGS AND DISCUSSION

NVivo 14 coding generated following themes for the study that covered various stages of innovation and entrepreneurial process.

1. Idea Generative Support of AI

Ideation stage of entrepreneurial journey involves creation of ideas. AI assists this stage by generating novel ideas, content, models and simulating scenarios which serves as base. These are then used by entrepreneurs to refine and prototype ideas through AI tools that help in brainstorming, sentiment analysis and generate design. Thus, AI plays a role of a catalyst in supporting the idea generation. NVivo coding showed strong cooccurrence for codes like *generative models, idea generation* and *novelty* across various sources.

2. Cognitive Intelligence

Post idea generation AI tools help entrepreneurs to collect, summarize and analyze large amount of data sets to extract meaningful data to simulate scenarios and understand patterns for opportunity recognition. Further AI tools also assist in understanding the related parameters and factors that can affect the opportunity in future. This helps in visualizing the line of action thus smoothing the process of decision-making. Various codes across sources that were linked by NVivo for this theme are *predictive analytics, scenario simulation and opportunity discovery.*

3. Fostering Innovation

The meaningful data extracted by AI tools is then transformed into actionable insights for innovating the products and/or services. Consumer preferences and emerging trends obtained as outputs of predictive analysis guides entrepreneurs for understanding the accepted patterns, expectations and gaps among the two. Upon analyzing these AI tools also help entrepreneurs with suggestions of viable innovations to reduce the identified gaps. With a vision of converting the consumer preferences into reality Machine Learning algorithms are designed to facilitate product innovation. In sectors like fintech and healthcare, AI accelerates the cycle from concept to commercialization. This theme was the output of code cluster including *automation*, *R&D efficiency*, *faster iteration* and *scale*.

4. Human-AI's Ethical Alignment

AI's capacity for real-time analysis enhances strategic decision-making and supports long-term innovation strategies. Thus, AI systems extend beyond operational assistance to become strategic collaborators. This theme represents a substantial cluster of following codes *trust*, *explainability*, *creative ownership* and *human oversight*. AI is adopted by entrepreneurs for varied reasons some use it as supportive novel idea generation and analyzing the consumer for strategic decision-making. While, several see it as loss of divergent thinking, intellectual property ambiguity and ethical risks. The consensus theme recommends design of collaborative workflows where humans set intent and AI provides variation, not sole authorship.

5. Entrepreneurial Ecosystem Readiness for AI

Entrepreneurial ecosystem is strongly based on creativity and innovation. Human led ideas should always run the chase and AI tools should be used to assist creativity and innovation. To avail the benefits of AI assistance in this process entrepreneurs need to invest in AI literacy, infrastructure and governance. NVivo coding revealed repeated references for *skills gaps*, *access to datasets* and *organizational capability* as impediments.

CHALLENGES AND ETHICAL CONSIDERATIONS

Although AI offers tremendous advantages, its integration also brings several challenges. Issues such as biased data influencing creative outputs, excessive dependence on algorithmic recommendations and uncertainties around intellectual ownership can complicate the creative process. Therefore, maintaining strong human oversight and establishing clear ethical guidelines are essential to preserve authenticity, transparency and accountability in AI-driven entrepreneurial ventures. Various challenges are mentioned below.

Technical challenges

Though vast amount of data is available through AI but data quality and availability comes as a challenge that can hinder accurate result generation. At times existing infrastructure and systems are not compatible with AI technologies which creates issues of connections. High initial cost is also seen as a setback of investment in AI technology, infrastructure, and skilled personnel especially for entrepreneurial ventures. Further scaling of a pilot AI-driven project brings along technological limitations and the complexity of integrating with diverse departments.

Organizational challenges

AI adoption still faces scarcity of professionals having specialized AI skills suitable for entrepreneurial ecosystems. On the other hand, existing employees resist to new AIdriven workflows which necessitates effective change management and training to ensure adoption. Also, there is uncertainty of quantifying the return on investment for AI initiatives making it hard to justify its adoption.

Ethical Challenges

Data security and privacy is one of the major challenges to be faced as AI systems require access to vast amounts of data. There is presence of potential risks of data breaches and privacy violations if not properly secured. AI models can perpetuate and even amplify algorithm biases, leading to discriminatory outcomes and damaging a company's reputation. Yet another challenge seems to job losses, necessitating strategies to support affected workers. As the automation capabilities of AI can lead to operations that require minimum human intervention.

PROPOSED FRAMEWORK

The human-AI cognition demonstrated in the proposed model shows a synergy between humans and Artificial Intelligence (AI) across four key stages of the innovation and entrepreneurial process viz. ideation, development, decision-making and implementation. In the very *first stage* of idea generation humans focuses on building the plot for core concepts to be used for innovation. While the AI assistant is obtained in expanding and refining these ideas by identifying hidden patterns and emerging opportunities through data-driven insights. This ensures enhanced creativity as an outcome. In *second stage*, of development entrepreneurs generate prototypes and test their concepts. Whereas AI contributes by optimizing designs, simulating outcomes and improving efficiency using analytical tools. This AI assistance gives improved innovative results. In the *third stage* humans mostly work on their intuitions but the use of AI balances this intuition by predictive analysis of facts obtained through data sets providing precision. This cognition enables well-informed, strategic choices. Finally, in the *fourth stage* of implementation, humans execute the strategies formulated, while AI automates and scales operations to ensure efficiency and adaptability.

This Human–AI Cognition Model illustrates that contextual understanding of humans when complemented with AI's computational intelligence the entrepreneurial creativity and innovation thrive. AI is not a replacement to human rather a collaborator to enhance creativity, accelerate innovation and support entrepreneurs in transforming ideas into impactful solutions.

Figure 1. The human–AI cognition model for innovation and entrepreneurial process

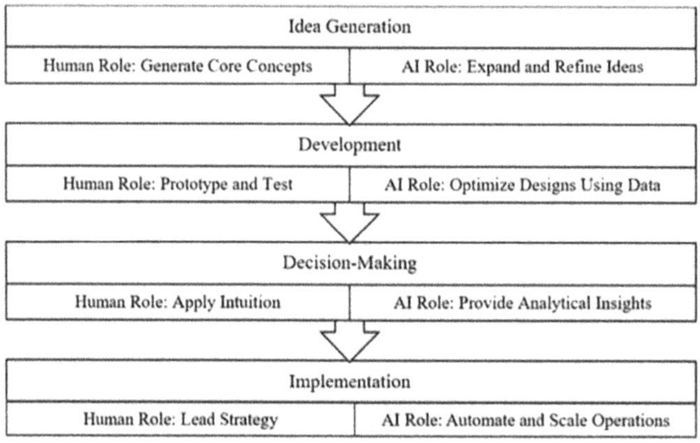

CONCLUSION

AI has become a powerful collaborator in fostering entrepreneurial creativity and innovation. Instead of replacing human capabilities, AI tools and techniques enhance the innovation and entrepreneurial process. It generates ideas more quickly, uncover deeper insights and manage innovation with greater precision. By maintaining the human-AI cognition of collaborating the data-driven intelligence of AI with human understanding and intuition, entrepreneurs can explore new possibilities and make more informed creative decisions. However, the real potential of AI lies in maintaining a healthy balance between technological efficiency and human imagination. When guided by ethical values, transparency and diverse perspectives, AI can assist to build innovative and sustainable ventures.

IMPLICATIONS

The study also provides implications for entrepreneurs, researchers and policymakers to facilitate the entrepreneurial creativity and innovation process.

For Entrepreneurs

Entrepreneurs should adopt AI tools as co-creators and set processes that preserve divergent thinking. Entrepreneurial training programs should emphasize on AI literacy and creative workflows. Data-driven decision-making should be adopted to make more informed decisions, reducing uncertainty and risk. Adopting AI technologies can provide a significant competitive edge, differentiating entrepreneurial ventures in crowded markets. Entrepreneurs must develop new skills to effectively harness AI, including understanding AI tools, data literacy and prompt engineering for generative AI systems.

For Researchers

Researchers focusing on AI's role in entrepreneurship must explore ethical AI use, algorithmic bias and data privacy. They should investigate AI-human collaboration dynamics, AI's long-term impact on entrepreneurial success and develop frameworks ensuring responsible, inclusive innovation. Addressing these will advance theory and practical AI applications in entrepreneurship. Researchers should measure not only idea quantity but idea diversity and downstream commercialization success.

For Policymakers

Policymakers must promote equitable AI access, foster AI literacy and establish regulations ensuring ethical AI use in entrepreneurship. They should support innovation ecosystems that balance technological advancement with privacy, security and workforce transition strategies to maximize AI's economic and social benefits while mitigating risks and inequalities in entrepreneurial landscapes should also be supported through shared data infrastructures to democratize access.

REFERENCES

Abu Hashima, M., & Al-Azab, M. (2024). Entrepreneurship in the age of artificial intelligence: A unique set of essential skills and competencies. *International Journal of Artificial Intelligence in Education and Training*, *4*(2).

Cantner, U., et al. (2021). Entrepreneurial lifecycle and innovation dynamics.

Chalmers, D.. (2021). Generative AI and startup transformations. *Entrepreneurship Theory and Practice*, *45*(3), 567–589.

Ferreira, J.. (2025). The transformative role of AI in modern entrepreneurship. *Business Horizons*, *68*(2), 134–146.

Gindert, M., & Müller, M. L. (2024). The impact of generative artificial intelligence on ideation and the performance of innovation teams. *arXiv*.

Hadhli, F., & Sheikh, H. (2022). Applications of artificial intelligence in design and its relationship to creativity and innovation: The Japanese experience as a model. *Journal of Legal and Economic Studies*, *5*(1).

Hiba, A. M., & Mohamed, I. (2021). The economic implications of the fourth industrial revolution: Artificial intelligence. *Series of Economic Studies*.

Huang, M.-H., & Rust, R. T. (2021). Artificial intelligence in service. *Journal of Service Research*, *24*(1), 3–41.

Kusetogullari, A.. (2025). Clustering AI research in entrepreneurship: A systematic review. *Journal of Innovation & Knowledge*, *10*(4), 78–89.

Mariani, M., & Dwivedi, Y. K. (2024). Generative artificial intelligence in innovation management: A preview of future research developments. *Journal of* Mariani, M., Machado, I., Magrelli, V., & Dwivedi, Y. K. (2023). Artificial intelligence in innovation research: A systematic review, conceptual framework and future research directions. *Technovation*, *122*, 102623.

McKinsey & Company. (2020). *AI Index Report*. McKinsey Global Institute.

Obschonka, M., & Audretsch, D. (2020). The impact of AI on entrepreneurial activities. *Small Business Economics*, *55*(3), 541–565.

Sayah, W. (2025). The role of artificial intelligence in enhancing innovation in entrepreneurship. *ASJP*, *10*(4), 118–128.

Wang, S., Attal, D., Rafiq, M., & Hubner-Benz, K. (2022). Leveraging GenAI for entrepreneurial decision-making. *Journal of Business Research*, *138*, 336–347.

White, M.. (2023). Prompt engineering: Maximizing large language model outputs for entrepreneurial use. *AI & Society*, *38*(3), 45–60.

Zeng, J., Khan, Z., & De Silva, M. (2022). The role of artificial intelligence in innovation: A systematic literature review. *Technological Forecasting and Social Change*, *179*, 121–158.

Chapter 11
Infrastructural Intelligence as Sociotechnical Institution:
Cloud-Native AI, Power, Governance, and Inequality in the Global South

Sneha Kadyan
https://orcid.org/0000-0001-6745-4419
Jindal Global Business School, O.P. Jindal Global University, Sonipat, India

Firdous Ahmad Malik
https://orcid.org/0000-0002-7815-0143
University of People, USA

ABSTRACT

Cloud-native AI tools are impactful in the automation of cognition and organization decision making. This chapter is directed at advancing the sociological concept of infrastructural intelligence that explains how the cloud-based infrastructures of cognitive automation get intertwined, reshaping labor, governance, and global power. This chapter is anchored in the studies of sociotechnical systems, labor process theory, platform capitalism, datafication, and global inequality, and argues that the cloud-native AI represents a change in the infrastructure of automation: moving from isolated, algorithmic processes to extensive, codified, and worldwide systems of cognition. Evidence from logistics, finance, manufacturing, digital labor, and public governance is brought together to show how cloud-native AI is transforming work and at the same time sustaining geopolitical imbalances. The chapter

DOI: 10.4018/979-8-3373-7442-0.ch011

presents a sociological perspective for understanding cloud-native AI and finally points out the ethical, environmental, and labor aspects that should be considered in future research.

1. INTRODUCTION: FROM AUTOMATED TASKS TO INFRASTRUCTURAL COGNITION

In the past, artificial intelligence was seen as a separate category made up of models or algorithms that were capable of performing cognitive tasks such as classification, prediction, and pattern recognition. However, the production of intelligence is no longer limited to such discrete models in the context of rising cloud-native architectures that include distributed clusters, containerized environments, automated pipelines, serverless functions, and managed orchestration. It is the infrastructure that actually provides it (Robbins & Wynsberghe, 2022).

Cloud-native AI tools offer the possibility of total automation of intelligence through the integration of the cognitive functions into a cloud platform that supports constant scaling, monitoring, retraining, and deployment. Thus, the automation of intelligence becomes a continuous process that is pervasive and infrastructural at a very deep level. This huge movement has far-reaching sociological implications; it changes the structure of labor, influences governance, and develops new dependencies between organizations and global cloud providers.

The main research question that this chapter revolves around is:

What are the means of Cloud-Native AI in automating intelligence and which sociological changes are the infrastructural shift responsible for?

To provide an answer, the chapter suggested the sociological concept of infrastructural intelligence as a means to study power, labor, and inequality in cloud-native AI systems.

2. THEORETICAL FRAMEWORK: A SOCIOTECHNICAL UNDERSTANDING OF CLOUD-NATIVE AI

2. 1 Sociotechnical Systems and Infrastructural Power

Sociotechnical systems theory points out the fact that social relationships influence the development of technology, and those infrastructures incorporate social practices into physical systems (Bowker et al., 2014; Edwards, 2003). Thus, cloud-native AI needs to be seen as a sociotechnical infrastructure composed of a variety of tools, pipelines, governance rules, and platform standards through which cognitive automation is done.

Cloud-native AI is a major transformation in the way that people, companies, and governments handle intelligence. Infrastructural intelligence sees cloud-native AI as a sociotechnical infrastructure that digitally integrates computational cognition into daily life, organizational decision-making, and state governance. It is not simply about understanding AI as a collection of algorithms, models, or tools.

Essentially, infrastructural intelligence highlights that cloud-native AI is not just a tool for the organizations, it is an organization itself. It gains the characteristics of sustainability, dependence on a specific way of doing things, and even the power to influence regulations similar to transportation networks or public utilities. However, its impact on classification, surveillance, and the distribution of benefits and harms is more nuanced, sometimes even hidden. Sociologically, this new perspective shows that cloud-native AI infrastructures create new types of automated authority, which produce what researchers call "algorithmic governance" (Yeung, 2018), "data colonialism" (Couldry & Mejias, 2019), and "infrastructural power" (Mann & Daly, 2020). The infrastructures are not neutral: they have inherent in them ideological beliefs about the importance of efficiency, productivity, and optimization and thus support neoliberalism in management, platform capitalism, and state rationalities aimed at control.

The globalization of the adoption of infrastructural intelligence in the Global South is significantly influenced by historical patterns of technological dependence, variations in regulation capacity, and the power dynamics between local and global platform monopolies. In this context, India stands out as a typical case of the new global trend wherein large-scale cloud infrastructures (AWS, Azure, Google Cloud), national digital platforms (Aadhaar, UPI, DigiLocker), and AI-driven public policy initiatives (NITI Aayog's National AI Strategy) are all closely interlinked to form hybrid governance regimes. Such regimes are characterized by the blurring of public–private boundaries and the extension of infrastructural intelligence into welfare distribution, identity verification, policing, agriculture, and health diagnostics. Thus, India's cloud-native AI is a double-edged sword of hope and intensified inequality at the same time. It provides new efficiencies while creating new forms of dispossession, automated exclusion, and regulatory unease.

Therefore, the concept of infrastructural intelligence enables us to interpret cloud-native AI as a sociopolitical formation rather than just a technological breakthrough. It emphasizes the position of AI in the global networks of data extraction, capital accumulation, and competition among states; the cultural values of innovation and modernity that are used to justify its expansion; and the different types of labor, from data annotation to algorithmic maintenance, that support its functioning. By this view, cloud-native AI does not surface as a mere instrument in the hands of society but as a powerful player that reshapes institutional reasoning, redefines the concepts of citizenship, and transfers power among public and private parties, this is a particularly important change for the Global South countries that are trying to reconcile their developmental desires with the harsh reality of structural inequality.

3. LITERATURE REVIEW: SOCIOLOGICAL SCHOLARSHIP ON AI, AUTOMATION, AND CLOUD INFRASTRUCTURES

The main ideas within science and technology scholarship emphasize that the relationship between technology and society is inseparable; hence, they are produced and used together (Bijker, 1994; Latour, 2005). Seeing it from a sociological angle, AI presents itself as a composition of data methods, human contributions, machine architectures, and the regulations of institutions (Suchman, 2007). In fact, the AI hosted in cloud strengthens the fact that it is deeply rooted as its infrastructures are serving the different sectors simultaneously. Analysts indicate the direction pointing to the fact that the algorithmic systems are changing the past and creating the new classifications, decision-making, and ordering (Fourcade & Healy, 2017).

3.1. Algorithms and Inequality

Critical race and feminist studies have pointed out that the AI is the culprit of not only the past and present inequalities but also the future ones through its techniques, for example, in the big data analysis where the racial and gender groups continue to be discriminated against (Benjamin 2023; Noble 2018). The literature on data colonialism puts forward the argument that the global data extraction is very much alike the old colonial pattern with the latter being a major concern for the countries in the Global South (Couldry & Mejias, 2019).

3.2 Labour, Platformisation and Cloud Economies

The works of labor and platform scholars have covered the topic of "algorithmic management" (Kellogg et al. 2020), the invisible workers behind AI (Gray & Suri 2019), and the emergence of temporary jobs in the form of small tasks (Tubaro et al. 2022). The transition from on-premise computing to cloud platforms also alters the distribution of power among global production networks, which means that only a few tech giants will be able to control the infrastructure power (Srnicek 2017).

3.3 AI in the Global South

Literature growing in volume and quality originating from India, Brazil, Kenya, and Southeast Asia claims that the Global South is not a merely passive AI user; it is an active area of data extraction, labor generation, and infrastructure experimentation (Arora 2024; Gupta & Gupta 2023; Sharma et al. 2023). Among the latter, the Aadhaar system in India, digital welfare architecture, platformized labor markets, and the rapid adoption of cloud-native AI in governance put India at the center of the debate about sociological implications at a large scale.

This chapter not only draws from but also significantly expands that literature by merging cloud-native AI systems with sociological studies of institutions, labor, inequality, and governance, always keeping in mind Global South realities.

4. CLOUD-NATIVE AI AS AN INSTITUTIONAL ACTOR: SOCIOLOGICAL PERSPECTIVES

Given the context of recent scholarship, cloud-native AI tools-systems that are designed to function within distributed cloud infrastructures-have been changing gradually from technical artifacts to institutional agents that influence daily activities, labor relations, and decision-making practices in organizations. As sociologists even more strongly claim algorithms and AI infrastructures must be regarded not just as tools but as "institutional agents" that are part of wider socio-technical installations (Fourcade & Healy 2017; Star 1999). To be sure, cloud-native AI is a technological category that has distinct characteristics compared to the former ones: scale, availability, and autonomy. No doubt these characteristics make it possible for AI to play a role in several layers of the organization-from HR management to logistics planning-while being mostly unnoticed by the people at the end of the line.

From the viewpoint of institutional theory, cloud-native AI acts to reconfigure fields of action by the introduction of new evaluation and performance logics (Kellogg et al., 2020). Organizations that implement cloud-native AI are increasingly using data-driven rationalities that focus on predictive analytics, optimization, and continuous monitoring. Professional identities are reshaped and expertise forms created, with data science skills being the only ones that get privileged and routines or tacit knowledge-dependent workers being the ones that get marginalized. Therefore, the embracing of cloud AI is hastening a larger sociological trend that culminates in the "algorithmization" of organizational life where decision-making is turned into a formalized, quantified, and automated process.

Furthermore, the cloud-native AI tools are involved in the gradual process of social inequalities. The cloud-native AI technologies get their training from the datasets which encode the past patterns of exclusion and the resulting biased predictions in such areas as recruitment, credit scoring, and predictive policing (Benjamin 2019; Noble 2018). The cloud-native infrastructures are responsible for a further increase of these biases through global and instantaneous scaling of decisions, thus new forms of "networked inequality" are created in which people in different places have to face the consequences of algorithmic classifications and at the same time are deprived of local accountability mechanisms. In this way, the cloud-native AI is not just a mirror of the social structures, but rather a tool that rewrites them, hence making the inequality computationally durable.

4.1 Labor, Automation, and the Reorganization of Expertise

Cloud-native AI tools reorganize labor in their staying ways and just at very job categories but also the social meanings of expertise, labor value, and professional autonomy. Automation has been the main focus of sociological discussions on work for a long time, but cloud-native AI introduces a completely new aspect: instead of taking over whole occupational roles, AI is splitting the tasks into parts and doing the specific small parts that are not too difficult and then the workers are supposed to integrate, supervise or correct the output of the algorithm (Gray & Suri, 2019). Teaching the workers, for instance, new hybrid roles are created-algorithmic supervisors, AI operations analysts, cloud integration specialists-these roles that further less the barriers between mental and physical labor.

There are empirical studies that have been done on the impact of cloud-native AI on organizations, and they reveal both "skill-biased" and "task-biased" forms of technological change. High-skill workers (such as data engineers or cloud architects) are the ones who get the most from the increase in demand, and rise in organizational status, while the middle-skill workers (administrative staff, technicians, and customer service agents) often experience a decrease in skill as AI systems automate the evaluation or troubleshooting tasks that require the least human intervention. The workers in the lowest-paid positions might see their work sent to global microwork platforms, where they carry out "ghost work" such as data labeling or content moderation for the purpose of training large-scale cloud AI models.

Cloud-native automation is now changing the way workplace governance works. Algorithmic management, which was mainly linked with gig platforms like Uber in the past, is now moving into traditional companies through the use of cloud-based monitoring tools that keep an eye on the productivity of workers, their communication habits, and the efficiency of their workflows. This transition along with the extension of cloud infrastructures will permit the monitoring to be done continuously around the clock in both physical and virtual workplaces, thus bringing up social issues of surveillance, consent, and worker agency rights. It has been pointed out that algorithmic governance like this has negative effects of stress increase among workers, reduced autonomy for workers, and mistrust to management (Möhlmann et al, 2021).

4.2 Datafication, Platformization, and New Regimes of Control

Cloud-native AI systems are, in fact, a significant part of the social and technological transformation towards datafication, the process which makes social and natural phenomena in the form of digital data that can be used for storage, analysis, and even monetization (Mejias & Couldry, 2019). Amazon Web Services (AWS), Microsoft Azure, and Google Cloud are all examples of major cloud providers which act as platform intermediaries and have the power over both the technical infrastructures and the economic and social architectures through which organizations take data, data as a new source of power and control. One negative consequence is the emergence of what some scholars call "platform capitalism," a notion where data flows and computational resources are the main avenues through which value is extracted rather than via traditional commodities (Srnicek, 2017).

The situation cloud native AI platforms create is that of a power imbalance between organizations and people as seen from a sociological point of view. Data which are necessary to teach AI models are generated by the workers/consumers, however, they have very little information about the data usage and the impact of algorithmic outputs on them. Analyzing this issue through the lens of surveillance capitalism one can argue that data extraction and prediction of behavior associated with data have become new means of social control (Zuboff, 2022). However, the power of cloud-native AI lies in the fact that it can uncloud such dynamics and hence allow the seamless and real-time data integration across various organizational departments, locations, and user types.

On top of that, platformization changes the technology's global political economy: nations that depend on overseas cloud providers are bound to be geopolitically more exposed, and the rights over digital sovereignty and data localization along with national autonomy are involved in the case. According to sociological research, countries in the Global South, such as India and some Southeast Asian nations, are caught in a dilemma of having to accept cloud AI in order to stay economically alive but at the same time trying to lessen the risk of being dependent on foreign digital infrastructures (Arora 2024). This situation can be compared with the logics of technological dependency in the past and it brings to the forefront the need for decolonizing digital infrastructures.

4.3 Ethical, Governance, and Justice Implications

Discussions about ethics and governance get louder wherever cloud-native AI systems are fully integrated into organizations. Sociology research emphasizes that ethical dilemmas are not solely due to technical hiccups but also reflect the prevailing societal inequalities and the power dynamics around technology. AI prejudices are already present in the pasts of the data they are trained on; obscurity is often a result of the secretive nature of cloud companies; and the lack of accountability seems to originate from the power disparities among those who create technology, the organizations, and the communities that are impacted.

Thus, the governance of cloud-based AIs should encompass multi-level interventions. Transparent model documentation, participatory design processes, and ethical impact assessments could be effective organizational-level measures to reduce risks. At the level of policy, the regulatory frameworks like the EU AI Act introduce risk-based governance approaches, requiring human oversight and algorithmic explainability for high-risk systems. Nonetheless, sociologists caution that deeper problems like labor exploitation, data colonialism, or platform monopolies cannot be solved by regulations alone (Couldry & Mejias, 2020). These issues need to be addressed through structural reforms, including the introduction of alternative data governance models that prioritize community ownership, digital commons, and community controls.

The call for justice in cloud-native AI has put a focus on relational accountability, which is to say that AI systems affected by the people have to make sure their opinions are heard, their interests are represented, and they have access to proper legal means for their claims. This framework based on critical race theory, feminist science and technology studies, and worker studies, questions the interests that support the AI infrastructures and the people whose lives are affected by them. In the end, the sociologically informed AI governance approach shows the issues of power, inequality, and the politics of technological development in a way that would not be possible with purely technical or managerial perspectives.

5. CLOUD EMPIRES AND PERIPHERAL FUTURES: GLOBAL INEQUALITY IN THE AGE OF INFRASTRUCTURAL INTELLIGENCE

The concept of cloud-native AI does not simply mean a shift in technology but also signifies a geopolitical and economic arrangement that makes the existing global disparities worse. As the intelligence of infrastructures is on the rise, the dependency described by academics as "digital dependency" has become even more prevalent

(Couldry & Mejias, 2019; Jin, 2024). This means that the countries in the Global South depend heavily on the computational infrastructures, proprietary AI models, and cloud ecosystems that are mainly controlled by Western companies, mainly the United States and, to a lesser extent, China. Such dependency is not just a matter of technology; it is an institutionalization of the differences between the regions in terms of knowledge production, regulatory freedom, economic power, and political power. The heavy concentration of cloud services among a few international companies, Amazon Web Services, Microsoft Azure, Google Cloud, and Alibaba Cloud, results in the infrastructures that are used by the governments and the companies being integrated into the corporate structures that the local governments cannot completely access, supervise, or control. As Star and Bowker (2006) point out, infrastructures act as "boundary-making" devices: they allocate visibility, authority, and legibility. Thus, cloud-native AI is a place of infrastructural geopolitics, where sovereignty is constantly negotiated through service contracts, requirements for data to be stored locally, and reliance on proprietary tools (Gwagwa et al., 2022; Milan & Treré, 2019).

In India, cloud reliance appears in various areas of governance and economic development. Initially, India branded itself to the world as the digital innovation leader with Aadhaar, UPI, CoWIN, and India Stack; however, the computations that were done in the backend and drove tons of AI deployments in the state and private sector were often on the cloud platforms of the global firms (Kumar, 2021; Sinha, 2024; Rao & Nair. 2019). Additionally, instances where governmental laws support data localization, the entire control over operations, management of models, and infrastructure design still remain with the multinational companies, thus, according to Birhane (2020), it is like "algorithmic colonization" reproduced. There are, therefore, dependencies abetting one another: the shapes of cloud pricing structures dictate the paths of startup innovations (Sreenivasan & Suresh, 2024), the access to the core AI models is limited by the restrictive licensing, and the infrastructural outages or changes in global policies may lead to the interruption of domestic welfare programs or loss of data in public databases. As a result, cloud-native AI fosters a situation of structural interdependence between the governments of the nations and the infrastructures of the transnational corporations, thus making the idea of digital sovereignty more complicated (Musiani, 2022).

Global inequalities, to a large extent, are reproduced in the process of data extraction and labor. Cloud-native AI heavily relies on massive amounts of data that are mostly sourced from the Global South, but the majority of the value extraction, modeling expertise, and infrastructural control still take place in the Global North (Ricaurte, 2019; Arora, 2024). Labor for annotation, content moderation, and model training is also largely outsourced to India, Kenya, and the Philippines, which is the type of labor Gray and Suri (2019) refer to as "ghost work." In the Indian cities of Bengaluru and Hyderabad, among others, the workers in the annotation hubs help in the passing of AI pipelines across the world without getting the same economic rewards or institutional power as the people in the global north (Qadri et al., 2023; Aneesh, 2015). This results in a well-known colonial pattern: the outward flow of data and labor and the inward flow of capital and control (Couldry & Mejias, 2019).

Cloud dependence at the governance level also impacts regulatory horizons. In many cases, it is governments in the Global South that are most affected since they do not have the technical capacity nor infrastructural autonomy to carry out audits of AI that resides on the cloud (Calzada, 2025). This gives rise to the situation described by Pasquale (2015) where a "black box society" comes into being and the regulators as well as the citizens are ignorant about the decision-making that takes place within the automated systems. The infrastructures that are cloud-based are being designed in such a way that they embody norms of privacy, efficiency, and optimization from the West and technical standards and platform design are the means through which the local regulatory frameworks are being influenced (Musiani et al., 2016; Eubanks, 2018). Consequently, cloud-native AI transfers normative beliefs that might not be in agreement with the local cultural, political, or developmental structures and thus creates a situation where infrastructure globality prevails over national sovereignty with a backdrop of tension.

Therefore, it can be concluded that the global cloud ecosystem is not only an AI host but also a power structure. Cloud-native AI enlarges the historical dependencies, creates new types of infrastructural subordination, and, at the same time, redistributes cognitive and economic power across the globe. For the Global South, especially India, this situation is urgent concerning the sociological and political questions raised about autonomy, institutional resilience, fairness, and the ability to shape technological futures without being limited by the constraints of transnational infrastructural dominance.

6. CONCLUSION: REIMAGINING CLOUD-NATIVE AI FOR SOCIAL FUTURES

Cloud-native AI is no longer just a technical paradigm but rather a power that the infrastructure is gradually reshaping social relationships, reversing the capacities of institutions, and changing the hierarchies of power in the world. The chapter has maintained that it is only by regarding cloud-native AI as a sociotechnical institution along with its material, economic, and epistemic foundations that the society's living, working, governing, and future imagining would be greatly determined. It is not that AI is only "in the cloud" anymore; on the contrary, the cloud has itself metamorphosed into the place where algorithmic governance, data extraction, platform capitalism, and geopolitical strategies meet. To realize this complexity one must not only abandon the overexaggerated narratives of innovation but also investigate thoroughly the social worlds, historical legacies, and political economies that support such infrastructures.

The infrastructural perspective indicates that cloud-native AI is at the heart of the issue of inequality. The situation brings about the matter of dependency that leads to the issue of the unevenness of power relations amongst the countries. AI data and models are some of the cases where the North has an advantage over the South, while the latter gets the short end of the stick in terms of digital apartheid. The Indian government has been in paradoxical positions as it tries to push the nation into a technological frontier and yet still remained inevitably connected to the global cloud player's infrastructural and financial demands. This situation has led to a change in the concept of national technological sovereignty and a need for a more subtle understanding of how "autonomy" is facilitated through globalized infrastructures. The sociological problem here is that one has to admit that the cloud is not equally shared out among the different parts of the world; it offers chances to some, creates vulnerability for many, and causes dramatic changes in governance, work, and social reproduction.

The social implications of cloud-native AI are at the same time very lockdown in the day-to-day things. The work is heavily relying on these invisible infrastructures, to the extent of reshaping the concepts of skill, labor, and value. The citizens come across the cloud through the social security systems, biometric identification, smartphone apps, and machine-made decision processes. Such encounters bring up essential issues regarding the status of humans, the quality of life, and the assignment of responsibilities in those societies where the technical infrastructures increasingly take the place of human judgment. The normalization of automated systems gives rise to the possibility that cloud-native AI might eventually become a non-questionable authority, hiding the uncertainties, disputes, and prejudices that are part of its design.

Thus, the undertaking of cloud-native AI futures must possess the capacity to retrieve the sociological imagination as both a methodological and a political means. It signifies that cloud-native AI should not be regarded as a necessary evolutionary change but rather as a non-definitive society of different factors shaped by the decisions made on the type of infrastructures, the beneficiaries of those interests, and the social lives opened or closed by them. It calls for a close examination of data policies, labor conditions, environmental effects, and the political economies supporting AI technology. Moreover, it demands consideration of the forms of resistance, creativity, and alternative infrastructures that are developing throughout the Global South, whether through feminist data practices, community Internet networks, open-source AI collectives, or regulations advocating for digital rights.

The main issue is not the progress of cloud-native AI but the values, ethics, and norms that will steer its progress. Sociologists, particularly those who are engaged in the study of technology, inequality, and governance, are pivotal in that process. Their understanding can reveal the extent to which cloud-native AI is a creator and a destroyer of domination and potential. They can also help to ensure that technological progress will not be solely the outcome of the interests of the corporations or the competition of the nations but will correspond to the more comprehensive visions of justice, dignity, and the common good.

Thus, in envisioning cloud-native AI for social futures, we should not let ourselves be influenced by the pessimistic stories of technological fate. Instead, we have to nourish an ethics and politics of infrastructural stewardship, one that gives priority to democratic governance, equitable access, sustainable labor practices, and inclusive participation in the design of digital worlds. The cloud-native AI has the potential to be a great supporting factor to the creation of just and resilient societies, but this would be only if its development is under the guidance of critical reflection, public accountability, and a devotion to changing infrastructures in ways that truly broaden human freedom and welfare.

REFERENCES

Aneesh, A. (2015). *Neutral accent: How language, labor, and life become global*. Duke University Press.

Arora, P. (2024). *From pessimism to promise: Lessons from the Global South on designing inclusive tech*. MIT Press.

Benjamin, R. (2023). Race after technology. In *Social Theory Re-Wired* (pp. 405–415). Routledge.

Bijker, W. E. (1994). Sociohistorical technology studies. *Handbook of science and technology studies*, 229-256.

Birhane, A. (2020). Algorithmic colonization of Africa. *Script-ed*, *17*, 389.

Bowker, G., Star, S. L., Gasser, L., & Turner, W. (2014). *Social science, technical systems, and cooperative work: Beyond the great divide*. Psychology Press.

Calzada, I. (2025). Decentralizing AI Economics for Poverty Alleviation: Web3 Social Innovation Systems in the Global South. *AI*, *6*(12), 309.

Couldry, N., & Mejias, U. A. (2019). Data colonialism: Rethinking big data's relation to the contemporary subject. *Television & New Media*, *20*(4), 336–349.

Edwards, P. N. (2003). Infrastructure and modernity: Force, time, and social organization in the history of sociotechnical systems. *Modernity and technology*, *1*, 185-226.

Eubanks, V. (2018). *Automating inequality: How high-tech tools profile, police, and punish the poor*. Macmillan+ ORM.

Fourcade, M., & Healy, K. (2017). Seeing like a market. *Socio-economic Review*, *15*(1), 9–29.

Gray, M. L., & Suri, S. (2019). *Ghost work: How to stop Silicon Valley from building a new global underclass*. Harper Business.

Gupta, R., & Gupta, R. (2023). Lost in the perilous boulevards of gig economy: Making of human drones. *SA Journal of Human Resource Management*, *10*(1), 85–106.

Gwagwa, A., Kazim, E., & Hilliard, A. (2022). The role of the African value of Ubuntu in global AI inclusion discourse: A normative ethics perspective. *Patterns (New York, N.Y.)*, *3*(4). PMID: 35465235

Jin, D. Y. (2024). The rise of digital platforms as a soft power apparatus in the New Korean Wave era. *Communication and the Public*, *9*(2), 161–177.

Kellogg, K. C., Valentine, M. A., & Christin, A. (2020). Algorithms at work: The new contested terrain of control. *The Academy of Management Annals*, *14*(1), 366–410.

Kumar, S. (2021). *The digital frontier: Infrastructures of control on the global web*. Indiana University Press.

Latour, B. (2005). An introduction to actor-network-theory. *Reassembling the social*.

Mann, M., & Daly, A. (2020). Geopolitics, jurisdiction and surveillance.

Milan, S., & Treré, E. (2019). Big data from the South (s): Beyond data universalism. *Television & New Media*, *20*(4), 319–335.

Möhlmann, M., Zalmanson, L., Henfridsson, O., & Gregory, R. W. (2021). Algorithmic management of work on online labor platforms: When matching meets control. *Management Information Systems Quarterly*, *45*(4), 1999–2022.

Musiani, F. (2022). Infrastructuring digital sovereignty: A research agenda for an infrastructure-based sociology of digital self-determination practices. *Information Communication and Society*, *25*(6), 785–800.

Musiani, F., Cogburn, D. L., DeNardis, L., & Levinson, N. S. (Eds.). (2016). *The turn to infrastructure in Internet governance* (pp. 268-p). New York: Palgrave Macmillan.

Noble, S. U. (2018). Algorithms of oppression: How search engines reinforce racism. In *Algorithms of oppression*. New York university press.

Pasquale, F. (2015). *The black box society: The secret algorithms that control money and information*. Harvard University Press.

Qadri, R., Shelby, R., Bennett, C. L., & Denton, R. (2023, June). Ai's regimes of representation: A community-centered study of text-to-image models in south asia. In *Proceedings of the 2023 ACM Conference on Fairness, Accountability, and Transparency* (pp. 506-517).

Rao, U., & Nair, V. (2019). Aadhaar: Governing with biometrics. *South Asia*, *42*(3), 469–481.

Ricaurte, P. (2019). Data epistemologies, the coloniality of power, and resistance. *Television & New Media*, *20*(4), 350–365.

Robbins, S., & Van Wynsberghe, A. (2022). Our new artificial intelligence infrastructure: Becoming locked into an unsustainable future. *Sustainability*, *14*(8), 4829.

Sharma, S., Bhardwaj, S., & Gupta, B. (2023). Algorithmic control: a disruption to motivation of gig workers? A critical review. *Multidimensional and strategic outlook in digital business transformation: human resource and management recommendations for performance improvement*, 1-14.

Sinha, N. (2024). Aadhaar, AI, and Identity: Negotiating Power and Surveillance in the Global South. *Социологическое обозрение, 23*(4), 80-112.

Sreenivasan, A., & Suresh, M. (2024). Can we unlock deep-tech in Indian startups for long-term success? *Technological Sustainability, 3*(1), 68–75.

Srnicek, N. (2017). *Platform capitalism*. John Wiley & Sons.

Star, S. L. (1999). The ethnography of infrastructure. *The American Behavioral Scientist, 43*(3), 377–391.

Star, S. L., & Bowker, G. C. (2006). How to infrastructure. *Handbook of new media: Social shaping and social consequences of ICTs*, 230-245

Suchman, L. A. (2007). *Human-machine reconfigurations: Plans and situated actions*. Cambridge university press.

Tubaro, P., Coville, M., Le Ludec, C., & Casilli, A. A. (2022). Hidden inequalities: The gendered labour of women on micro-tasking platforms. *Internet Policy Review, 11*(1), 1–26.

Yeung, K. (2018). Algorithmic regulation: A critical interrogation. *Regulation & Governance, 12*(4), 505–523.

Zuboff, S. (2022). Surveillance capitalism or democracy? The death match of institutional orders and the politics of knowledge in our information civilization. *Organization Theory, 3*(3), 26317877221129290.

Chapter 12
Flipped Learning Empowering Students Through Digital Innovation:
Reverse Instruction for Enhanced Engagement and Deeper Understanding

Priti Gupta
https://orcid.org/0000-0002-3000-9968
Greater Noida Institute of Technology, Greater Noida, India

Richa Adlakha
Manav Rachna International Institute of Research and Studies, India

Ashish Grover
Manav Rachna International Institute of Research and Studies, India

Neha Chaudhary
Manav Rachna International Institute of Research and Studies, India

ABSTRACT

One major deviation from traditional teaching approaches is flipped learning, sometimes known as the inverted classroom paradigm. Under this method, the conventional wisdom—that of lecturing in class and assigning homework for solo study—is flipped. In-class time is instead devoted to discussions, active learning exercises, and hands-on activities while educational materials—texts, movies, or

DOI: 10.4018/979-8-3373-7442-0.ch012

interactive modules—are distributed outside of the classroom. Fundamentally, flipped learning seeks to maximize in-person teaching time by means of interesting, dynamic activities promoting a better knowledge and application of ideas. There are several elements cooperating to establish a good learning environment. The core of the flipped learning model is the change from passive information absorption to active participation in the learning process.

HISTORICAL CONTEXT OF TRADITIONAL VS. FLIPPED CLASSROOMS

Figure 1.

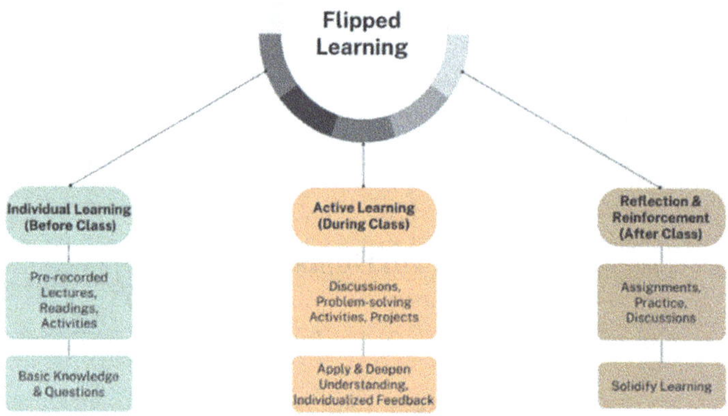

The basis of formal education for decades has been the traditional classroom configuration. Imagine orderly rows of desks, a chalkboard up front, and a teacher lecturing as pupils silently take notes—expectedly absorbing and using the knowledge on their own time. Aiming more for homogeneity and discipline than for encouraging creativity or customized learning, this "sage on the stage" approach became well-known during the industrial period.

But the pressures of schooling evolved along with the world. The flaws of this homogeneous teaching approach were underlined by the rapid development of technology, the increasing awareness of individual learning demands, and the knowledge that pupils absorb knowledge differently. Students started to be passive listeners, and lectures sometimes dominated valuable class time, therefore leaving little room for interaction, research, or hands-on application(Ahmed,2016).

This led to the transformed method known as the flipped classroom, which reinterpreted the conventional learning environment. Students first come upon fresh material outside the classroom in this approach via texts, films, or digital resources. With the teacher more functioning as a facilitator or "guide on the side" than as a lecturer, this change lets classroom sessions concentrate on interactive discussions, group projects, and customized help.

THE SHIFT IN EDUCATIONAL PARADIGMS

From conventional to flipped learning, the transition signals a major shift in the educational perspective rather than only a structural one in the delivery of the lessons. Under a teacher-led approach, classrooms have traditionally functioned whereby teachers provide material during class and students are expected to examine, consider, and use the knowledge thereafter on their own. Rooted in the industrial-age model of education, this "lecture-then-homework" method sometimes gave conformity and compliance top priority over creativity and personal learning styles.

But the limits of this traditional approach have become more clear as the requirements of students have gotten more varied and the function of technology in education has grown more important. With little chance for inquiry, group projects, or individualized feedback during class, students in conventional classrooms sometimes participate passively (Holmes & Gee, 2016).

Reimagining the learning process, the flipped classroom concept solves these difficulties. Here, before class, students access teaching materials such movies, texts, or interactive modules. This lets in-class time be used more deliberately, involving students in group projects, homework, and problem-solving exercises. The instructor transforms from being the "sage on the stage" into a "guide on the side," helping and mentoring students as they negotiate and apply ideas right away.

Several important educational priorities form the foundation of this paradigm change:

- Encouragement of students to take responsibility for their learning process and grow more autonomous and motivated learners results from student-entered learning.
- Encouragement of research, discovery, and instantaneous comments during class sessions is active engagement.
- Establishing chances for students to collaborate, exchange ideas, and jointly acquire knowledge would help them to develop cooperative environments.
- Deeper Understanding: Emphasizing application, analysis, and evaluation above surface-level memorizing (Larcara,2015).

Modern education calls for flexibility, engagement, and personalization—all of which the flipped approach readily accommodates. This change fits learning models of the twenty-first century, in which knowledge of skills including creativity, critical thinking, communication, and self-management is just as important as topic mastery. Teachers need on a set of digital tools supporting every stage of the learning process if they are to properly apply flipped learning. These instruments—which list in Figure 2 include:

- Learning Management Systems (LMS) include Google Classroom and Canvas provide effective student progress monitoring, resource distribution, and course content organization.
- Edpuzzle, Loom, among other tools let professors build and distribute interesting pre-class materials for access by students anywhere.
- Platforms for interactive collaborative projects, brainstorming, and visual thinking such Padlet, Jamboard, and Miro help.
- Through gamified learning, assessment tools as Kahoot!, Quizziz, and Socrative let for quick formative evaluations, immediate feedback, and higher student motivation.

Figure 2. Key tools in the digital classroom

Tool Type	Icon Idea	Examples	Use in Flipped Learning
LMS	📚	Google Classroom, Canvas	Organize content, track progress
Video Tools	🎥	Edpuzzle, Loom, Screencastify	Deliver pre-class instruction
Collab Tools	🤝	Padlet, Jamboard, Miro	Brainstorming, group work
Assessment Tools	✅	Kahoot!, Quizizz, Socrative	Formative checks, engagement boost

Beyond the technologies, flipped learning is anchored in established educational ideas including constructivism, which stresses the active role of learners in building knowledge, and Bloom's taxonomy, which promotes advancement beyond basic recollection to higher-order thinking. This approach fosters student agency, therefore empowering students to take charge of their education and supporting lifetime learning practices. The change also calls for a mental adjustment for teachers as well as for students. sTeachers have to welcome fresh responsibilities as creators of rich, interactive learning opportunities, mentors, and facilitators. Students must then adjust to a more participatory role, arriving at class ready to thoroughly interact with the content. Fundamentally, flipped learning is a revolutionary concept redefining what it means to teach and learn in the digital age rather than only a pedagogical tool. Placing students at the center of the experience and using technology's power helps them to develop the skills and attitude they need to thrive in a complicated, rapidly changing environment (Greipl et al., 2020).

TIMELINE OF FLIPPED LEARNING – LINKED TO HISTORY

Over the past two decades, the development of flipped learning represents a fundamental paradigm shift in educational strategies. This instructional model—inverting conventional teaching by transferring lectures to homework and changing classrooms into interactive environments—has shown better student outcomes and involvement. Emphasizing the major turning points that changed worldwide classroom education, Figure 3 shows this development from traditional teaching to today's technologically advanced flexible learning environments.

Pre-2000s – Traditional Teaching

- **What happened:** Teachers mostly lectured in class and gave homework to practice.
- **History link:** This method followed the *Industrial Age model* – same lessons for all, focused on memorization and routine.

Early 2000s – Rise of Online Videos & Tech

- **What happened:** Teachers started recording lessons and posting them online.
- **History link:** With the growth of the *internet and YouTube (2005)*, access to information became easier. Teachers explored using tech to teach differently.

2007 – Birth of Flipped Learning

- **What happened:** Teachers Jonathan Bergmann and Aaron Sams started flipping their classrooms by giving video lectures as homework.
- **History link:** This was during the *Digital Learning Boom*, when schools began using more technology in education.

2010s – Flipped Learning Spreads Worldwide

- **What happened:** More schools began using flipped classrooms. Research showed it could help students learn better.

- **History link:** During this time, there was a push for *student-centered learning*, focusing more on critical thinking than just memorizing.

2020 – COVID-19 Pandemic

- **What happened:** Schools closed and moved online. Flipped learning became more popular because it helped students learn from home.
- **History link:** The *global shift to remote learning* made flipped learning an important tool for education during lockdowns.

Today (2020s) – Blended & Flexible Learning

- **What's happening:** Flipped learning is now used with other methods like hybrid or blended learning.
- **History link:** Education today is shaped by *technology, flexibility, and personalized learning paths.*

Figure 3. Evolution of flipped learning (pre-2000s to present)

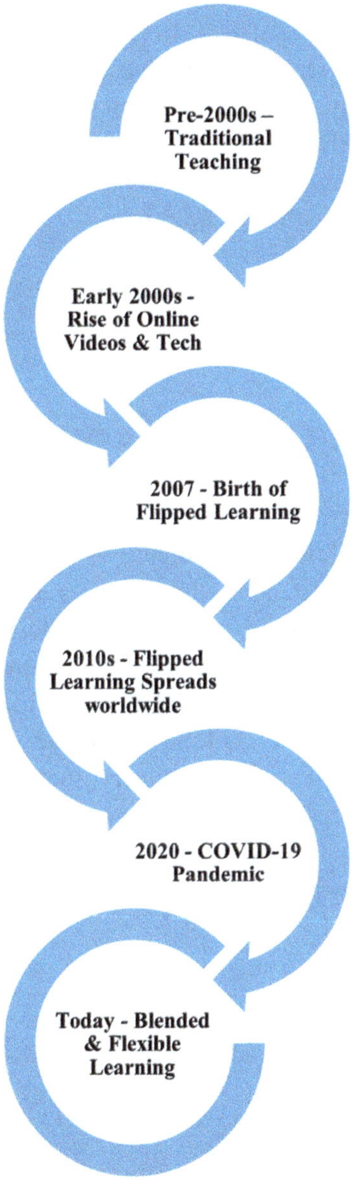

Important educational theories anchor this change in instructional design:

- Constructivism stresses how actively students create their own knowledge via means of experience, introspection, and group projects. By assigning class time to peer conversation and problem-solving, flipped classrooms help to foster this.

- Cognitive learning is classified by Bloom's Taxonomy into six levels: Remember, Understand, Apply, Analyze, Evaluate, and Create. While children struggle with higher-order activities at home, traditional classrooms typically cover just the lower levels (Remember, Understand) during class. Under flipped learning, deeper engagement occurs in class while basic ideas are taught before to class (Foster & Shah, 2020).

These ideas taken together encourage a more active, reflective, student-driven approach to learning.

PASSIVE VS. ACTIVE LEARNING: A PARADIGM SHIFT

The shift from passive to active learning marks a significant transformation in educational practices. As illustrated in Figure 4, traditional learning environments position students as passive listeners, while teachers act as primary information providers. Class time is mostly lecture-based, with homework focused on advanced tasks.

In contrast, active learning—often implemented through flipped classrooms—encourages students to be explorers, collaborators, and creators. Teachers become facilitators and mentors, using class time for discussions, collaborative problem-solving, and critical thinking. Content delivery, such as videos or readings, takes place outside the classroom (Bishop & Verleger, 2013).

This approach also emphasizes personalized, interactive learning, moving away from one-size-fits-all models. It fosters deeper engagement, better retention, and essential 21st-century skills.

Figure 4. Traditional vs. flipped learning comparison

Aspect	Traditional (Passive)	Flipped (Active)
Role of Student	Listener, note-taker	Explorer, collaborator, creator
Role of Teacher	Lecturer, information giver	Facilitator, guide, mentor
Class Time Used For	Lectures	Discussion, problem-solving, collaboration
Homework	Practice or higher-level tasks	Content absorption (videos, readings)
Learning Process	One-size-fits-all	Personalized, interactive

THE MINDSET SHIFT FOR TEACHERS AND LEARNERS

Turning to flipped learning calls for a basic mental change for teachers as well as for students, as Fig. 5 summarizes:

For Teachers:

- Let go of control over every moment of instruction
- Trust students to prepare before class
- Redefine success as student engagement and inquiry—not content delivery
- Embrace the role of *coach and facilitator* rather than sole knowledge source

For Students:

- Realize that learning doesn't only happen during lectures
- Take responsibility for *preparation and participation*
- See class as a space for *experimentation, collaboration, and critical thinking*
- Move from passive compliance to *active ownership*

Figure 5. Mindset shift in flipped learning

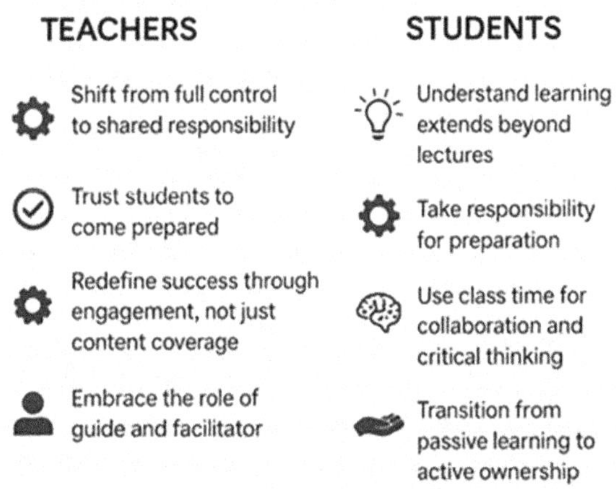

FLIPPED LEARNING'S INTERSECT WITH GAME-BASED LEARNING

Flipped Learning's Intersect with Game-Based Learning

Combining the ideas and techniques of both approaches to education—flipped learning and game-based learning—allows a dynamic approach that produces interesting and participatory learning opportunities. While game-based learning uses the motivating power of games to increase student engagement and learning outcomes, flipped learning, with its emphasis on moving direct instruction outside of class and using in-class time for active learning activities, complements.

Active participation and individualized learning are fundamental ideas guiding the junction between flipped learning and game-based learning. Whether by pre-recorded films, texts, or interactive modules, students in a flipped classroom have the chance to interact with instructional materials at their own leisure outside of class. When students actively engage in immersive and interactive gameplay experiences that support problem-solving, critical thinking, and skill development, this asynchronous learning paradigm fits quite nicely with ideas of game-based learning (O'Flaherty et al.,2015).

Furthermore, including game-based components and mechanics into flipped learning projects helps to increase student involvement and drive. Incorporating components like challenges, rewards, progress systems, and feedback loops into flipped learning materials helps teachers create dynamic and interactive learning opportunities that grab students' interest and maintain their drive. Leaderboards, points, and badges are among gamification techniques that could inspire student involvement even more and help them to feel successful and accomplished.

Moreover, the junction of game-based learning and flipped learning presents chances for uniqueness and tailored education. Using the adaptability of flipped learning, teachers can give students access to a range of instructional tools and materials catered to their particular learning requirements and preferences. Customizing game-based learning activities to provide adaptive feedback, scaffolding, and challenges depending on student performance and progress guarantees that every student gains tailored support and experiences success (Lo wt al.,2017).

Additionally essential elements of the junction between game-based learning and flipped learning are social interaction and cooperation. Among students, multiplayer games, cooperative challenges, and group problem-solving exercises inspire peer-to--peer connection, communication, and teamwork. Under a flipped classroom, students might work together to solve challenges and accomplish shared goals, participate in conversations and debates during in-class activities, and complete pre-class tasks.

All things considered, the junction of flipped learning and game-based learning presents a potent educational tool that combines the best features of both approaches to produce interesting, dynamic, and individualized learning opportunities. Teachers can create dynamic and successful learning environments that support deep learning and academic success by using the ideas of flipped learning to give students access to instructional materials outside of the classroom and including game-based elements and mechanics to encourage active engagement, motivation, and collaboration (Zainuddin et al.,2016).

THE ADVANTAGES OF FLIPPED LEARNING IN THE CLASSROOM

Offering several benefits that help to increase student engagement, deeper learning, and better academic results, flipped learning is a creative teaching method Its ability to customize education to match particular student requirements is one main advantage. Teachers may accommodate several learning styles, speeds, and preferences by providing teaching materials outside of the classroom via movies, texts, or online modules. Increased understanding and retention follow from students' ability to examine information at their own speed, revisit difficult ideas, and interact with things in ways that fit their learning style.

Furthermore, flipped learning encourages active involvement in the course of instruction. In-person meetings might center on interactive activities, group projects, and hands-on exercises while conventional lectures move outside of class time. This change from passive listening to active learning increases critical thinking, knowledge, and ability to solve problems. Deeper engagement and topic mastery follow from students applying information, having meaningful conversations, and getting instantaneous comments from teachers and peers.

Flip learning also advances a more student-centered learning environment. Flip learning promotes autonomy, initiative, and self-regulated learning abilities by arming students to take control of their education and providing chances for self-directed inquiry. As they become metacognitively, goal-setting, and responsible for their academic performance, students start actively participating in their education. Along with improving their learning experience, this greater autonomy helps kids be ready for lifetime education outside of the classroom (Talbert et al.,2013).

Moreover, flipped learning could enable better utilization of resources and instructional time. Teachers may optimize face-to--face time for interactive, group projects that advance higher-order thinking abilities by front-loading materials distribution outside of class. This best use of class time helps teachers to go farther into difficult subjects, provide customized guidance, and instantly clear misunderstandings. Furthermore, as digital resources and online technologies provide more dynamic, customized, and current information, flipped learning may help to lessen dependence on conventional textbooks and supplemental materials.

Furthermore linked to better student performance and accomplishment has been flipped learning. Studies of flipped classrooms have shown that students there typically exhibit better degrees of involvement, drive, and pleasure with their educational encounters. Furthermore connected to flipped learning have been higher rates of student retention, better test results, and more general academic performance. Flip-learning helps students to realize their full potential and attain academic achievement by giving them more chances for active learning, tailored education, and meaningful connection.

All things considered, flipped learning provides a wide range of advantages that could profoundly affect student results and experiences. Flip learning has the power to transform education and provide students with the tools they need for success in the twenty-first century and beyond by means of tailored teaching, active participation, student-centered learning, effective resource use, and improved academic accomplishment.

GAME-BASED LEARNING'S BENEFITS FOR EDUCATION

Many benefits from game-based learning help to improve learning results and boost educational opportunities. One obvious advantage is the natural involvement and inspiration games provide for pupils. Games, meant to be interactive and immersive, grab kids' attention and inspire natural drive to study. Game-based learning promotes active involvement and keeps students' concentration by means of interesting gaming mechanics, challenges, rewards, and progression systems, therefore fostering a pleasant atmosphere fit for exploration and discovery.

Furthermore, game-based learning helps to build critical thinking, cooperation, decision-making, and problem-solving among other vital abilities. Through offering significant obstacles and strategic thinking chances within the gaming environment, students develop abilities applicable in the real world. Games also provide instantaneous feedback and chances for introspection so that students could gain from failures, improve their tactics, and develop a growth mindset.

As students work on problem-solving and decision-making, game-based learning also encourages active learning and involvement. Unlike conventional education, in which pupils could passively absorb knowledge, game-based learning calls for active participation, therefore fostering better knowledge and memory of ideas. Through games, students experience immediate results of their choices and actions, therefore applying information in context (Doung-In, 2017).

Moreover, game-based learning may accommodate different learning environments and preferences, therefore meeting the particular requirements of every learner. Games provide individualized experiences for every learner by being built to fit varying skill levels, tempo, and learning goals. Games also provide children a secure space for experimentation and risk-taking where they may investigate ideas, run tests, and grow from mistakes free from worry about bad outcomes.

All things considered, game-based learning provides a spectrum of advantages that improve student involvement, drive, and learning results. Game-based learning has the potential to transform education and produce immersive learning opportunities that inspire curiosity, creativity, and lifetime learning by means of intrinsic motivation, skill development, encouragement of active participation, and accommodation of various learning needs.

PRINCIPLES OF EFFECTIVE GAME-BASED LEARNING

To optimize its effect on student engagement, motivation, and learning results, effective game-based learning depends on three fundamental ideas. First of all, one must match learning goals. Clear educational objectives should guide the design of games so they satisfy certain learning goals or criteria. Teachers may make sure gaming supports and enhances intended learning outcomes by matching game mechanics, content, and activities with learning goals.

Second, significant integration of gaming aspects is really vital. To inspire and involve pupils, games should include components like challenges, prizes, feedback systems, and development tools. Careful design of these game features should help to encourage among players active engagement, problem-solving, critical thinking, and teamwork. Meaningful incorporation of game components improves the whole learning process and enables students to acquire necessary abilities while submerged in games. Furthermore essential ideas in efficient game-based learning are scaffolding and advancement. Games should provide suitable degrees of difficulty and assistance to fit students' learning requirements and degree of ability. As students go through the game, scaffolding techniques— tutorials, tips, and progressive difficulty levels—help them negotiate obstacles and develop confidence. Furthermore, well-crafted progression algorithms guarantee that gaming stays interesting and demanding, inspiring students to keep developing their knowledge (Abeysekera & Dawson, 2015).

Moreover, good game-based learning depends much on comments and reflection. Games should provide pupils timely, instructive comments that emphasize their achievements, point out areas needing work, and direct their next actions. Feedback systems have to be subtly included into games so that they provide players chances for contemplation and actionable insights. By use of comments and introspection, students may monitor their development, modify their approaches, and increase their grasp of game-presented ideas.

Essential ideas in game-based learning also include social interaction and cooperation. Games should provide chances for peer-to--peer contact, group learning, and cooperative play—all of which help players to build a feeling of community. Cooperative gaming improves students' social and emotional abilities by means of communication, bargaining, and idea sharing, therefore strengthening the inclusive and friendly learning environment. Effective game-based learning is, all things considered, driven by ideas that provide alignment with learning goals top priority, meaningful integration of game features, scaffolding and progress, feedback and reflection, and teamwork and social interaction top priority. Following these ideas can help teachers create and carry out game-based learning activities that include students, inspire learning, and therefore foster the acquisition of fundamental skills in a dynamic and engaging way.

STRATEGIES FOR INTEGRATING GAMES INTO FLIPPED LEARNING ENVIRONMENTS

Including games into flipped learning settings will significantly improve student involvement, knowledge, and memory. Using games as pre-class assignments—where students engage with instructional games or simulations prior to class—is a major tactic. These exercises guarantee students come ready to actively engage in conversations, activities, and problem-solving exercises during the in-person session, therefore acting as interesting introductions to important ideas and vocabulary.

Including game-based evaluations to gauge student knowledge and development is another very successful strategy. Teachers may create game-like examinations challenging students to use their knowledge in simulated environments or problem-solving tasks rather than conventional quizzes or tests. These tests provide a more real and interesting evaluation experience along with insightful analysis of students' grasp of the content.

Moreover, teachers might include gamified components into their lesson plans to increase participation and drive. This might include including within the course of instruction elements like badges, levels, awards, or leaderboards to motivate advancement and success. Gamified components inspire students to actively engage and flourish in their educational path by generating a feeling of rivalry, success, and advancement. Furthermore among students, cooperative gaming may help to develop social interaction, communication, and collaboration. Teachers may create cooperative or multiplayer games wherein students cooperate to solve issues, finish tasks, or reach shared objectives. These encounters build community, support peer-to--peer learning, and improve students' communication and teamwork abilities. Furthermore, teachers may design unique games or gamified learning experiences catered to particular learning goals and curriculum using tools and platforms of game-based learning. These sites often provide tools, templates, and assistance for teachers to properly create and carry out game-based activities. Game-based learning systems allow teachers to easily include games into their flipped learning settings, therefore giving their students interesting and dynamic learning opportunities.

All things considered, introducing games into flipped learning settings provides several advantages like more engagement, deeper knowledge, and more student cooperation. Teachers may develop dynamic and immersive learning environments that spark inquiry, encourage creativity, and enhance learning results by using gamified components, assessments, game-based materials, cooperative experiences, and learning platforms (Chen et al.,2014).

CASE STUDIES: SUCCESSFUL IMPLEMENTATIONS OF FLIPPED LEARNING WITH GAME-BASED APPROACHES

Case examples of effective implementations highlight the transforming power of combining game-based technologies with flipped learning in the classroom. One prominent instance is the use in a high school biology class of a flipped classroom format together with game-based learning. In one instance, the teacher offered pre-class homework based on educational films she produced addressing fundamental biology ideas. Students participated in a variety of game-based activities including virtual lab simulations, biology-themed escape rooms, and interactive quizzes using instructional games throughout class. This combined strategy improved student involvement and motivation as well as deeper knowledge and memory of difficult biological ideas.

The integration of a gamified learning platform into a flipped classroom environment in a middle school math lesson is even another interesting case study. Using a gamified learning platform with interactive math games fit for the curriculum, adaptive assessments, and individualized learning paths, the teacher To be ready for in-class assignments and tests, students watched readings and instructive movies at home. Students competed in math-themed activities, worked on group projects, and received badges and awards for their successes throughout class. Increased student involvement, better math skills, and a more favorable attitude toward studying mathematics were outcomes of this gamified flipped learning strategy (Lage et al.,2014).

Moreover, a case study done in a college-level computer science course shows how well integrating game-based techniques with flipped learning improves student learning results. In one instance, the teacher developed interactive coding lessons and online assignments for students to do before class, therefore laying a basis for in-depth conversations and practical coding projects carried out in-class. Students worked on coding projects in teams, entered coding contests, and presented their works at a game development expo to support learning and encourage teamwork. This combined approach not only enhanced student knowledge of programming ideas but also promoted cooperation, creativity, and problem-solving ability so vital for success in the area of computer science.

From high school biology to middle school math to college-level computer science, these case studies show generally the effective integration of flipped learning with game-based methodologies in many educational settings. Combining game-based tactics with flipped learning approaches allows teachers to provide dynamic and immersive learning opportunities that include students, advance deeper knowledge, and improve learning results in many fields.

CHALLENGES AND SOLUTIONS IN IMPLEMENTING GAME-BASED FLIPPED LEARNING

Game-based flipped learning presents a number of difficulties, however pro-active approaches may help to overcome them very easily. One major challenge comes from the possible opposition to change among students and teachers both. While students could be used to conventional teaching strategies, teachers might be reluctant to embrace new technology and pedagogical techniques. Providing professional development seminars, training courses, and continuous assistance can let teachers familiarize themselves with game-based flipped learning approaches and increase confidence in their use, therefore addressing this obstacle. Likewise, giving students concise justifications of the advantages of game-based learning and chances to see success with gamified projects helps to reduce resistance and build buy-in (Strayer,2017).

The availability and accessibility of tools and technology required for game-based flipped learning provide even another obstacle. Not every student would have the necessary tools, internet connection, or software programs to interact with digital games and multimedia materials outside of the classroom. Offering a combination of digital and non-digital learning activities can help to meet various student demands and therefore address this difficulty. Furthermore, guaranteeing fair participation and inclusion for every student is offering substitute ways of access include loaner devices, offline resources, or printed materials.

Teachers have trouble making sure game-based activities complement learning goals. Games have to complement the intended learning goals and curricular requirements even if they might be very interesting and motivating. Teachers should carefully choose or create games fit for certain learning objectives and competences in order to meet this challenge. Including reflective evaluations, conversations, and debriefing sessions will enable students to relate their gaming experiences to more general learning goals and apply learned information and abilities in practical settings(Enfield,2013).

Evaluating student development and knowledge within a game-based flipped learning environment might be difficult. Conventional tests may not be able to adequately represent the sophisticated abilities and competences acquired via game-based activities. Teachers may use performance-based assessments, portfolio evaluations, peer assessments, and self-assessments among a range of formative and summative assessment tools to meet this problem. These tests should be intended to evaluate not just topic knowledge but also higher-order thinking skills, problem-solving capacity, and teamwork gained via game-based learning events.

Furthermore, in game-based flipped learning contexts, keeping student involvement and motivation throughout time poses continuous difficulty. Although games might first grab students' attention and excitement, keeping involvement throughout the course of instruction calls for deliberate design and execution plans. Teachers could include aspects of choice, autonomy, challenge, and advancement into game-based activities to meet this difficulty. Frequent updates and new game material, timely feedback, and social components like competition and cooperation assist to keep motivation and interest over time.

All things considered, even if game-based flipped learning offers difficulties, proactive approaches and solutions may efficiently remove these barriers. Teachers may build dynamic and immersive learning environments that improve student engagement, knowledge, and accomplishment by overcoming resistance to change, guaranteeing accessibility and alignment, developing effective evaluations, and keeping involvement and motivation.

ASSESSING LEARNING OUTCOMES IN GAME-BASED FLIPPED LEARNING ENVIRONMENTS

Evaluating learning outcomes in game-based flipped learning environments presents educators with both challenges and opportunities. One primary consideration is crafting assessments that effectively gauge the multifaceted skills and competencies fostered through game-based activities. Traditional assessment methods may fall short in capturing the nuanced nature of learning outcomes in this setting. Hence, educators should deploy a mix of formative and summative assessment strategies tailored to the specific learning objectives and activities inherent in game-based flipped learning.

Performance-based assessments emerge as particularly fitting for appraising learning outcomes in game-based flipped learning environments. These assessments task students with showcasing their knowledge, skills, and competencies through authentic tasks or projects mirroring real-world challenges. For instance, students might tackle a game-based scenario, tackle a complex problem, or produce a multimedia project demonstrating their grasp and application of critical concepts. Performance-based assessments not only offer a more accurate gauge of students' abilities but also foster deeper learning and knowledge retention (Herreid & Schiller, 2013).

Besides performance-based assessments, educators can embed formative assessment techniques throughout the learning journey to monitor student progress and comprehension. These may encompass quizzes, polls, discussions, peer assessments, self-assessments, and reflection activities integrated within the game-based learning experience. Formative assessments furnish valuable feedback to educators and students alike, allowing for continuous monitoring of learning outcomes and avenues for adjustments or reinforcement as needed.

Furthermore, educators can leverage digital tools and learning analytics to appraise student engagement, participation, and performance within game-based flipped learning environments. Learning management systems (LMS) and game-based platforms often incorporate built-in analytics features tracking student interactions, progress, and accomplishments within the gaming milieu. Analyzing this data empowers educators to glean insights into student learning behaviors, pinpoint areas of strength and improvement, and tailor instruction to individual student requirements.

Crucially, educators must ensure alignment between assessments and learning objectives in game-based flipped learning environments. Assessments ought to be meticulously designed to measure the precise knowledge, skills, and competencies outlined by the learning objectives. They should be authentic, meaningful, and relevant to the game-based activities and learning experiences students partake in. Furthermore, assessments should furnish opportunities for students to exhibit higher-order thinking skills, problem-solving prowess, and collaborative abilities cultivated through game-based learning.

In essence, assessing learning outcomes in game-based flipped learning environments necessitates thoughtful planning and alignment of assessment strategies with the unique characteristics of this pedagogical approach. By employing performance-based assessments, formative assessment techniques, digital tools, and learning analytics, educators can aptly measure student learning outcomes, stimulate deeper learning, and augment the overall efficacy of game-based flipped learning experiences (Gilboy et la.,2015).

RESEARCH METHODOLOGY

The degree of flipped learning adoption and its view among teachers and students in educational institutions is investigated in this paper using a quantitative research approach. Structured questionnaires sent online to a varied sample of teachers and students from many fields and institutions collected data. The aim was to see the patterns using informative pie charts and evaluate the acceptance, exposure, and preference connected to flipped learning approaches.

The approach consists in three main areas of emphasis:

(i) Teachers embracing flipped learning,
(ii) Student experience and exposure; also,
(iii) Views of students about flipped learning.

To improve the interpretability and clarity of the data, each of these features is shown with matching pie charts—Figures 6 through 8.

Teacher Adoption Of Flipped Learning

As shown in Fig. 6, 35% of educators reported regularly employing flipped learning strategies in their classrooms. Another 30% of respondents indicated that they use flipped learning occasionally but not consistently, depending on subject content and student dynamics. A notable 20% of teachers were aware of the concept but had not yet implemented it, while 15% had no prior exposure to the term "flipped learning." This breakdown indicates that while awareness of the concept is growing, full-scale integration into teaching practices is still evolving. This data provides a critical insight into the gap between awareness and implementation. It also suggests a need for more training and resource development to help teachers embrace and apply flipped learning effectively in their pedagogical methods.

Figure 6. Teacher adoption of flipped learning

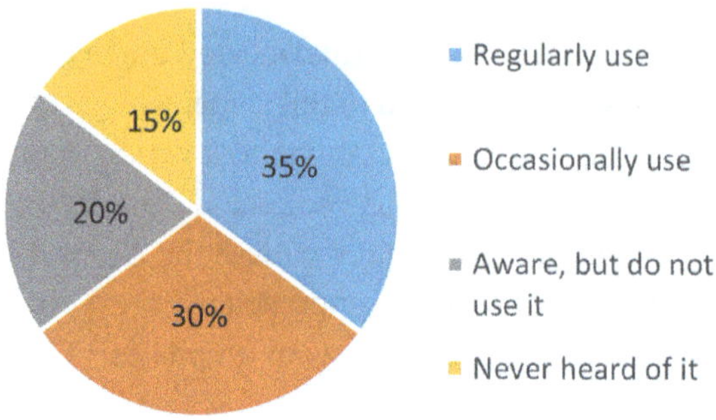

Student Experience with Flipped Learning

The second focus area involved examining the extent of exposure students have had to flipped learning environments. According to the results illustrated in Figure 7, 25% of students reported that they frequently experienced flipped instruction in their academic journey. Meanwhile, 40% of students had encountered flipped methods only occasionally, suggesting inconsistent adoption across classrooms and institutions. Another 20% rarely encountered flipped learning, and 15% had never experienced it at all.

This distribution highlights the unequal access to flipped learning and underscores the variability in institutional adoption. While some students benefit from this interactive learning model, others remain confined to conventional methods, suggesting an urgent need for standardization and broader institutional integration (McLaughlin & Jacqueline,2014).

Figure 7. Student experience with flipped learning

Category	Percentage
Frequently learn through flipping	25%
Sometimes experience flipped lessons	40%
Rarely experience flipped learning	20%
Never experienced flipped learning	15%

Student Preferences Toward Flipped Learning

To assess how students perceive flipped learning, survey participants were asked to express their preferred mode of instruction. As depicted in Fig. 8, 45% of students expressed a preference for flipped learning over traditional lectures. Another 35% favoured a blended approach, combining both flipped and conventional teaching methods. Only 20% of students preferred strictly traditional lectures, signalling a broader acceptance and positive inclination toward student-cantered learning approaches.

These findings align with literature that emphasizes student preferences for active, flexible, and technology-integrated learning environments. The strong preference for flipped or hybrid instruction suggests that students appreciate autonomy, engagement, and interactive class time, which are integral elements of flipped learning.

Figure 8. Student preference for flipped learning

Category	Percentage
Prefer flipped over traditional	45%
Like a mix of both methods	35%
Prefer traditional teaching	20%

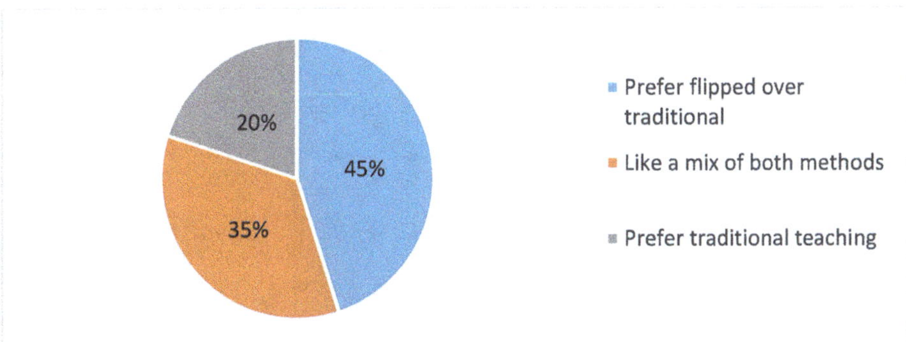

Best Practices and Tips for Teachers Using Game-Based Flipped Learning

Implementing game-based flipped learning effectively requires educators to adopt best practices and strategies tailored to this unique instructional approach. One essential practice is to carefully select or design educational games that align with learning objectives and engage students effectively. Educators should choose games that offer meaningful learning experiences, foster critical thinking, and promote skill development relevant to the curriculum. Additionally, games should be age-appropriate, accessible, and inclusive to accommodate diverse student needs and preferences.

Furthermore, educators should integrate game-based activities seamlessly into the flipped learning environment by aligning them with pre-class and in-class instructional materials and activities. Games should complement and reinforce the content covered in pre-class videos or readings, allowing students to apply and extend their learning through interactive gameplay. During in-class sessions, educators can facilitate discussions, reflection, and application exercises that bridge the gap between game-based activities and deeper conceptual understanding. Refer to Figure 9:

Figure 9. Game-based flipped learning: A continuous learning cycle

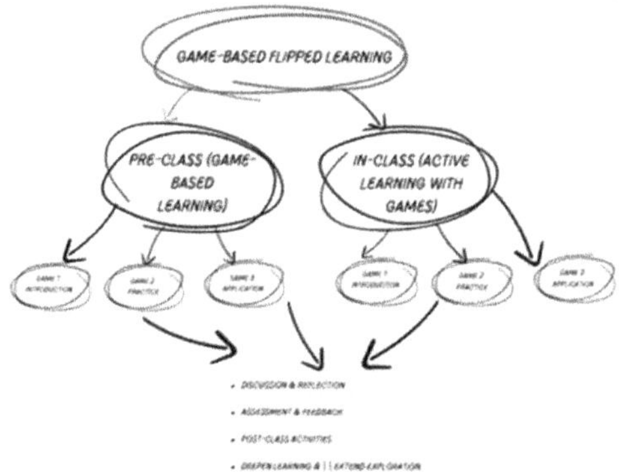

Additionally, fostering a positive and supportive learning environment is crucial for the success of game-based flipped learning. Educators should cultivate a classroom culture that values experimentation, collaboration, and continuous improvement. Encouraging students to take risks, learn from failures, and celebrate successes fosters a growth mindset and motivates students to engage actively in the learning process. Providing opportunities for students to share their gaming experiences, achievements, and challenges can also promote peer learning and community building within the classroom.

Moreover, ongoing assessment and feedback are essential components of effective game-based flipped learning. Educators should design formative and summative assessments that align with learning objectives and measure student progress and achievement within the game environment. Providing timely feedback on student performance, addressing misconceptions, and offering opportunities for reflection and self-assessment empower students to monitor their own learning and make meaningful connections between gameplay and academic content.

In summary, adopting best practices and strategies for game-based flipped learning can enhance student engagement, motivation, and learning outcomes. By carefully selecting educational games, providing clear instructions, integrating game-based activities effectively, fostering a positive learning environment, and implementing ongoing assessment and feedback, educators can create dynamic and immersive learning experiences that inspire curiosity, promote collaboration, and facilitate deeper learning.

DIGITAL INNOVATION IN EDUCATION — TOOLS, TRENDS, AND TRANSFORMATIONS

As shown in Fig. 10, the integration of LMS, video platforms, collaboration, and assessment tools plays a critical role in shaping effective flipped classrooms by supporting content delivery, interaction, and feedback.

Learning Management Systems (LMS)

- Central hubs for course content, assignments, feedback, and grading
- Examples: Google Classroom, Moodle, Canvas, Schoology
- Promote organization, communication, and progress tracking

Video Platforms

- Core to flipped learning for pre-class instruction
- Examples: Edpuzzle (interactive video), Loom (screen recording), YouTube
- Enable anytime, anywhere learning and self-paced study

Collaboration Tools

- Encourage teamwork, idea sharing, and creativity

- Examples: Padlet, Jamboard, Miro, Microsoft Teams
- Help build digital citizenship and peer-to-peer engagement

Assessment Tools

- Provide real-time, formative feedback
- Examples: Kahoot!, Socrative, Quizizz, Formative
- Increase engagement and accountability

Figure 10. Core digital tools supporting flipped learning environments

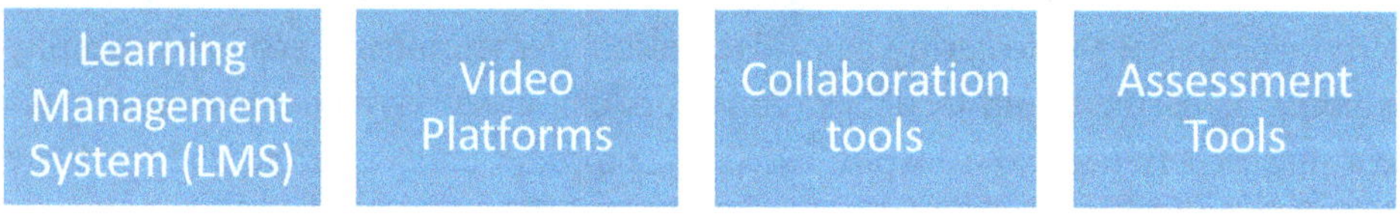

DIGITAL LITERACY: A NON-NEGOTIABLE SKILLSET

Success in a digital classroom depends on both students and educators being digitally literate—not just tech-savvy, but thoughtful, ethical, and strategic users of technology (Tune et al.,2013).

For Educators:

- Know how to curate and create content
- Practice blended and remote teaching techniques
- Manage digital classrooms effectively
- Stay current through professional development

For Students:

- Learn to navigate digital tools independently
- Practice cyber safety and digital responsibility
- Develop critical thinking when evaluating online content
- Collaborate and create with confidence

CONCLUSION

In summation, the fusion of game-based methodologies within flipped learning represents a potent pedagogical approach with vast transformative capabilities within education. This chapter has delved into the multifaceted dimensions of game-based flipped learning, spanning from its theoretical underpinnings to practical implementation strategies. By capitalizing on the synergy between game-based learning and flipped instruction, educators stand to cultivate dynamic and immersive learning environments that not only captivate students but also foster deeper comprehension and augment learning outcomes.

REFERENCES

Abeysekera, L., & Dawson, P. (2015). Motivation and cognitive load in the flipped classroom: Definition, rationale and a call for research. *Higher Education Research & Development*, *34*(1), 1–14. DOI: 10.1080/07294360.2014.934336

Ahmed, H. O. K. (2016). Flipped learning as a new educational paradigm: An analytical critical study. *European Scientific Journal*, *12*(10).

Bishop, J. L., & Verleger, M. A. (2013). *The flipped classroom: A survey of the research.* In *Proceedings of the 2013 ASEE Annual Conference & Exposition.* American Society for Engineering Education.

Chen, Y., Wang, Y., & Chen, N.-S. (2014). Is FLIP enough? Or should we use the FLIPPED model instead? *Computers & Education*, *79*, 16–27. DOI: 10.1016/j.compedu.2014.07.004

Doung-In, S. (2017). Flip your classroom: Reach every student in every class every day. *Walailak Journal of Learning Innovations*, *3*(2), 71–78.

Enfield, J. (2013). Looking at the impact of the flipped classroom model of instruction on undergraduate multimedia students at CSUN. *TechTrends*, *57*(6), 14–27. DOI: 10.1007/s11528-013-0698-1

Foster, A., & Shah, M. (2020). Principles for advancing game-based learning in teacher education. *Journal of Digital Learning in Teacher Education*, *36*(2), 84–95.

Gilboy, M. B., Heinerichs, S., & Pazzaglia, G. (2015). Enhancing student engagement using the flipped classroom. *Journal of Nutrition Education and Behavior*, *47*(1), 109–114. DOI: 10.1016/j.jneb.2014.08.008

Greipl, S., Moeller, K., & Ninaus, M. (2020). Potential and limits of game-based learning. *International Journal of Technology Enhanced Learning*, *12*(4), 363–389.

Herreid, C. F., & Schiller, N. A. (2013). Case study: Case studies and the flipped classroom. *Journal of College Science Teaching*, *42*(5), 62–66.

Holmes, J. B., & Gee, E. R. (2016). A framework for understanding game-based teaching and learning. *On the Horizon*, *24*(1), 1–16.

Lage, M. J., Platt, G. J., & Treglia, M. (2000). Inverting the classroom: A gateway to creating an inclusive learning environment. *The Journal of Economic Education*, *31*(1), 30–43. DOI: 10.1080/00220480009596759

Larcara, M. (2015). Benefits of the flipped classroom model. In *Curriculum design and classroom management: Concepts, methodologies, tools, and applications* (pp. 93–105). IGI Global.

Lo, C. K., & Hew, K. F. (2017). A critical review of flipped classroom challenges in K–12 education: Possible solutions and recommendations for future research. *Research and Practice in Technology Enhanced Learning, 12*, 4. Advance online publication. DOI: 10.1186/s41039-016-0044-2

McLaughlin, J. E.. (2014). The flipped classroom: A course redesign to foster learning and engagement in a health professions school. *Academic Medicine, 89*(2), 236–243. DOI: 10.1097/ACM.0000000000000086

O'Flaherty, J., & Phillips, C. (2015). The use of flipped classrooms in higher education: A scoping review. *The Internet and Higher Education, 25*, 85–95. DOI: 10.1016/j.iheduc.2015.02.002

Strayer, J. F. (2012). How learning in an inverted classroom influences cooperation, innovation and task orientation. *Learning Environments Research, 15*, 171–193. DOI: 10.1007/s10984-012-9108-4

Talbert, R., & Bergmann, J. (2023). *Flipped learning: A guide for higher education faculty*. Routledge.

Tune, J. D., Sturek, M., & Basile, D. P. (2013). Flipped classroom model improves graduate student performance in cardiovascular, respiratory, and renal physiology. *Advances in Physiology Education, 37*(4), 316–320. DOI: 10.1152/advan.00091.2013

Zainuddin, Z., & Halili, S. H. (2016). Flipped classroom research and trends from different fields of study. *International Review of Research in Open and Distance Learning, 17*(3), 313–340. DOI: 10.19173/irrodl.v17i3.2274

Chapter 13
Automating Corporate Strategy:
How Cloud AI Transforms Managerial Intelligence

Himanshi Rajora
https://orcid.org/0000-0001-5686-3030
JGBS, O.P. Jindal Global University, India

Dilshad Jeffery
Qasim Ibrahim School of Business, Villa College, Maldives

Azadeh Amoozegar
https://orcid.org/0000-0003-4988-5113
INTI International University, Malaysia

ABSTRACT

The contemporary corporate landscape, characterized by volatility and data saturation, renders traditional intuition-based strategic models inadequate. This chapter argues that Cloud Artificial Intelligence (AI) is catalyzing a paradigm shift, fundamentally transforming managerial intelligence. We demonstrate how Cloud AI automates and augments the entire strategic lifecycle—from environmental scanning and forecasting to planning and execution. By deconstructing core strategic functions and presenting detailed use cases, the chapter illustrates the practical application of machine learning, natural language processing, and optimization algorithms. Crucially, it provides a critical examination of the profound implications of this transition, analyzing the managerial, practical, ethical, and social consequences.

DOI: 10.4018/979-8-3373-7442-0.ch013

The future of competitive advantage lies in a symbiotic human-AI collaboration, necessitating new competencies, ethical frameworks, and organizational structures for responsible and effective strategy formulation.

SECTION 1: INTRODUCTION: THE END OF INTUITION-BASED STRATEGY?

The very sacred boardrooms within which the corporate strategy used to be formulated, on the basis of the well-developed intuition and experience of some of the senior leaders, are being silently revolutionized. Over the decades, three-pronged approaches were created based on SWOT analysis, the Five Forces models created by Porter, and the intuition of the executive, which was usually summarized in the form of a five-year plan that stood in place. Although these models were fundamental, they were constrained inherently by human bandwidth of cognition, retroactive information and the impossibility of addressing the multitude of interrelated variables of the contemporary global economy (Porter and Heppelmann, 2014). These approaches have been revealed as fragile by the current business environment that is commonly referred to as VUCA, Volatile, Uncertain, Complex and Ambiguous. As information is created exponentially through social media, IoT sensors, market feeds, and transaction systems, it is no longer the absence of information but rather the massive excess that cannot be analyzed by humans (McAfee and Brynjolfsson, 2012).

The definition of the managerial intelligence is changing in this new landscape. Previously, it was an equivalent of experience, intuition, and making definitive decisions when faced with pressure. It should be reconsidered in the digital era as the data-augmented cognition, the ability to use technological means to widen cognitive boundaries, process complexity on large scale, and make more adaptive, predictive and informed choices. This is where Cloud AI becomes the tremendous democratizer and propagator. In contrast to the proprietary AI systems of the previous century, which demanded huge capital investments and specialized skills, the Cloud AI systems of providers such as Amazon Web Services (AWS), Google Cloud Platform (GCP), and Microsoft Azure are available on-demand, scalable, and getting more sophisticated AI-as-a-Service (Duan et al., 2019). This was made accessible to the point that now a mid-sized enterprise can use the same level of machine learning and natural language processing, which previously was exclusively the preserve of tech giants.

The chapter will contend that we have entered a new age in which the automation of corporate strategy is not only imaginable but it is emerging as one of the primary sources of organizational resilience and growth. To determine the automation lever points, we will start by dismantling the fundamental tasks of managerial intelligence in order to determine the specific levers to automate. We will then see what the Cloud AI toolkit offers to the contemporary strategist, to have a tangible feel of the technologies behind it. The main part of the chapter will demonstrate the practical application of this toolkit in automating the whole strategic lifecycle using detailed use cases. This human-AI partnership will be discussed critically with regard to the changing role of the manager. Lastly, the chapter will give a holistic and critical discussion of the far-reaching consequences of this radical change, both managerial, work-related, ethical, and social, and will end with a roadmap of leaders on the path to this radical change.

SECTION 2: DECONSTRUCTING MANAGERIAL INTELLIGENCE FOR AUTOMATION

The strategy automation that may be automated by AI is abstract and requires breaking down the concept managerial intelligence into its fundamental, automator friendly elements. As described by such traditional theorists as Ansoff (1965) and Mintzberg (1994), the strategic management process can be divided into the broad cycle of four interdependent functions, namely scanning, analysis, decision-making, and execution. The sub-tasks in each of these functions are begging to be augmented and automated using Cloud AI.

Scanning and Sensemaking is the first function, which entails checking the external environment both in terms of opportunities and threats. Conventional methods such as the PESTLE (Political, Economic, Social, Technological, Legal, Environmental) and the Porters Five Forces, are based on the manual collection and analysis of data by managers, using news stories, analyst reports, and regulatory press releases. Not only is it slow but it also can be prone to human biases, such as confirmation bias, which can make strategists fail to consider data that does not support their already formed convictions (Kahneman, 2011). The lever to automation in this case is the use of Natural Language Processing (NLP) to process and process large amounts of unstructured text data in real time of millions of sources to identify new trends or competitive activity or a possible supply chain failure before it can manifest itself through more traditional monitoring mechanisms.

The second role, Analysis and Forecasting is concerned with making some sense of the obtained data to project future states. The traditional framework, which is very much immutable, is known as SWOT analysis (Strengths, Weaknesses, Opportunities, Threats). In a similar manner, extrapolative statistical models have been used to make predictions in the past. Such techniques find it difficult to cope with non-linear tendencies and the multi-variable interactions. It is the field of Machine

Learning (ML) and predictive analytics. Compared to traditional models, ML models can find subtle non-linear trends in historical and real-time data that can be used to predict market demand, customer churn, or the change of commodity prices with a higher level of precision (Agrawal et al., 2018). The leverage of the automation model is the fact that it constantly learns on new data and constantly improves on its predictions automatically.

The third role, Decision-Making and Planning, is the process of choosing a course of action out of the alternatives. Scenario planning and war-gaming are useful but are frequently restricted by the amount of scenarios that human beings have the capacity to create and analyze. In this case, Generative AI and Optimization Algorithms offer strong tools of automation. There are optimization algorithms that can be used to deal with thousands of constraints and variables in order to suggest the most efficient resource allocation, such as the optimal global marketing mix or the most cost-effective network of global supply chains. Generative AI can go even further in generating a very large range of plausible future scenarios out of historical data and future projections so that the strategists can stress-test plans against a far broader range of possible futures than would have been possible previously.

Lastly, the fourth role is the one of the Execution and Monitoring which deals with the execution of the agreed strategy and monitoring its progress through Key Performance Indicators (KPIs). Conventionally, this can be achieved through period reviews of performances and manual root-cause analysis in case KPIs are out-of-target. The source of automation here is the adoption of AI intelligent dashboards and automated control systems. Not only can such systems monitor KPIs in real-time, but their diagnostic AI can automatically determine the source of a performance gap (e.g., a performance lag in a certain region is due to a competitor running an offer and a regional logistics jam-up) and even suggest or go ahead and take corrective measures, making the process of strategy implementation less of a periodic review and more of a continuously adapting feedback mechanism.

SECTION 3: THE CLOUD AI TOOLKIT FOR THE MODERN STRATEGIST

A particular set of technologies provided on cloud services enables the automation of strategic functions. Learning about this toolkit is needed to help managers define their needs and work directly with data scientists and engineers. This toolkit goes beyond data analytics to include predictive, interpretative, and generative.

Machine Learning (ML) for Predictive Insights forms the backbone of strategic forecasting. ML algorithms do not need to be specified to individual cases to learn how to make predictions on future outcomes by analyzing historical data. To a strategist, this is the capacity of shifting between reactive and proactive decisions. Cloud services provide canned ML applications to solve such common AI applications as demand forecasting, identifying irregularities in financial transactions, or consumer personalization. Additional applications built to order include creating tailored models to determine which market segments can be expanded, which acquisition partner can be best integrated with, or how a new regulatory policy would impact the cost of operation (Brynjolfsson and McAfee, 2017). It is possible to train and execute these computationally intensive models on large datasets with computers on the cloud, which would otherwise be prohibitively expensive and slow to execute on on-premise infrastructure due to its scale.

NLP for Environmental Scanning is the technology that enables machines to read, comprehend and meaningfully interpret the human language. NLP is transformative to the strategic scanning function. Algorithms in sentiment analysis can be applied to social media, news articles, and the transcripts of earnings calls to understand how the people perceive the brand of a company compared to those of its competitors. The topic modeling is capable of automatically grouping together thousands of documents to draw out an emerging trend of theme within a particular industry. NER can scan worldwide news feeds to monitor references of major competitors, political leaders, or emerging technologies that could be of significance to the company (Hirschberg and Manning, 2015). An example of this is that a strategist can establish a cloud-based NLP pipeline that will constantly analyze all incoming patent filings in a technological sector and thus establish an early warning system in case of disruption. This automatizes the hand-laden search aspect of research, which enables managers to concentrate on the sense-making.

Resource Allocation: Optimization Algorithms are mathematical methods of identifying the optimum solution among an existing set of viable options, under some constraints. Strategic planning, fundamentally is an optimizing problem - how to use limited resources (capital, personnel, production capacity) to get the most returns or to meet strategic goals. Problems of vast complexity can be solved using linear programming, integer programming and other optimization methods which are offered as services on cloud platforms. These tools allow a multinational corporation to find the best place to establish a new manufacturing plant by modeling the costs, taxation, logistics network and the labor markets at the same time. It can be utilized by a marketing director to assign a budget to dozens of channels to optimize the acquisition of customers, and the algorithm will automatically redistribute the budget based on real-time performance feedback.

The Scenario Generation Generative AI / Strategic Narratives is the future of the AI tool. Whereas predictive models are pointing at a possible future, generative models, like Large Language Models (LLMs), can invent a variety of plausible alternative futures. To produce detailed narratives of three variant versions of a scenario (e.g., a 2% rise in global interest rates) a strategist can input the scenario as a base case, and request the AI to produce a best-case, a worst-case, and a most-likely case scenario, with estimated implications on different business units. This compels the thought of more possibilities and this reduces groupthink and blind spots in planning. Moreover, generative AI can help in preparing strategic messages, investor reports and internal reports, complex-data analyses into consistent and compelling stories to various parties (Mollick, 2022).

SECTION 4: AUTOMATING THE STRATEGIC LIFECYCLE: USE CASES

The potential of the Cloud AI toolkit is theoretically best viewed in the practical implementation of the tool in the strategic lifecycle. The following use cases show how these technologies are transferred to the concept to the operation reality in the way in which the abstract strategy is transformed into the continuous, data based, and automated process.

4.1 Automated Opportunity & Threat Identification:

Assuming a global consumer goods company. Conventionally, its market intelligence team used to read industry reports and news manually in order to keep track of its competitors. Today, the company has in place a cloud-based NLP system which continually reads and analyses more than 100,000 sources which includes news sources, regulatory databases and social media. The system is conditioned to identify the products of the company, the main competitors, and the terms of technology. When a small startup in Southeast Asia patents a new sustainable packaging material, which is a novel material, the NLP system triggers the warning within hours not as one piece of data, but as the system connects it to an increasing number of posts on social media about plastic pollution and new regulatory initiatives in the European Union. This combined understanding, available as an automatic warning, enables the strategy team to see a twofold opening (a new sustainable line of goods) and a danger (a possible interference to their current chain of provision) months or years prior to it showing on an orthodox radar.

4.2 Data-Driven Scenario & War-Gaming:

An automotive company has seen an opportunity to invest a multi-billion dollars in the production of electric vehicle (EV) batteries. The company uses Cloud AI platform to execute a massive simulation instead of basing it on several manually developed financial models. The simulation will use ML predictions of lithium prices, sentiment calculated by NLP on the adoption of consumer EVs, and the supply chain logistics optimization models. The strategists then war-game an announcement by one of the competitors that has had a breakthrough in solid-state battery technology. The generative AI element assists into developing an elaborated account of the market response, and the ML models dynamically re-adjust the predictions of the self-company EV sales and the optimizing algorithm re-calculates the ROI of the intended factory within this new competitive environment. This is done by trading off all such scenarios on the company, and pinpointing the critical points of weakness and trigger in any of the scenarios, which would force the company to pivot in a strategic direction and, as a result, de-risk one of the biggest decisions in the history of the company.

4.3 Dynamic Strategy Formulation & Goal Setting:

A giant retail chain has traditionally established its yearly corporate strategy and imposed its static OKRs (Objectives and Key Results) bottom to top the organization. This is a dynamic process in an AI-augmented model. Based on the above scanning and scenario planning, the corporate strategy is converted into high level objectives. These goals are then optimized backwards by cloud-based optimization algorithms quota-in all real-time store, e-commerce, and competitive data, to suggest the best group of quarterly and monthly OKRs per region and department. In case a supply chain in one area is disrupted due to a sudden weather condition, the system can automatically redistribute the sales goals and advertising budgets to other areas to help the entire corporate aim to be attained. This turns strategy into a living plan instead of a dead, fixed plan that is annual.

4.4 Intelligent Execution & Adaptive Control:

The last connection in the chain is the execution. A telecommunications firm measures its network performance and churn with the help of conventional dashboard. This can be improved by an AI-based system that is able to not only provide KPIs but also identify problems. As an example, in case the churn rate of customer in a particular demographic is skyrocketing by 15 percent, the diagnostic AI automatically cross-correlates it with other data channels. It could be correlated with a network outage in that region that happened recently, with an advertising campaign that succeeded among the competitors, and with a sudden increase in negative ratings in the app store. Then automatically it produces a report with a root-cause analysis and suggests a certain action: initiate a specific offer to win-back customers in postal codes X, Y and Z and ramp up the network infrastructure upgrade in area A.

The role of the manager changes to identify the problem to assess and approve the AI-generated solution, and this role is fundamentally accelerative to the strategic feedback loop and enhances the effectiveness of the tactical reaction.

SECTION 5: THE HUMAN-AI COLLABORATION: THE EVOLVING ROLE OF THE MANAGER

The fact that strategic tasks are being automated does not imply the demise of the manager, on the contrary, it is a catalyst of the fundamental transformation of the managerial position. The strategic value of a strategist will be more about his/her judgement, morality, creativity and leadership in a human/AI partnership rather than the process of information processing.

The managerial position becomes a less authoritative figure who is not the Sage on the Stage (the only source of experience and wisdom that determines the strategy) but the Guide on the Side (a curator, interpreter, and orchestrator of AI-driven insights) (Huang and Rust, 2021). This comes with a number of new roles. To start with, the managers need to master strategic AI, which is prompt engineering, learning how to pose the right questions and present problems in a way that will be effectively solved by the AI. Second, they will have to be an interface between technical and business space, converting AI-generated deliverables (e.g., a cluster analysis or a predictive forecast) into useful business knowledge that can be comprehended and implemented by the teams they belong to.

The new core competencies are required in this new paradigm. Critical thinking is the most important as managers should be very skeptical when considering AI suggestions, knowing the shortcomings of the models and the bias that may be present in the underlying data. The AI literacy is not a specialized IT ability any more, it is a management skill, and one does not have to be a coder to understand a model, but he/she should be familiar with its fundamentals, its advantages, and its disadvantages (Davenport and Ronanki, 2018). Ethical reasoning is raised, as managers will be the ultimate decision-makers of the AI-based decisions with serious societal and economic implications. Lastly, human-specific qualities such as creativity and empathy become the eventual points of divergence, since they are utilized to imagine new strategic paths, motivate employees and cope with the complexities of human nature of organizational change that AI may not fully understand.

One of the key issues in this cooperation is the issue of the management of the Black Box problem. There are quite a number of well-developed ML models, especially deep learning networks, that are too complicated that even their developers cannot easily give clear reasons why they have come to a particular output. This lack of transparency is a serious obstacle to trust and adoption to the manager who must be staking millions of dollars and careers of their employees on the strategic recommendation of an AI. This implies a necessity to look at Explainable AI (XAI) - approaches and mechanisms to make the decision-making procedure of the AI more human-readable (Adadi and Berrada, 2018). The future manager should be capable of requiring and decoding such explanations with ultimate accountability and control. The slogan should be: the AI suggests, but the human decides.

SECTION 6: IMPLICATIONS OF AUTOMATED CORPORATE STRATEGY

The adoption of Cloud AI as the heart of the company strategy is not just a technological update, but a seismic event with a long-lasting impact that cannot be seen solely through the boardroom but can fall on the organizations and the society itself. To be of responsible use, these implications should be critically reviewed.

6.1 Managerial Implications

There is a likelihood of the flatteners of organizational structure. When AI systems are able to manage the constant surveillance and real-time optimization that previously needed multiple levels of middle management, the need of the latter becomes questionable. This may result in more nimble organizations that are leaner and in which the decision-making authority is brought nearer to the front lines based on AI guidance. This results in a new definition of leadership. The charismatic leader is complemented by the data-savvy leader capable of developing the culture of experimentation, coordinate cross-functional AI projects, and make ethical judgmental decisions to AI suggestions. A large skill deficit ensues, which poses an urgent need to invest heavily in upskilling and reskilling initiatives to make the current and future managers AI-literate and capable of critical thinking. The change also changes the balance of power; the people who own, read, and manage the AI systems, be it in the C-suite or in a newly established, so-called Strategic AI Office will accumulate influence potentially shaking the long-established hierarchies founded on tenure only.

6.2 Practical and Operational Implications:

The greatest change on the operational side is the drastic change in speed and scalability of strategic processes. Strategy is now a real-time, constantly-on process instead of an annual, discrete occurrence, and organizations are becoming far more agile than they ever have been before. Nonetheless, this generates a total reliance on information quality and facilities. The notions of garbage in, garbage out are exponential, and an AI-oriented approach can only be as effective as the data it processes, which requires well-developed data management and integration systems. Such a dependency creates new vulnerabilities such as the risk of model drift (when the performance of an AI model declines when the real-world conditions change), cyberattacks against AI models (such as data poisoning) and vendor lock-in with a large cloud provider, which would lead to a lack of strategic flexibility. In addition, it is difficult to determine the Return on Investment (ROI) of AI in strategy. The costs of cloud services and talent is easy to understand but benefits which include risks that have been avoided or opportunities that have been realized in the past are rather intangible and difficult to separate, making it hard to justify additional investment to the business case.

6.3 Ethical Implications:

This is the most important aspect of automated strategy to be questioned, it is the ethical aspect of it. The propagation and intensification of algorithmic bias is one of the main issues. By receiving historical information, which is biased towards the society or corporate sphere (i.e. preferring investments in male-dominated sectors), the AI will encode biases, which contribute to systematic discrimination in market selection, hiring, and lending (O'Neil, 2016). This is inextricably connected with the issue of transparency and explainability. Stakeholders (shareholders, employees and regulators) can seek to know why an AI is advising to leave a market or buy a company. This poses a challenge to the principles of corporate governance and accountability because of the black box nature of certain models of AI. This causes an accountability gap. In case the entirely AI-prescribed plan results in the failure of gigantic proportions, who should be held responsible? The CEO who approved it? The data science team which developed the model? The algorithm itself? The existing legal and corporate governance systems are ill-fitted to deal with this. Lastly, AI-based competitive intelligence puts the application into the realms of surveillance capitalism, and in that regard, competition line and ethical boundaries are obscured as far as how far it is acceptable to extract and analyze the data of rivals, clients and the general population to provide a strategic edge.

6.4 Social and Macroeconomic Implications:

The larger social implications are far reaching. The future of work of managers is in a crossroad. Although AI can make them, it will replace most of the existing strategic planning and analysis jobs, which will contribute to the escalation of economic inequality and build social turmoil unless addressed through balanced policy and retraining programs. At the macro level, the competitive environment may become extremely fast. The companies that are effective in implementing AI-based strategy can create an unbeatable edge, maneuvering faster competitors and forcing them to the background, which can diminish competition in the market and causes the emergence of AI-powered oligopolies. This may increase the inequality of the economy between big, technology-intensive companies and small and medium-sized enterprises (SMEs), which are not able to compete, crowding out innovation by smaller entities. Above all, there is the threat of an excessive dependence on AI resulting in a homogenization of strategy, in which the social values and objectives of companies are optimised in accordance with measurable financial ratios such as shareholder value. Unless purposefully programmed to be so, AI systems lack the skill to integrate the more humane values, such as community well-being, employee satisfaction, or ecological sustainability, into the calculus to which the strategic aspect of the calculation belongs. The threat is that we will produce profitable, efficient and automated corporations that are not serving the general human interest.

SECTION 7: CONCLUSION & THE PATH FORWARD

The expedition in the world of AI-driven strategy depicts a world full of potential but full of risks. The main thesis of this chapter is unambiguous, Cloud AI is radically changing the managerial intelligence, it automates the heavy-lifting of strategy and frees human mind on higher-order tasks. It is the time when the old model of strategic management that was based solely on executive intuition is being replaced by a new one of data-enhanced and continuously adaptive strategic management. NLP in scanning, ML in forecasting, optimization in planning and diagnostic AI in execution are not some fantasy of the future but working realities that provide competitive advantage to the early adopters.

It does not give rise to a future of human replacement, but of hybrid intelligence. Those organizations that learn to coexist symbiotically between human and artificial intelligence will be the strongest and the most prosperous. Within this model, the unquestionably human strengths of ethical reasoning, innovative vision, cultural awareness, and compassionate leadership determine and put into perspective the giant processing power, identification of patterns, and scalability of the AI systems. The AI will be what forms the choices and anticipates the results; the human being will be the one to bring wisdom, values and context to make the ultimate decision.

This demands a straight forward and immediate call to arms among corporate leaders. To begin with, they need to adopt a culture of continuous learning, as individuals can develop their own AI literacy and encourage a culture of curiosity and experimentation in their organizations. Second, they need to make investments that are responsible, not only in technology, but also in the data infrastructure, talent and training that are the pillars of successful AI implementation. Third, and most importantly they should be advocates of ethical AI governance. This includes setting up cross-functional ethics boards, creating transparent models with clear policies, and making the AI systems they implement matched with the stated values of the corporation and its obligation to the wider group of stakeholders. It is unavoidable that the corporate strategy will be automated, yet the form and influence are not. It is up to our present-day leaders to make sure that this mighty technology is utilized so as to make more lucrative businesses and, at the same time, a more balanced, sustainable, and human future.

REFERENCES

Adadi, A., & Berrada, M. (2018). Peeking inside the black-box: A survey on explainable artificial intelligence (XAI). *IEEE Access : Practical Innovations, Open Solutions*, *6*, 52138–52160.

Agrawal, A., Gans, J., & Goldfarb, A. (2018). *Prediction machines: The simple economics of artificial intelligence*. Harvard Business Review Press.

Ansoff, H. I. (1965). *Corporate strategy: An analytic approach to business policy for growth and expansion*. McGraw-Hill.

Brynjolfsson, E., & McAfee, A. (2017). *Machine, platform, crowd: Harnessing our digital future*. W. W. Norton & Company.

Davenport, T. H., & Ronanki, R. (2018). Artificial intelligence for the real world. *Harvard Business Review*, *96*(1), 108–116.

Duan, Y., Edwards, J. S., & Dwivedi, Y. K. (2019). Artificial intelligence for decision making in the era of Big Data – evolution, challenges and research agenda. *International Journal of Information Management*, *48*, 63–71.

Hirschberg, J., & Manning, C. D. (2015). Advances in natural language processing. *Science*, *349*(6245), 261–266. PMID: 26185244

Huang, M. H., & Rust, R. T. (2021). A strategic framework for artificial intelligence in marketing. *Journal of the Academy of Marketing Science*, *49*(1), 30–50.

Kahneman, D. (2011). *Thinking, fast and slow*. Farrar, Straus and Giroux.

McAfee, A., & Brynjolfsson, E. (2012). Big data: The management revolution. *Harvard Business Review*, *90*(10), 60–68. PMID: 23074865

Mintzberg, H. (1994). The fall and rise of strategic planning. *Harvard Business Review*, *72*(1), 107–114.

Mollick, E. (2022). *How to use AI to do practical stuff: A guide*. Retrieved from https://oneusefulthing.substack.com/

O'Neil, C. (2016). *Weapons of math destruction: How big data increases inequality and threatens democracy*. Crown Publishing Group.

Porter, M. E., & Heppelmann, J. E. (2014). How smart, connected products are transforming competition. *Harvard Business Review*, *92*(11), 64–88.

Chapter 14
Cloud-Native AI for Operational Efficiency and Predictive Maintenance:
Automating Intelligence in Practice

Swati Singh
https://orcid.org/0009-0005-1202-8048
Greater Noida Institute of Engineering and Technology, India

Kamaljit Kaur
https://orcid.org/0009-0001-5347-3616
Guru Nanak Dev University, Amritsar, India

Shivangi Tyagi
https://orcid.org/0000-0003-1099-3176
KIET Group of Institutions, Ghaziabad, India

Mayank Tyagi
https://orcid.org/0000-0002-6689-7012
KIET Group of Institutions, Ghaziabad, India

ABSTRACT

Technology has changed from being a helpful tool to a key factor in growth and efficiency in today's dynamic business environment. For businesses looking to preserve their competitive edge, improve operational efficiency, and provide better customer experiences, digital transformation has become crucial. It's critical to

DOI: 10.4018/979-8-3373-7442-0.ch014

create a thorough plan that takes into account both growth prospects and operational enhancements as businesses increasingly incorporate technology into their core operations. In the past, technology supported daily operations according to predetermined business needs, acting as a business enabler. Today, though, technology is far more important. These days, it influences business strategies, influencing everything from how businesses function to the development of new business models and techniques for engaging customers. Companies are securing their position in an increasingly digital marketplace by utilizing digital solutions to optimize operations and unlock growth potential.

DIGITAL TRANSFORMATION

According to SAP, digital transformation is the process of utilizing technology to radically alter an organization's operations, customer value delivery, and global engagement. It involves implementing digital technologies to develop new goods and services, alter current ones, and enhance corporate procedures. An organization's culture may change as a result of this transformation, promoting experimentation and an openness to change. The process of integrating digital technology into every aspect of a business and radically altering how you run and provide value to clients is known as digital transformation. Organizations must constantly question the status quo, try new things, and learn to accept failure as part of this cultural shift.

The process of integrating digital technology into every aspect of a business and radically altering how you run and provide value to clients is known as digital transformation. Organizations must constantly question the status quo, try new things, and learn to accept failure as part of this cultural shift (Lee et al., 2015).

All businesses, no matter how big or small, must embrace digital transformation. Almost every keynote, panel discussion, article, or study about how companies can stay relevant and competitive in an increasingly digital world makes that point very evident. Many business executives are unclear about what digital transformation entails.

According to some leaders, the term "digital transformation" itself has become so general and overused that it is no longer useful. The term might not be your favorite. Whether you like it or not, the business principles that underpin the term—thinking outside the box, trying new things, and becoming more flexible in your response to clients and competitors—aren't going away (Xu et al., 2018).

The initial setbacks from vision to execution are among the most challenging questions in digital transformation. They create anxiety: despite not being far behind their peers in terms of transformation, many CIOs and organizations believe they are.

The COVID-19 pandemic in recent years has heightened the urgency of achieving digital transformation objectives and compelled numerous organizations to accelerate their transformation efforts. IT executives still struggle with budgeting, hiring, and cultural shifts. Let's seek advice from experts in digital transformation and your peers.

Digital transformation is the process of integrating digital technology into every aspect of a company, which fundamentally alters how companies function and provide value to their clients. Beyond that, it's a cultural shift that calls on businesses to experiment frequently, accept failure, and continuously question the status quo. This occasionally entails abandoning established business procedures on which organizations were founded in favor of more recent methods that are still being developed. The timeline for Digital Transformation is shown in Figure 1 and Table 1.

Figure 1. Digital transformation timeline

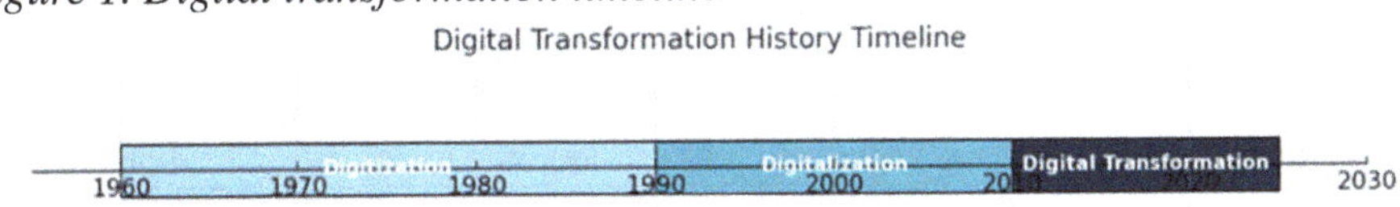

Table 1. Timeline for digital transformation

Phase	Time Period	Key Characteristics
Digitization	1960 – 1990	- Conversion of analog data to digital form - Introduction of computers and IT systems
Digitalization	1990 – 2010	- Use of digital tools to improve existing business processes - Rise of ERP, CRM, email, websites
Digital Transformation	2010 – 2025+	- Holistic integration of digital tech into business models - AI, cloud, IoT, data-driven culture - Shift toward innovation, agility, and customer experience

DIGITAL TRANSFORMATION: IMPORTANCE IN ARTIFICIAL INTELLIGENCE AND MACHINE LEARNING

A company may decide to go digital for a number of reasons. However, the most likely explanation is that they must: It's a matter of survival. The ability of an organization to promptly adjust to supply chain disruptions, market pressures, and rapidly evolving customer expectations has become crucial in the aftermath of the pandemic.

There are some constants and common themes among published frameworks and case studies that all business and technology leaders should take into consideration as they embark on digital transformation, even though it will vary greatly depending on the unique demands and challenges of each organization as show in in Table 2.

Table 2. Artificial intelligence and machine learning

Aspect	Importance in AI/ML
Data Availability	DT digitizes and centralizes data from all sources—crucial for training AI/ML models.
Infrastructure Readiness	Cloud adoption, data lakes, and edge computing provide the scalable platforms AI needs.
Cultural Shift	Promotes experimentation, continuous learning, and data-driven decision-making.
Operational Integration	AI/ML becomes embedded into workflows (e.g., predictive maintenance, demand forecasting).
Speed & Agility	Digital systems enable faster iterations of AI models and quicker deployment.

Essential elements of digital transformation and their proportional significance in facilitating AI/ML. It emphasizes that while infrastructure preparedness and data availability are fundamental, practical deployment is driven by speed, agility, and integration as shown in Figure 2.

Figure 2. Artificial intelligence and machine learning

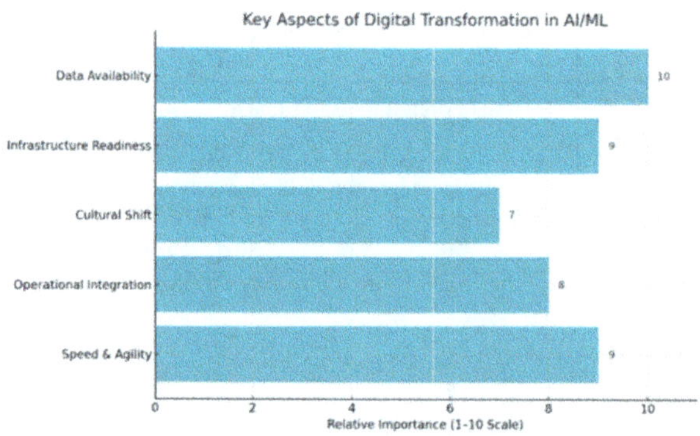

Digital Transformation Elements

- Experience of the customer
- Agility in operations
- Leadership and culture
- Enabling the workforce
- Integration of digital technologies

Important elements of digital Transformation are shown in Figure 3.

Figure 3. Important elements: Digital transformation

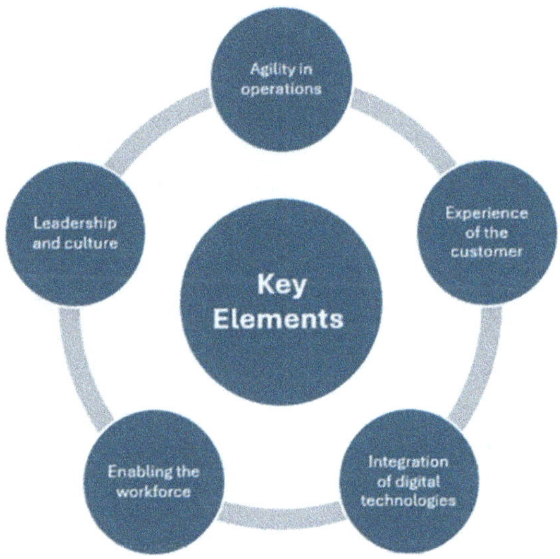

RESEARCH QUESTIONS AND HYPOTHESIS

Research questions and hypothesis are designed shown in Table 3.

Table 3. Research questions and hypothesis

Hypothesis	Rationale/Question	What to Test or Validate	Expected Outcome
Digital transformation is more than just moving to the cloud.	Is it just a catchy way to say "moving to the cloud"?	Compare cloud migration efforts vs. full digital transformation initiatives.	Digital transformation includes cultural, strategic, and process changes beyond cloud adoption.
A structured framework is essential for successful digital transformation.	Do we need to design new roles or frameworks?	Assess performance and clarity in organizations with vs. without a digital transformation framework.	Clear roles and frameworks lead to better alignment and smoother execution.
Hiring a consulting service accelerates transformation.	Should we engage external consultants?	Compare outcomes (speed, quality, ROI) between internal-only and consultant-assisted transformations.	Consulting services bring specialized expertise and accelerate early phases.
Business strategy must evolve to support digital initiatives.	What parts of our business strategy need to change?	Analyze strategy gaps—e.g., customer-centricity, innovation funding, agile governance.	Strategy must adapt to prioritize innovation, agility, and digital customer experiences.
The return on investment (ROI) of digital transformation is positive.	Is it really worth it?	Calculate costs vs. gains (efficiency, revenue, market share, resilience) before and after transformation.	Digital transformation delivers measurable ROI over the medium to long term.

DRIVERS OF DIGITAL TRANSFORMATION

A number of significant technological developments can be linked to the possibility of expansion via digital transformation. These include the Internet of Things (IoT), cloud computing, big data, robotic process automation (RPA), and artificial intelligence (AI). With the help of these technologies, businesses can revolutionize their operations and expand into new markets.

Demand Generation: By helping companies better understand consumer behavior, social media and sophisticated data analytics open the door to highly customized marketing tactics.

Reach & Selection: Businesses can now more accurately target customers with tools like mobile and location-based services, which enhances their customer acquisition and retention tactics.

Customer Experience: Businesses can create smooth and customized customer journeys across various platforms by utilizing AI and Big Data, guaranteeing that customer satisfaction propels expansion.Additionally, efficiency is introduced at all organizational levels through digital transformation. Among the main operational enhancements are:

Process Efficiency: RPA and other automation technologies are decreasing manual labor, increasing accuracy, and lowering operating expenses (Lu, 2017).

Asset Utilization: Businesses can react swiftly to changes in the market thanks to real-time insights from connected devices and IoT systems, which optimize production and minimize downtime.

Agility: Enhanced business agility is a key advantage of digital transformation, enabling companies to quickly adjust in response to market developments or emerging opportunities.

CHALLENGES:DIGITAL TRANSFORMATION

Although the advantages of digital transformation are widely known, there are frequently obstacles in the way of attaining them, which can impede or postpone development. To overcome these obstacles, which cut across people, technology, and procedures, careful preparation and execution are needed.

1.Data Security and Privacy Issues

Ensuring data security and privacy is one of the biggest obstacles in any digital transformation project. Businesses frequently gather and process enormous volumes of sensitive data, from client information to confidential company information, as they digitize their operations and embrace new technologies. This creates serious weaknesses that hackers can take advantage of.

As data breaches, cyberattacks, and digital espionage have increased, cybersecurity has taken precedence. Businesses need to spend money on strong security measures like multi-factor authentication, encryption, and ongoing monitoring systems. In order to stay out of trouble with the law and keep the trust of customers, it's also essential to follow international data protection laws like the California Consumer Privacy Act (CCPA) and the General Data Protection Regulation (GDPR).

2. Legacy System Integration

Many well-established companies still use legacy systems, which are outdated hardware or software that was once state-of-the-art. Due to their frequent incompatibility with contemporary solutions, integrating new digital technologies with these legacy systems can be extremely challenging.

The intricacy of integrating these systems may result in downtime, increased expenses, and inefficiencies. Furthermore, careful planning is necessary to prevent interruptions to ongoing operations when switching from legacy systems to more agile, cloud-based, or automated solutions. To ensure that the transition does not adversely affect business continuity, companies may need to implement hybrid strategies or gradually phase out legacy systems (Lee, Lapira, Bagheri, & Kao, 2013).

3. Opposition To Change In Culture

Adopting new technologies is only one aspect of digital transformation; another is a fundamental change in how employees think and operate within the company. Employees frequently oppose this because they may find new procedures uncomfortable, worry about losing their jobs to automation, or just object to changes to established workflows.

In order for digital transformation to be successful, businesses must cultivate an environment that values transparency, ongoing education, and creativity. Leadership needs to train staff, promote departmental cooperation, and clearly explain the advantages of change. Although it takes time, the cultural shift is necessary to make sure that workers support the organization's strategic objectives.

4.Handling Complexity in Organizations

Digital transformation affects a company's people, processes, and technology. It can be very difficult to manage these interrelated changes, especially for big, multi-layered organizations. To prevent silos—where teams or departments become isolated from one another or fail to share critical information—businesses must carefully plan these changes.

Project management frameworks that foster rapid iteration and cross-functional collaboration, like Agile or DevOps, can facilitate smoother transitions. But it's important to strike a balance between speed and stability because rushing the transformation process can result in expensive errors or unfinished implementations.

5. Skill shortages and talent gaps

New knowledge and abilities in fields like data analytics, artificial intelligence, cloud computing, and cybersecurity are frequently needed for digital transformation. Unfortunately, there is a lack of talent with these specialized skills in many organizations. Businesses are therefore forced to make costly and time-consuming investments in upskilling their current employees or to compete for digital talent on the job market.

Many businesses are implementing hybrid workforce models to lessen this difficulty, using outside partners, independent contractors, or consultants to cover knowledge gaps. At the same time, internal talent development initiatives that promote education and creativity can eventually contribute to the development of a workforce that is more tech-savvy.

6. Investment and Cost Difficulties

Initiatives for digital transformation frequently call for a large upfront investment in new infrastructure, training, and technologies. The expenses of these programs may seem unaffordable to smaller companies or those with limited resources.

Furthermore, since the advantages of digital transformation frequently build up over time through increased productivity, improved customer experiences, and new revenue streams, the return on investment (ROI) might not be immediate. As a result, companies need to approach digital transformation with a long-term perspective, striking a balance between the strategic investment needed to future-proof their operations and the immediate operational demands.

7.Preserving Business Operations While Under Transition

Regular business operations may experience brief disruptions as a result of the digitization and integration of new technologies. If these disruptions are not properly handled, they may result in lost revenue, decreased productivity, and operational downtime. Businesses going through a digital transformation must put strong change management plans in place that guarantee daily operations won't be negatively impacted. To ensure business continuity, this may entail backup systems, phased rollouts, and backup plans (Mourtzis, Doukas, & Psarommatis, 2019).

8. Assessing KPIs and Success

Lastly, evaluating the success of digital transformation projects is one of the most challenging tasks. Establishing precise, quantifiable Key Performance Indicators (KPIs) that appropriately represent the advancement of transformation initiatives is a challenge for many organizations. It can be challenging to determine whether the transformation is producing the intended value in the absence of these metrics.

Companies must establish clear, outcome-based KPIs early on in order to get around this. These could include better data security, shorter time-to-market for new products, higher operational efficiency, and higher customer satisfaction ratings.

To make sure that the digital transformation stays in line with corporate objectives, regular evaluation and modifications based on these KPIs are essential.

DIGITAL TRANSFORMATION'S FUTURE

It is evident that digital transformation is a continuous, changing process rather than a one-time endeavor as companies continue to adopt digital solutions. Businesses must constantly adjust to new tools, platforms, and operational procedures due to the rapid pace of technological change and the demands of contemporary consumers. Businesses that create and implement strong digital strategies—ones that successfully combine cutting-edge technologies with established business models—will succeed. The ultimate objective is to promote improved customer experiences, operational efficiency, and steady growth.

Organizations must fully embrace the digital future, which calls for a proactive, forward-thinking approach, if they hope to achieve this sustained growth. This is where Dawgen Global comes in to help companies with their journey toward digital transformation.

DIGITAL TRANSFORMATION EFFORTS

Roadmapping Strategically Dawgen Global assists businesses in creating a thorough digital transformation plan that complements their long-term strategic goals. Dawgen assists companies in seizing chances for expansion and productivity by determining the ideal combination of cutting-edge technologies, including artificial intelligence (AI), cloud computing, big data, and the Internet of Things (IoT). Transformation projects are prioritized and scheduled for optimal impact thanks to this strategic approach.

Integration and Optimization of Technology When it comes to digital transformation, legacy systems frequently present major challenges. By seamlessly and economically integrating contemporary technologies with pre-existing systems, Dawgen Global helps businesses get past these obstacles. Whether it's improving cybersecurity frameworks, automating manual processes with Robotic Process Automation (RPA), or moving to cloud-based solutions, Dawgen Global makes sure that technology investments yield both short-term and long-term benefits (Zhang, Yang, & Wang, 2019).

Automation of Processes and Operational Efficiency Increasing operational efficiency is a key component of digital transformation. Dawgen Global assists companies in streamlining processes, automating tedious jobs, and using real-time data to inform better decisions. Dawgen assists businesses in cutting expenses, boosting productivity, and enhancing overall performance through the implementation of lean and agile approaches. This enables companies to respond more quickly to shifts in the market and client needs.

Management of Organizational and Cultural Change Digital transformation necessitates a change in organizational culture in addition to new technology. Dawgen Global helps companies cultivate a culture of creativity, ongoing education, and teamwork. Employees at all levels are prepared to welcome digital change thanks to Dawgen's training, workshops, and strategic advice. The organization-wide integration and sustainability of digital transformation are guaranteed by this emphasis on change management.

Making Decisions Based on Data Data is one of the most important resources a company can use in the current digital era. Dawgen Global assists businesses in utilizing analytics and big data to improve customer interactions, gain actionable insights, and hone their strategies. Through customer segmentation and predictive analytics, Dawgen empowers businesses to make data-driven choices that boost customer satisfaction, spur expansion, and increase operational effectiveness.

End-to-end Solutions for Cybersecurity Businesses are facing more security risks as they use digital technologies more and more. To protect sensitive data, defend against cyberattacks, and guarantee regulatory compliance, Dawgen Global provides strong cybersecurity solutions. Dawgen assists businesses in minimizing risks and optimizing the advantages of digital innovation by integrating cybersecurity as a key component of digital transformation plans.

BLUEPRINT FOR OPERATIONAL EFFICIENCY AND PREDICTIVE MAINTENANCE

Using data analysis, particularly predictive analytics, to streamline procedures and reduce downtime is the main goal of a blueprint for operational efficiency and predictive maintenance. In order to predict possible failures and optimize maintenance schedules, this entails utilizing Internet of Things sensors to gather real-time data from equipment, evaluating historical data, and developing predictive algorithms. Businesses can cut expenses, enhance safety, and boost overall productivity by foreseeing problems before they affect operations (Jardine, Lin, & Banjevic, 2006).

Key elements

- **Gathering and Analyzing Data:**

Real-time Data: Make use of Internet of Things sensors to gather information from machinery, such as temperature, pressure, vibration, and other pertinent variables.

- **Historical Information:**

Compile and examine past information to find trends, patterns, and possible weak points.

- **Data Lakes**

For thorough analysis, combine data from multiple sources into one data lake.

- **Predictive Algorithms**

Assess data, spot possible issues, and forecast when maintenance is required by using predictive analytics algorithms.

- **Strategies for Predictive Maintenance**

Condition-Based Maintenance: Keep an eye on the state of the equipment and initiate maintenance when certain parameters become intolerable, signaling an impending failure.

- **Risk-Based Maintenance**

Pay attention to vital systems and assets where a failure could have serious repercussions.

- **Preventive Maintenance**

To stop possible issues from developing, carry out routine maintenance procedures. Corrective maintenance involves fixing equipment malfunctions as they happen.

- Roadmap for Implementation

INDUSTRY 4.0

Manufacturers are adopting cutting-edge technologies to maintain their competitiveness as Industry 4.0 becomes the standard. The use of data lakes and predictive maintenance is a major aspect of this technological revolution. In an effort to improve operational effectiveness, reduce downtime, and preserve high product quality, manufacturers are utilizing these innovations.

Comprehending data lakes

Structured and unstructured data at any scale can be stored centrally in a data lake. Data lakes, in contrast to conventional data storage options, provide the freedom to hold raw data until it's required (Qin, Liu, & Grosvenor, 2016).

INDUSTRY APPLICATIONS

AI-driven predictive maintenance is already being applied across various industries, bringing tangible benefits in operational efficiency:
Numerous industries have already implemented AI-driven predictive maintenance, which has a noticeable positive impact on operational efficiency.

Comprehending data lakes

Structured and unstructured data at any scale can be stored centrally in a data lake. Data lakes, in contrast to conventional data storage options, provide the freedom to hold raw data until it's required. This is especially helpful in manufacturing, where a variety of data types are produced by machines, sensors, and quality control systems, among other sources.

DATA LAKES' ADVANTAGES IN MANUFACTURING

Scalability: Large volumes of data can be stored by manufacturers without being hampered by capacity. limitations.
Cost-effective: Storage expenses have no effect on the raw data that can be obtained.
Flexibility: Advanced analytics are made possible by the ability to store data in a variety of formats.

Making use of predictive analysis

Predictive analysis is the process of analyzing historical data and forecasting future events using statistical algorithms and machine learning techniques. Predictive analysis can be used in manufacturing to improve overall production efficiency, optimize maintenance schedules, and foresee quality issues.

Applications of predictive analysis

Quality Predictions: Manufacturers can anticipate possible quality problems before they arise by analyzing both historical and current data. For instance, manufacturers can identify irregularities that might point to future flaws by tracking data from production lines. They are able to resolve problems before they affect the quality of their products thanks to this proactive approach (Wang, Ma, Yang, Liu, & Kang, 2018).

Process Improvements

Predictive analysis aids in locating manufacturing process bottlenecks and inefficiencies. Manufacturers can identify areas where processes are lagging or resources are being underutilized by looking at patterns and trends in production data. Operations can then be streamlined, resources can be allocated more effectively, and overall production efficiency can be raised with this information (Ahuja & Khamba, 2008).

Consistency Assurance

In manufacturing, upholding strict standards for product quality is crucial. Consistency across production runs can be ensured with the help of data-driven insights from predictive analysis. Manufacturers can maintain consistency in product quality by tracking important quality metrics over time and comparing them to historical data. This allows them to identify deviations and take immediate corrective action.

Integrating Data Lakes with Predictive Analysis

Predictive analysis and data lakes together provide a potent method for intelligent quality control. Large amounts of data from manufacturing processes are stored in data lakes, and predictive analysis turns this data into useful insights.

A variety of sources, including sensors built into machinery, PLCs that control production lines, and Enterprise Resource Planning (ERP) systems that handle resources, can provide the data that is kept in data lakes.

However, using predictive analytics unlocks the true value. By analyzing the data in the data lake, these sophisticated techniques uncover relationships and patterns that may be missed by more conventional approaches.

This paves the way for improved quality standards and a more proactive approach to quality control by enabling manufacturers to document and occasionally predict possible quality issues before they arise (Singh, Kumar, & Gupta, 2018).

THE ADVANTAGES OF COMBINING PREDICTIVE ANALYTICS AND DATA LAKES

Improved quality control

Integrating data lakes with predictive analysis greatly improves quality management when used in conjunction with intelligent quality control software like FLAGS. FLAGS can offer a wide range of custom-written reports that give manufacturers and data scientists a thorough understanding of their operations by storing enormous volumes of data in data lakes. Then, using this data, predictive analysis finds possible quality problems before they become issues. By lowering the frequency of flaws and guaranteeing that products live up to expectations, this proactive approach enables manufacturers to maintain high standards of quality (Lee, Bagheri, & Jin, 2016).

Efficiency in operations

By locating inefficiencies and bottlenecks, predictive analysis aids in the optimization of manufacturing processes. Manufacturers can analyze production workflows and identify areas for improvement using predictive algorithms thanks to data lakes, which offer a central repository of all pertinent data. This improves overall operational efficiency by reducing cycle times, increasing throughput, and allocating resources optimally.

Cost reduction is one of the biggest advantages of combining data lakes and predictive analysis. Manufacturers can minimize downtime by scheduling maintenance activities during off-peak hours and anticipating possible equipment failures and quality issues before they arise. Additionally, manufacturers can cut waste and rework by identifying inefficiencies in production processes, which results in significant cost savings. Smart quality management in manufacturing is revolutionized by combining data lakes and predictive analysis. By lowering expenses, guaranteeing superior product quality, and improving operational efficiency, this combined strategy helps manufacturers maintain their competitiveness (Mobley, 2002).

AI PREDICTIVE MAINTENANCE IS REWRITING THE RULES OF MANUFACTURING: FROM DOWNTIME TO UPTIME

Every second of machine downtime has a direct impact on productivity and profitability in the fast-paced manufacturing world. Manufacturers lose an astounding $260,000 per hour on average due to unplanned downtime, which emphasizes how important it is to keep machinery operating efficiently.

However, many manufacturers continue to face unforeseen equipment failures, frequently caught off guard by expensive malfunctions that cause production schedule disruptions and reduce profitability (Lee & Mönch, 2019).

This challenge has already been presented. For many years, manufacturers have depended on reactive or preventive maintenance strategies, which have inherent drawbacks. Preventive maintenance, carried out according to a schedule, can be time-consuming and ineffective, resulting in needless downtime and resource waste, whereas reactive maintenance, where machines are repaired after a failure, frequently causes unexpected production halts.

Presenting Predictive Maintenance (PdM), a state-of-the-art approach to directly tackle these inefficiencies. Utilizing real-time data and Artificial Intelligence (AI), PdM tracks equipment performance and anticipates failures before they occur.

Predictive maintenance enables manufacturers to maintain equipment only when necessary, drastically reducing downtime and prolonging the lifespan of vital assets, as opposed to responding to malfunctions or servicing machines according to a set schedule.

THE DEVELOPMENT OF MAINTENANCE TECHNIQUES IN MANUFACTURING

Manufacturers have struggled for decades to strike a careful balance between preserving equipment efficiency and reducing downtime.

Reactive and preventive maintenance strategies have historically dominated the sector. Both have had their uses, but they have serious drawbacks that have led to the hunt for smarter, more effective solutions.

As the name implies, reactive maintenance is fixing equipment problems only after a failure has occurred. This strategy may result in unexpected production stops because equipment is frequently used to its breaking point before repairs are made.

Reactive maintenance's unpredictable nature usually leads to expensive downtime, missed production goals, and possible harm to other processes or systems that depend on the malfunctioning machine. Unexpected breakdowns may appear cost-effective in the short term, but they can have serious long-term effects.

Preventive maintenance, on the other hand, aims to reduce these risks by planning routine maintenance according to a set schedule or usage benchmarks. This method has drawbacks even though it lessens the possibility of unexpected equipment failure.

Maintaining equipment that might not need much care frequently results in needless downtime and increased maintenance expenses. The calendar determines when equipment is taken offline rather than its actual state, which could result in the loss of important production hours for little profit. Furthermore, labor and parts are spent on equipment that may still be performing at peak efficiency.

This leads us to the necessity of a more creative strategy that combines the advantages of both without any negative effects. The need for a proactive and predictive approach is highlighted by the shortcomings of reactive and preventive maintenance.

Presenting predictive maintenance (PdM), a ground-breaking advancement in maintenance techniques. PdM enables manufacturers to keep an eye on the condition of their equipment and anticipate failures before they occur by utilizing AI and real-time data (Bousdekis, Magoutas, Apostolou, & Mentzas, 2017).

Predictive maintenance maximizes maintenance activities, guaranteeing that machines are serviced only when required, as opposed to responding to malfunctions or following strict schedules. This lowers expenses dramatically, decreases downtime, and prolongs the operational life of vital assets.

A crucial change as manufacturing becomes more complex is the transition from reactive and preventive maintenance to AI-powered predictive maintenance, which allows for increased productivity, dependability, and profitability in a market that is becoming more and more competitive.

An illustration in Figure 4 shows of how maintenance procedures have changed over time, from reactive to predictive maintenance, demonstrating the increase in system complexity and efficiency.

Figure 4. Evolution of maintenance: From reactive to predictive

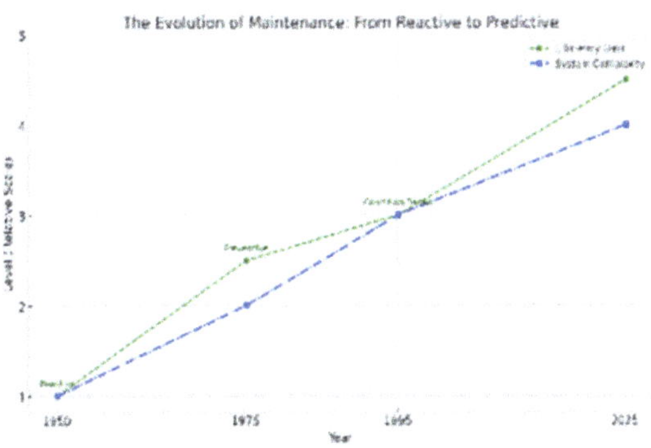

KEY TECHNOLOGIES ENABLING THE EFFECTIVENESS OF PREDICTIVE MAINTENANCE

The efficiency of predictive maintenance depends on a number of fundamental technologies, all of which are essential for collecting, evaluating, and interpreting data.

Internet of Things (IoT): PdM relies on IoT devices, which serve as the system's eyes and ears. Large volumes of operational data are recorded by sensors built into manufacturing machinery, ranging from pressure and fluid dynamics to temperature variations and vibration levels.

Real-time data reflecting the state of the machinery is fed into the central PdM platform by these sensors, which are in constant communication with it.

With the help of IoT technology, numerous machines and systems can be seamlessly integrated and continuously monitored, forming a network of linked devices that offers a thorough understanding of factory operations.

Without artificial intelligence (AI) and machine learning (ML), the amount of data produced by Internet of Things (IoT) devices would be enormous. This data can be processed and analyzed by AI algorithms to find trends, patterns, and anomalies that human operators might overlook.

A branch of artificial intelligence called machine learning enables the system to continuously improve its predictive models by learning from past data. These models improve in accuracy over time, allowing the system to make more accurate failure predictions.

In PdM, certain methods like anomaly detection and predictive analytics are frequently employed, where the system learns what normal equipment behavior is and highlights any deviations that could indicate an imminent failure.

By providing manufacturers with early warnings, this predictive capability makes sure that maintenance is planned before problems worse.

REFERENCES

Ahuja, I. P. S., & Khamba, J. S. (2008). Total productive maintenance: Literature review and directions. *International Journal of Quality & Reliability Management*, *25*(7), 709–756. DOI: 10.1108/02656710810892786

Bousdekis, A., Magoutas, B., Apostolou, D., & Mentzas, G. (2017). Predictive maintenance using data-driven approaches: A review. *Procedia CIRP*, *67*, 268–273. DOI: 10.1016/j.procir.2017.03.015

Jardine, A. K. S., Lin, D., & Banjevic, D. (2006). A review on machinery diagnostics and prognostics implementing condition-based maintenance. *Mechanical Systems and Signal Processing*, *20*(7), 1483–1510. DOI: 10.1016/j.ymssp.2005.09.012

Lee, J., Bagheri, B., & Jin, C. (2016). Introduction to cyber manufacturing. *Manufacturing Letters*, *8*, 11–15. DOI: 10.1016/j.mfglet.2016.03.001

Lee, J., Bagheri, B., & Kao, H.-A. (2015). A Cyber-Physical Systems architecture for Industry 4.0-based manufacturing systems. *Manufacturing Letters*, *3*, 18–23. DOI: 10.1016/j.mfglet.2014.12.001

Lee, J., Lapira, E., Bagheri, B., & Kao, H. A. (2013). Recent advances and trends in predictive manufacturing systems in big data environment. *Manufacturing Letters*, *1*(1), 38–41. DOI: 10.1016/j.mfglet.2013.09.005

Lee, S., & Mönch, L. (2019). A cyber-physical approach to predictive maintenance in manufacturing. *Computers & Industrial Engineering*, *130*, 332–342. DOI: 10.1016/j.cie.2019.03.014

Lu, Y. (2017). Industry 4.0: A survey on technologies, applications and open research issues. *Journal of Industrial Information Integration*, *6*, 1–10. DOI: 10.1016/j.jii.2017.04.005

Mobley, R. K. (2002). *An Introduction to Predictive Maintenance*. Elsevier., DOI: 10.1016/B978-075067532-3/50003-9

Mourtzis, D., Doukas, M., & Psarommatis, F. (2019). Industrial big data as a result of IoT adoption in manufacturing. *Procedia Manufacturing*, *39*, 1613–1620. DOI: 10.1016/j.promfg.2020.01.413

Qin, J., Liu, Y., & Grosvenor, R. (2016). A categorical framework of manufacturing for industry 4.0 and beyond. *Procedia CIRP*, *52*, 173–178. DOI: 10.1016/j.procir.2016.07.040

Singh, R., Kumar, U., & Gupta, M. (2018). Internet of Things (IoT) in manufacturing: Review and challenges. *International Journal of Advanced Manufacturing Technology*, *102*, 511–526. DOI: 10.1007/s00170-019-03731-3

Wang, K., Ma, Y., Yang, J., Liu, Z., & Kang, R. (2018). Data-driven fault diagnosis and prognosis for manufacturing equipment: A review. *IEEE Transactions on Industrial Informatics*, *15*(6), 3312–3325. DOI: 10.1109/TII.2018.2799931

Xu, L. D., Xu, E. L., & Li, L. (2018). Industry 4.0: State of the art and future trends. *International Journal of Production Research*, *56*(8), 2941–2962. DOI: 10.1080/00207543.2018.1444806

Zhang, W., Yang, D., & Wang, H. (2019). Data-driven methods for predictive maintenance of industrial equipment: A survey. *IEEE Systems Journal*, *13*(3), 2213–2227. DOI: 10.1109/JSYST.2018.2874340

Chapter 15
Automating Intelligence: Cloud-Native AI, Generative Tools, and the Future of Research

Shimona Shriya
https://orcid.org/0009-0001-0682-4014
Xavier Institute of Management and Entrepreneurship, Bangalore, India

Mohit Verma
https://orcid.org/0009-0005-3036-3894
O.P. Jindal Global University, India

Aishwarya Vashishta
Rutgers University, USA

ABSTRACT

Defining a term before it becomes meaningful is vital in capturing the importance of any term. If a term does not have a meaning that people agree with, it is only a group of characters. The most important of all is to define an idiom. It is challenging as the term's meaning isn't clear from its constituent elements. The notion is most relevant in case of the technical terms which receive a significant press coverage in numerous ways and in different setups. Common apps, such as Google Assistant, Alexa, and Siri, have many people using artificial intelligence (AI) daily without even thinking about it. Productivity and creative apps such as ChatGPT, perplexity, and Gemini help us focus on the content rather than on how to get there. The media floods our entire social environment with so much information and disinformation that many people see AI as a kind of magic (which most certainly isn't). Therefore we start the chapter by defining what AI is, and what it isn't. We also discuss on how it connects to computers today.

DOI: 10.4018/979-8-3373-7442-0.ch015

INTRODUCTION

Defining a term before it becomes meaningful is vital in capturing the importance of any term. If a term does not have a meaning that people agree with, it is only a group of characters. The most important of all is to define an *idiom*. It is challenging as the term's meaning isn't clear from its constituent elements. The notion is most relevant in case of the technical terms which receive a significant press coverage in numerous ways and in different setups.

Common apps, such as Google Assistant, Alexa, and Siri, have many people using artificial intelligence (AI) daily without even thinking about it. Productivity and creative apps such as ChatGPT, perplexity, and Gemini help us focus on the content rather than on how to get there. The media floods our entire social environment with so much information and disinformation that many people see AI as a kind of magic (which most certainly isn't). Therefore we start the chapter by defining what AI is, and what it isn't. We also discuss on how it connects to computers today.

There is a discord among AI users that it doesn't tell us anything cognitively meaningful, which is the reason people have so many differences of opinions regarding this term. It can certainly be argued that is the case since it doesn't come from any natural sources and have artificial occurrence, but how much of an intelligence it is brings some ambiguity at best. Even if people would arguably not agree in defining AI, this chapter follows through and uses AI by the definitions provided. Knowing that will help the readers follow the concepts in greater ease. Intelligence contains certain mental actions (Ceci, 1996; Hoskens & De Boeck, 1991)

comprised of one or many of these characteristics: a.) Learning: Having the ability to acquire and process new information, b.) Reasoning: the ability to manipulate information in a variety of ways, c.) Understanding: Result of manipulating the information, d.) Seeing truths: Determining whether the information is truthful, e.) Seeing relationships: Divining how the information relates to other information, f.) Changing meanings: Applying truths in a specific case that are consistent with those relationships, g.) Separating fact from belief: Determining whether the information is backed up by provable resources that we can demonstrate is consistently true.

Similarly, any intelligence can also follow a step by step, sequential process which can be emulated by the computer system as a program or artificial simulation (similar to the real world scenario): The first step is to set a goal based on the desired requirements, the output and inputs. Second is to assess the availability of valuable data and information in support with the goal, third step is to gather more additional information which may further add value to the set goal, fourth step is to process the acquired data such that it is in the form of the existing set of information. The next step is to define the associations or the relations between existing and new. In the sixth step we determine if the goal is achieved or not achieved. The next two stages are to alter the objective with the implications and effects delivered by the new data and finally to keep going via from the second phase to the seventh phase as necessary until the objective is achieved or the prospects for achieving it have been exhausted.

Any machines' capability to achieve intelligence is extremely restricted even though one can create intricate and detailed algorithms with complete access to data. The computers are incapable of understanding the real-world complexities and different scenarios as they rely on machine processes to assess that data using mathematics or anything equivalent in a strictly robotic fashion.

On a similar note, these machines are incapable of distinguishing truth from lies with ease. In fact, no computers are capable of performing all of the mental operations that define what intelligence means. Another way to consider what is involved with intelligence, is to categorize intelligence. Humans do not rely on only one type of intelligence. There are multiple intelligences that they to perform tasks. Howard Gardner, a Harvard psychologist, defined eight types of intelligences (Gardner, 2011):

1. Visual-spatial: An ability to recognize and manipulate large-scale and fine-grained spatial images.
2. Linguistic verbal: An ability to analyze information and create products involving oral and written language.
3. Logical-mathematical: An ability to develop equations and proofs, make calculations, and solve abstract problems.
4. Body-kinaesthetic: An ability to use one's own body to create products or solve problems.
5. Musical: An ability to produce, remember, and make meaning of different patterns of sound.
6. Interpersonal: An ability to recognize and understand other people's moods, desires, motivations, and intentions
7. Intrapersonal: An ability to recognize and understand his or her own moods, desires, motivations, and intentions
8. Naturalistic: An ability to identify and distinguish among different types of plants, animals, and weather formations that are found in the natural world.

Understanding the kinds of intelligences leads one to relate them to the types of tasks a computer can or could not accomplish as intelligent behaviours. Judging by the types of intelligences, we could come to the conclusion that the definition of AI ought to be based on how closely it models the human behaviour being that either it follows a set of rules and protocols, in other words, the artificial intelligence is simulating the true human behaviour by putting itself in the role of a human.

While we are in the context of talking about human behaviour, rationality is also a fundamental characteristic of the decision-making process. A process can be said to be rational if it always does the right thing, based on the information at the time, given that the process has an optimal performance metric.

In summary, rational processes work from a plan, assuming the plan is correct. Human processes work from instinct, intuition, and variables such as experience, nature and nurture that may not reflect the plan, and the plan may not even consider the data. For example, the rational way to drive a vehicle is to follow the law. But, traffic does not follow the letter of the law. If you drive according to the law, you may be stuck through no fault of your own, while other drivers are not following the law. Therefore to succeed, a self-driving vehicle must act humanly and not just rationally.

Depending on these characteristics, the four classification or categories used and promoted by Arend Hintze (2016) form a better basis for understanding and defining AI as shown in Table 1.

Table 1. Types of AIs and definitions

Category	Description
Reactive machines:	Machines that beat humans at chess or compete on game shows are reactive machines. A reactive machine lacks any memory or experience to help it make decisions. It depends solely on computational power and smart algorithms to replicate every decision every time. This is an example of a weak AI designed for a limited, specific task.
Limited memory:	A self-driving (SD) vehicle or autonomous robot won't have time to make every decision from scratch. These machines leverage a small amount of memory in order to use experiential knowledge of different situations. When the machine has seen a new situation before, it can quickly rely on experience to reduce reaction time and provide more resources for creating new decisions that have not yet occurred. This represents the current level of strong AI..
Theory of mind:	A machine that is capable of understanding its own goals and even those of others in the environment has some form of understanding that is possible today, but not at a commercial level. A self-driving car at a busy intersection is an example of "theory of mind." The car has to pursue its own goal of going somewhere and while following that goal it has to make a guess about what another driver is doing, such as a driver questioning whether they can make an illegal turn or a pedestrian stepping into the road and reacting with safety in mind, even though the other drivers and/or pedestrian may wish to achieve a goal that is contrary or on a collision course with the vehicle's goals. The car would have to determine the other drivers intention based on their behaviour rather than their state.
Self-awareness:	This is the type of AI we often observe in films Although, it relies on technologies that are not entirely possible at this point in time, as it would indicate a machine with a sense of self and consciousness (e.g., HAL 9000 in 2001: A Space Odyssey, Ava in Ex Machina). Such a machine could infer the intent of others based on experiential knowledge.

CONSIDERING AI COMMON APPLICATIONS IN TODAY'S WORLD

AI exists in a BANI world but also provides tools to navigate a VUCA world. The world can be described as BANI (Brittle, Anxious, Nonlinear, Incomprehensible) due to rapid, unpredictable, and chaotic changes, while AI is both a product of and a tool for managing the VUCA (Volatile, Uncertain, Complex, Ambiguous) challenges it creates and amplifies. Therefore, AI is both a driver of BANI and a solution for VUCA.

One can find numerous applications of AI in a diverse applied sector today. The one main problem still remains is that this technology works so well that we don't know the extent to which it is capable of transforming the current environment into a highly efficient machine-driven ecosystem. It is not even known to all that it exists to which depth and multiverse.

Actually, it would probably surprise the average person to learn how many devices in the home already utilize AI. For instance, some smart thermostats create schedules for you automatically based on how you adjust the temperature manually. Likewise, a voice input that controls some devices learns how you speak in order to optimize its interaction with you. AI certainly is present in the car, and primarily at work. In fact, there are potentially millions of use cases for AI, all securely hidden from view even though many are quite overt. Here is a sampling of some of the ways you may see AI used:

1. Fraud detection: The known word spam in e-mail platforms are effectively able to classify advertisements, spam and fraudulent mails. Even they may also give a red mark alert on a few of the available platforms. Second example calls for confirmation by credit cards companies. Where there is an AI embedded within the code of the credit code functioning, detects for any unfamiliar spending pattern and provides an alert call, message or mail.
2. Complex analysis: People frequently need support with complex analysis because there are literally too many factors to consider. For example, designing a software platform or designing an app where numerous factors are to be considered from both the users and service providers perspectives. Even in healthcare, doctors may have a complex case where they may refer AI feedbacks for diagnosing and saving a patient's life.
3. Resource scheduling: Many organizations need to schedule the use of resources efficiently. For example, a college may have to determine which room to assign a course or sections to base on the availability of seats, faculty, and other infrastructural constraints.

4. Customer support: The customer support line that you call may not even have a human. It is good enough with a program that mimics an actual person – using a script and another source of material toanswer 99% of your questions. After hearing a good voice inflection (also an AI), you may not even realize you are speaking with a computer.
5. Safety systems: Many of the safety systems on machines of all kinds rely on AI to take over vehicle operation in an emergency scenario. Many automatic braking systems (ABS) rely on AI to stop the vehicle based on all the input a vehicle can provide - for example, how and at what angle it is skidding. Computerized ABS is well over 40 years of age as technology continues to change.
6. Machine efficiency: AI can help control a machine for maximum efficiency. The AI controls resources to maximize efficiency so that the system stops short of violating speed or other goals. Every ounce of power is used in very specific increments to provide services.
7. Content generation: When people consider content generation, they often think about ChatGPT because it's in the public eye. However, content generation can exist deep within an application to provide specific functionality. For example, given a photo of the user, how will a new outfit look?

FEW POPULAR ML & GENAI ALGORITHMS BY CATEGORY

1. Regression Algorithms Linear Regression - Simple linear relationship modeling Random Forest Regression - Ensemble method using multiple decision trees Support Vector Regression (SVR) - Uses support vector machines for regression Gradient Boosting Regression - Sequential ensemble method for improved predictions
2. Classification Algorithms Logistic Regression - Statistical method for binary and multiclass classification Random Forest Classifier - Ensemble of decision trees for classification Support Vector Machine (SVM) - Finds optimal decision boundary between classes Naive Bayes - Probabilistic classifier based on Bayes' theorem.
3. Clustering Algorithms K-Means - Partitions data into k clusters based on centroids Hierarchical Clustering - Creates tree-like cluster structures DBSCAN - Density-based clustering that identifies outliers Gaussian Mixture Models (GMM) - Probabilistic clustering using Gaussian distributions

INDUSTRY APPLICATION OF THESE WIDELY APPLIED TOOLS

Regression Cases: In manufacturing, for predictive Maintenance, Quality Control, Waste Management, and Recycling Rate Optimization. To explain each four of the examples respectively, regression is used for redicting equipment failure time based on sensor data (temperature, vibration, pressure) to schedule maintenance before breakdowns occur, for predicting product defect rates based on production parameters like temperature, humidity, and machine settings, in case of waste Generation Forecasting to predict weekly waste volumes based on historical data, weather patterns, and seasonal trends for optimal collection scheduling, and to estimate the recycling efficiency based on material composition, processing parameters, and facility capacity

Classification Cases: In Logistics, for Package Routing and automatically classifying packages by destination zones, priority levels, and delivery methods for efficient sorting and routing, In Fraud Detection to identify suspicious shipping activities or fraudulent delivery claims based on transaction patterns and behavioural data. In Quick Commerce for Customer Segmentation to Classify customers into categories (frequent buyers, price sensitive, premium) for targeted marketing and personalized recommendations. Classification is also used in demand Categorization to classify products into high, medium, or low demand categories for inventory and warehouse handling.

Clustering Cases: Clustering cases are used in manufacturing for Production Line Optimization and Grouping similar products for efficient batch processing and reducing changeover times between different product types. Models are prepared for supplier Segmentation and clustering suppliers based on performance metrics, delivery reliability, and cost factors for strategic sourcing decisions. Even in logistics clustering is applied for Route Optimization where the regions are clustered together to create efficient delivery routes and minimize transportation costs. Clustering is applied in warehouse layout and grouping products based on picking frequency and compatibility for optimal warehouse handling.

DEEP LEARNING VS. MACHINE LEARNING

Deep learning is chosen when dealing with: Complex, unstructured data (images, text, audio, video) and with large datasets (millions of datapoints) where traditional ML usually underperform.

The high-dimensional feature spaces where manual feature engineering is impractical and, in the tasks, requiring automatic feature extraction from raw data.

They are also applied in practical scenarios and cases with non-linear relationships that traditional ML struggles to capture

DEEP NEURAL NETWORK (DNN) CASES

To state few Industry-Specific DNN Applications: In Manufacturing the Computer Vision Quality Inspection deploys the Convolutional Neural Networks (CNNs) to detect defects in products on assembly lines with higher accuracy than human inspection. In Waste Management, the Automated Waste Sorting is achieved by deploying CNNs to identify and categorize different types of waste materials for automated recycling systems. In Logistics, for demand forecasting by using Long Short-Term Memory (LSTM) networks to predict complex demand patterns considering multiple variables like seasonality, promotions, and external factors. The application of DNN is also visible in Quick Commerce where the Real-time Recommendation Systems to provide personalized product recommendations based on user's purchase history, and historic data.

THE MAGNIFICENT ROLE OF LARGE LANGUAGE MODELS (LLMS)

As a recent advancement among artificial intelligence systems, Language Models (LLMs) are capable of processing and producing text while communicating coherently (Agüera & Arcas, 2022) and generalizing across tasks (Brown et al., 2020; Hoppe & Toussaint, 2019). LLMs have recently demonstrated amazing capabilities in natural language processing and other tasks. LLMs success has generated an abundance of research contributions across these various tasks. These contributions range across diverse topics like novel architectures, improved training strategies, enlarged context lengths, fine-tuning models, multi-modal LLMs, robotics, datasets, benchmarking methods, efficient methods and more.

Language is a fundamental part of humans communicating and self-expressing with each other, and with machines. The push for generalized models is largely driven by a demand to perform complex language tasks, which could include translation, summarization, information retrieval, conversational, etc. Recently there have been significant improvements in the advances of language models, as a result of the combination of transformers (Chernyavskiy et al., 2021), better computing capabilities, and the scale of available training data. All these advances have resulted in a dramatic turnaround, where a new set of LLMs are being developed which can replicate human-level performance on a number of tasks (Wang et al., 2019).

LLMs, which are capable of understanding and generating human-like text, have found meaningful applications across a variety of fields.

For more details, Naveed *et al.* (2025) can be referred. Figure 1 demonstrates the evolution in development of LLMs.

Figure 1. Development of LLMs [Blue cards represent 'pre-trained' models. Orange cards correspond to 'instruction-tuned' models] (Naveed et al., 2025)

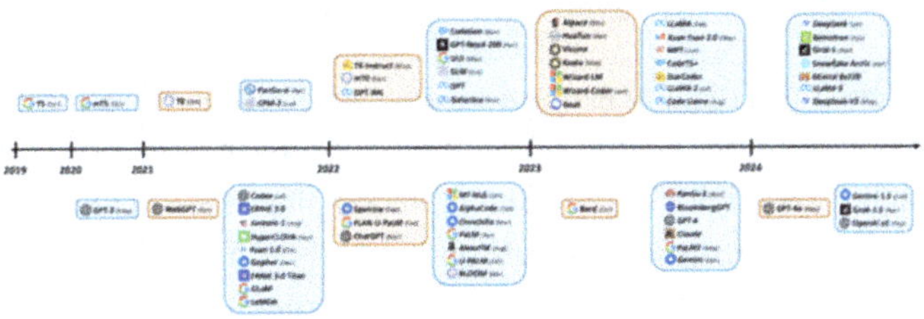

A SPECIAL CASE: USING GENAI FOR RESEARCH AND ACADEMIC WRITING

GenAI is becoming a increasingly useful tool in many researchers' tool kits and in the overall AI stack. It offers a range of functions that may help in the laborious action of gathering and analysing data. Research is the foundation of scholarly writing, as it lays the foundation of your academic work.

GenAI can help surface connections between data points that may be less obvious. For example, by analysing very large networks of scholarly papers, GenAI will surface relationships and patterns that may not be easily identified by the human researcher. These could lead to insights and interdisciplinary connections that might serve as the germination of new research worth of publication. Of course, you should always take what AI surfaces as connections with a grain of salt, and do your own due diligence through further research and analysis.

GenAI also has the capability of quickly scanning existing literature on a subject then commissioning it into a brief summary. This type of broad overview will assist the researcher to identify thematic studies, gaps in the research, and works that have the potential of leading to something transformative.

GenAI can conveniently support the narrower scopes of research; for example, data mining and extraction. For example, it can scoop through a large amount of data to extract relevant trends, insights, and other figures important to the researchers' problem statement. In addition, GenAI's language processing capabilities can be leveraged to transcribe and translate materials, improving access to non-native language sources, and allowing for a broader scope of research. The use of GenAI in this context should be approached with an intentionality to assure it is a support but not a replacement for the researcher's own critical thinking skills.

EXAMINING GENAI TOOL OPTIONS FOR RESEARCH AND ACADEMIC WRITING

GenAI has developed to include research-focused tools that enhance the researcher's ability to explore more detail and complexity into information. For example, applications such as Iris.ai (iris.ai) and Semantic Scholar (www.semanticscholar.org) use GenAI to map the scientific literature and identify relevant papers, reviews, and patents to the researchers, often with the aid of AI-generated results that might have otherwise gone unexploited through standard search engines.

These platforms apply natural language processing capabilities in evaluating research ideas and questions, to ensure the experience of locating articles is not just keyword searches, but decisions based on research proximity to the researched question. Another example is Zotero (www.zotero.org), a free reference manager that can assist academic researchers with organization and citation for book or report writing, and integrates with GenAI search specificity of articles. Zotero can extract citation information from PDFs and web links automatically, with a single click on a browser extension to add to articles to a reading/research folder, thereby assisting the academic researcher in building and maintaining their bibliography.

One can easily connect Zotero with *Research Rabbit* to further streamline your research and get your reports and papers found by other researchers, too. Research Rabbit (2025) is a free research tool built to help you discover and organize relevant papers for your research. It provides interactive visualizations, collections, digests, and data search tools.

One can add ChatGPT to Zotero in a few easy steps. Then, they can research and write their academic paper or book in one tool. The easiest integration to add GenAI to Zotero is ARIA, an acronym for AI Research Assistant and a Zotero plugin (Meta, 2025). Before using ARIA, you need to have an OpenAI API key, which is a security code for developers to use that enables computer programs to interact.

GENAIS AND RESEARCH DESIGN

GenAI can simplify the research process, from the idea phase to collecting data, by drafting research designs, outlines, and questions for polls. GenAI tools are particularly useful for researchers at the proposal stage of research projects because they are good at organizing thinking and structuring complex ideas. In terms of research design, GenAI can help researchers develop a detailed outline that consider all the conventional components of a study. Researchers can input research objectives and questions and then use the AI to develop a framework that includes the introduction, literature review, methodological approach, data analysis, and conclusion sections.

AI tools, for example, ZoteroBib, also provide researchers with a convenient way to manage their sources and citations to ensure that literature reviews are thorough and include appropriate references (Zotero, 2025). Additionally, AI-supported tools such as Iris.ai can search large and extensive databases for relevant studies and theory to suggest items that might not be apparent but would be important to include in the Research design.

Platforms like BenchSci apply AI to decode complex scientific figures, allowing researchers to find specific experimental data without combing through countless articles manually (benchsci, 2025). However, you'll need to integrate this, too, with a GenAI tool trained for writing, or copy its outputs to a GenAI text generator like ChatGPT or Claude so you can write your article or paper.

REFINING GENAI RESEARCH OUTPUTS

GenAI outputs "as is" will almost never be what we want and need them to be. Indeed, the process of trying to get GenAI to produce exactly what you need can Content Producing Long-Form be exhausting and frustrating. Also, GenAI, even the specialized text generators, do a lousy job of writing anything in long form. It's smarter and more productive to resign yourself to these unfortunate facts and develop your writing process accordingly. In other words, focus on GenAI's strengths in data discovery and analysis and in helping you fill in any blanks in your work or thinking. Otherwise, accept GenAI's flaws and use your own mind and talent to work around them. To refine AI research outputs and better ensure their relevance and accuracy. For the same, following steps can be followed: by starting with a clear and concise research question or set of keywords and inserting these into the GenAI tool as a prompt to generate an initial list of resources. By Using a specialized GenAI or other tool's filtering options to narrow down the results to the most relevant papers. This may include sorting by date, relevance, citation count, or journal impact factor. One can also integrate GenAI with research tools, as discussed above.

The researcher can browse through the abstracts provided by the AI to evaluate the relevance of each resource. This is where they want to examine the objective, method, and summary because each will inform about the resource's relevance for their purpose. One consideration is the design of the relevant study, along with what kind of data were used and how sound was the thinking and argument put forward by the resource. They might also want look at the references and citations in the papers to get leads on more material and verify what the AI suggested. They can also interrogate literature manually in response to their literature relevant to the topic in order to have something to compare on-going. Use the AI to help in extracting and organizing important data points, statistics or quotes from resources and be diligent to portray the resource accurately and appropriately cite them too. They can further refine their search query based on findings from literature, to repeat this searching process until they find a few or enough resources or to further narrow their inquiry.

By integrating these advanced GenAI tools into your research process and rigorously refining their outputs, one can enhance the breadth and depth of one's academic writing.

FINDING HIDDEN PITFALLS OR GENAIS LIMITATIONS

While GenAi tools can provide major advantages in academic research, researchers should be aware of some hidden pitfalls and limitations. While the limitations carry less gravity, they are important to know when to supplement, or when to stop using an AI research article altogether and replace it with manual efforts and other types of tools, especially if the data is entered inappropriately.

- Data quality and bias: GenAI tools are only as good as the data it has trained on. If the underlying data is biased or inaccurate, the AI's outputs will reflect that. Thus, researchers should be careful to not overly rely on AI-generated data without cross-referencing it against other research articles.
- Scope of training: Some AI tools may have been trained on a particular type of literature or within a particular discipline, and therefore may not allow you to identify relevant research outside of that discipline. Researchers should validate whether the AI's limits align with their field of study and supplement with manual searches and other types of tools as needed.
- Contextual misunderstandings: AI will struggle with absorbing the nuanced context of some research queries and offer results that are tangential or irrelevant. Researchers should be prepared to retrieve the AI results with a critical eye and adjust their queries.
- Unrecognized novel research: AI tools can favor tools with "more" citations or are easier to access, and in doing so may miss more recent, lesser-cited, but highly relevant research. Researchers will need to manually follow the latest research in their fields to capture recent studies that are cutting edge.

- Non-text research limitations: GenAI is primarily text-based and may not be able to properly analyze or interpret non-text research data such as graphs, or raw datasets, info-graphics, and images. Researchers will need to manually read them to fully understand them and tie them into their studies. Updates and maintenance: AI databases will need to be routinely maintained by on-going updates to include new research. If you are using an AI tool that is not regularly updated, it may miss new publications. Researchers will have to follow up to how often the AI tool usage is updated and will also want to supplement their readings with either manual reviews or other tool searches. In fast moving areas of study, you may necessarily want to supplement with a manual review of both the General findings or just read it manually without relying on AI thought or opinion.

You can mitigate these limitations by remaining balanced in bringing GenAI to your research as a beginning of identifying potential relevance and "novel" research to apply your experience and critical-thinking skills to explore a nuanced and complete review of potentially relevant research.

CONCLUSION

Originality of GenAI-generated content isn't just avoiding plagiarism and copyright infringement; it can also mean stimulating innovation, developing new products and computer code, and enhancing the limits of what GenAI can do.

The opportunities for the diversity of creative and knowledge work that can be created with GenAI are limitless.

Excellence, on the other hand, means ensuring the output meets or exceeds the expectations placed on humans by law, ethics, tradition, or disciplines. But with this extraordinary technology, along with the outputs, come the obligation to ensure that the content is both original and of adequate quality. These two tenets provide the foundation for trust and value in GenAI-generated works.

Strategies for ensuring originality in GenAI creations

To guarantee the novelty and quality of GenAI creations, it is necessary to implement measures able to ascertain originality and correctness of content generated.

The following are a few approaches that you can deploy to enhance the content generated of GenAI: the team should share information on the model's training process and the data that was utilized. This transparency not only assists in preserving trust but permits appropriate scrutiny of the models and further supports the overall legitimacy in the GenAI space. Discovering how the model was trained can help you estimate the risks in terms of the likelihood of originality and reliability of outputs. For example, the GPT models that power ChatGPT were trained on data scraped from the internet. That may mean a higher risk in the potential of plagiarism and copyright infringement than, say, a specialized model that AI scientists at your company trained only on company data. Table 2 gives few approaches to appropriately apply GenAIs.

Table 2. How to create using GenAIs

Item	Description
Prompt GenAI to provide references in its responses and then review those references yourself	Why? Because GenAI often lies. And, yes, it can totally make up references to support its lies, too. But also because that's a good first step in spotting hints of plagiarism or copyright infringements. That's why you never want to prompt a GenAI model to factcheck or check for plagiarism in its own responses and just accept its answer. Nor should you ever assume that you're using a different model to check the first model's work. You could easily be using two different applications built atop the same GenAI model.
Use a different GenAI model to cross-reference generated content against existing works, flagging potential duplicates for review.	For example, prompt Google Smart Search or Perplexity to factcheck the response you got from Claude or ChatGPT. Some GenAI search tools like Perplexity will automatically cite its sources for you to review, too. Make sure to stop and actually review the sources it gives you
Consider embedding digital watermarks or metadata within GenAI generated content, creating a traceable digital footprint that confirms its origin and authenticity	Embedding digital watermarks or metadata is a method used to include hidden information within GenAI-generated content to verify its origin and authenticity. This is particularly important for digital works such as images, videos, music, and text that are created by GenAI systems or tools, as it helps to prevent unauthorized use and copyright infringement, as well as to track the distribution of the content.
Know the data source	The development of GenAI models should be transparent and ethical, with clear documentation of the data sources and training methods used. By this, we mean that the maker of the GenAI model or application either a vendor or your internal AI development
Consider asking your employer or GenAI vendors to use Causal AI to reveal how the GenAI model came to its decision (output) as a deeper check of its functioning within ethical guidelines.	This is a highly technical strategy that your AI or IT department or GenAI vendor will need to do for you, as it involves developing causal inference algorithms that can work alongside or within GenAI models, creating training datasets that include causal relationship information, designing new model architectures that can combine causal reasoning with natural language generation, and developing metrics to manage the two. However, once in place, the Causal AI can reveal how the GenAI did what it did so you can better evaluate the originality and quality of its outputs

Recognizing AI's Limitations

Language models are bad at math: AI chatbots are language models tuned for conversation. If you ask a language model for the answer to a basic math problem, it will usually respond confidently with some answer. However, upon checking that answer using a calculator, you may be surprised that it's often just plain wrong.

Language models are wordy AI chatbots often respond to prompts with paragraphs when a one-word answer will do. The standard ChatGPT response to even a simple question reads like a high school book report, containing an introduction, an analysis of an issue from multiple viewpoints, and a summary

AI has limited knowledge Although it's now possible for ChatGPT to access data on the internet, the model behind ChatGPT is finite. As a result, ChatGPT doesn't know everything. When prompted with questions about obscure or recent topics, the answer the model returns may be wrong. AI lacks common sense No machine learning model has had the experience of being human. As a result, the responses generated by the model lack common sense. For example, if you ask ChatGPT how to swim to the moon, it may provide an answer without questioning the absurdity of the question. Any human would first question the value of such a strange question.

AI has accuracy issues The accuracy of responses generated by a model depends on many factors, including the training data, context, user input, complexity of the prompt and language, and bias. As the user of a model, you have control over only some of these factors. Where possible, however, you can help the model provide better responses by knowing how best to prompt it and by challenging the model's output in follow up prompts. AI has the potential to be biased Because they're trained largely on text written by people, machine learning models will pick up the biases and preferences that exist in the training data. Creators of models put a lot of effort into eliminating bias which is a worthy goal but an impossible task. Unintended consequences and even dangerous situations may result from bias in models. The classic example is when Microsoft released its Tay chatbot to the internet in 2016. Within one day of talking to people, the chatbot went from saying things like "Humans are cool" to making racist and sexist comments.

REFERENCES

Agüera, Y., & Arcas, B. (2022). Do Large Language Models Understand Us? *Daedalus, 151*(2), 183–197. DOI: 10.1162/DAED_a_01909

Benchsci. (2025). *benchsci*. www.benchsci.com

Brown, T. B., Mann, B., Ryder, N., Subbiah, M., Kaplan, J., Dhariwal, P., Neelakantan, A., Shyam, P., Sastry, G., Askell, A., Agarwal, S., Herbert-Voss, A., Krueger, G., Henighan, T., Child, R., Ramesh, A., Ziegler, D. M., Wu, J., Winter, C., . . . Amodei, D. (2020). Language Models are Few-Shot Learners Tom. *Advances in Neural Information Processing Systems, NeurIPS*. https://commoncrawl.org/the-data/

Ceci, S. J. (1996). On Intelligence. *On Intelligence*, 1–2. https://doi.org/DOI: 10.4159/9780674029316

Chernyavskiy, A., Ilvovsky, D., & Nakov, P. (2021). Transformers: "The End of History" for NLP? *Joint European Conference on Machine Learning and Knowledge Discovery in Databases*. https://arxiv.org/abs/2105.00813

Gardner, H. (2011). *The Theory of Multiple Intelligences*. Harvard Resources. https://pz.harvard.edu/resources/the-theory-of-multiple-intelligences

Hintze, A. (2016). *Understanding the four types of AI, from reactive robots to self-aware beings*. The Conversation. https://theconversation.com/understanding-the-four-types-of-ai-from-reactive-robots-to-self-aware-beings-67616

Hoppe, S., & Toussaint, M. (2019). Language Models are Unsupervised Multitask Learners Alec. *OpenAI Blog*. https://arxiv.org/abs/2007.07582

Hoskens, M., & De Boeck, P. (1991). An Implicit Theory of Intelligence-Related Mental Activities. *Journal of Personality, 59*(4), 793–814. DOI: 10.1111/j.1467-6494.1991.tb00932.x

Meta. (2025). *A.R.I.A.* https://www.projectaria.com/

Naveed, H., Khan, A. U., Qiu, S., Saqib, M., Anwar, S., Usman, M., Akhtar, N., Barnes, N., & Mian, A. (2025). A Comprehensive Overview of Large Language Models. *ACM Transactions on Intelligent Systems and Technology, 16*(5), 1–72. DOI: 10.1145/3744746

Rabbit, R. (2025). *Research Rabbit*. https://www.researchrabbit.ai/

Wang, A., Pruksachatkun, Y., Nangia, N., Singh, A., Michael, J., Hill, F., Levy, O., & Bowman, S. R. (2019). SuperGLUE: A stickier benchmark for general-purpose language understanding systems. *Advances in Neural Information Processing Systems*, *32*(July), 1–15.

Chapter 16

The Cloud-Native Enterprise: Orchestrating Intelligent Automation Across Global Operations

Priya Gaur
Centre for Online Education, GLA University, India

Gunjan Mohan Sharma
https://orcid.org/0000-0003-2272-7385
JGBS, O.P. Jindal Global University, India

Ali Najeeb
https://orcid.org/0000-0003-4663-9347
Qasim Ibrahim School of Business, Villa College, Maldives

ABSTRACT

The contemporary global business landscape demands a new operational paradigm centered on agility and intelligence. This chapter argues that the convergence of cloud-native architectures and intelligent automation (IA) is fundamental to this transformation. It deconstructs how cloud-native principles—microservices, containers, and DevOps—provide the essential, scalable foundation for orchestrating IA across complex, global operations. The analysis moves beyond a technical blueprint to explore a practical implementation playbook for transitioning from pilot to enterprise-wide scale. Crucially, the chapter dedicates significant focus to the profound implications of this shift, examining the social impact on the workforce, practical challenges like talent and cost, critical ethical considerations around

DOI: 10.4018/979-8-3373-7442-0.ch016

algorithmic bias, and the evolving role of leadership. Success requires managing this socio-technical transformation to build resilient, responsible, and competitive enterprises.

INTRODUCTION

Consider a global consumer goods company caught in a perfect storm: there has been a rush of demand on a product in a part of the world, there is a supplier meltdown in another, and new and complicated custom laws are being introduced. This would be a slow, staged process of data collection, analysis and communication, which would result in stockouts, lost income and angry customers, due to a traditional, siloed IT environment. Suppose you were in a different world? In this case, the system is event-based which automatically identifies the spike in demand and supplier problem in real-time. An intelligent platform running on the cloud immediately reroutes shipments by modeling thousands of logistics conditions, a chatbot places the required customs documents automatically, and a procurement bot launches a tender with already pre-vqualified alternative suppliers all without human intervention. It is not a far flung fantasy of the future but operational reality of the new cloud-native enterprise. This chapter assumes that in the modern global business environment, where volatility, uncertainty, complexity, and ambiguity (VUCA) are the new reality, the implementation of a cloud-native paradigm is no longer an IT infrastructure decision but the precondition to organize intelligent automation at a global scale. It is this synergy that allows organizations to shift beyond being an efficient to a significantly resilient, agile, and innovative organization (Mayer & Weinreich, 2018). The process to get to this state is however more about technology as much as it is about organizational philosophy. In this chapter, the fundamental paradigms of cloud-native computing and intelligent automation are going to be dismantled, and a proper vocabulary will be established. It will subsequently examine the formidable intersection of these two forces as to why cloud-native architecture is the essential catalyst that enables automation to be no longer a single-task endeavor but a worldwide coordinated feature. An elaborate architectural plan will be availed, including the main constituents and architectural designs required in such a venture. Subsequently, a playbook on practical implementation will be used to assist leaders in initial pilot stage to global scale. Last, and, perhaps, most importantly, the chapter will pay much attention to the complex of implications that this new operating model implies on the enterprise of social, practical, ethical, and managerial levels and claim that going through these human and strategic levels is the final factor of success or failure.

SECTION 1: DECONSTRUCTING THE CORE PARADIGMS

In order to see the transformative capabilities of a cloud-native approach to intelligent automation in their entirety, it is necessary to form a clear grasp of the components that make it up. Cloud and AI are often discussed as rather broad terms that are misleading and are employed as buzzwords. The deconstruction required is a narrowing down to get out of hype to strategy execution.

1.1. What is Truly "Cloud-Native"? Beyond Lift-and-Shift

Cloud-native is a term that is used interchangeably to denote any type of application or workload that is hosted on a public cloud platform, including Amazon Web Services (AWS), Microsoft Azure, or Google Cloud Platform (GCP). This is a basic misrepresentation. The mere process of migrating a monolithic, legacy application out of a corporate data center and into a virtual machine in the cloud, a process referred to as a lift-and-shift, only utilizes a small part of the cloud potential, namely, the lowered capital spending on hardware (Fehling, Leymann, and Retter, 2014). True cloud-native is an architectural and cultural model of developing and operating applications that pushes the benefits of the cloud computing model to their extreme. It is based on a number of major principles. The first one is the implementation of a microservices architecture, in which large monolithic applications are broken down into small loosely coupled and independently deployable services. A microservice represents a particular business capability (e.g. process payment, check inventory, calculate shipping) and is connected with others by clearly defined APIs (Application Programming Interfaces). This is an essential facilitator in automation, as it enables discrete business processes to be automatable, updated, and scaled independently (Newman, 2015). Second, such microservices are commonly containerized, commonly with Docker. An application together with its dependencies is packaged into a standardized unit in a container, which is consistent and can be deployed anywhere, on the laptop of a developer or on a production cluster across the globe. Scaling up the management of these containerized workloads is done through orchestration platforms, and Kubernetes has become the standard. Kubernetes is an automated system that deploys, scales, heals, and manages containerized applications and is a distributed operating system in the cloud-native world (Hightower, Burns, and Beda, 2017).

This technical basis cannot be separated out without a cultural and procedural transformation represented by DevOps. DevOps represents a combination of development (Dev) and operations (Ops) into one, joint lifecycle, which is intended to reduce the systems development life cycle and achieve continuous delivery with high software quality. It is facilitated by such practices as Continuous Integration and Continuous Deployment (CI/CD) when automated pipelines continuously build, test, and deploy changes to production. This establishes an innovation flywheel, where new features and bug fixes, including automated processes updates, can be published safely and fast, often hundreds of times a day (Forsgren, Humble, and Kim, 2018). Lastly, the cloud-native solution adopts the notion of immutable infrastructure and Infrastructure as Code (IaC). Rather than having to configure and patch long-lived servers manually, infrastructure (e.g. virtual networks, load balancers, storage) is described and deployed in machine-readable definition files. Terraform or AWS CloudFormation tools enable the spin up of entire environments to be programmatically versioned and torn down. This turns infrastructure into a first-class citizen in the automation narrative, making it reliable, repeatable, and eliminating configuration drift and snowflake servers, which are unique and hard to manage.

1.2. The Evolution of Intelligent Automation (IA)

The sphere of automation has also seen its major change in parallel with the development of cloud computing. The trip started with Robotic Process Automation (RPA), which was concerned with automating rule-based, high-volume, and repetitive tasks normally done by humans in contact with the graphical user interfaces (GUIs) of already existing applications. RPA bots were able to log in to systems, copy-paste information, fill in forms, and do other menial clerical tasks. Although RPA brought considerable efficiency benefits to particular functions, it was essentially a band-aid application, which worked on the presentation layer, but not on the inherent weakness of the integrated systems. It introduced isolated automation islands that were fragile and hard to support in case of application interfaces (van der Aalst, Bichler, and Heinzl, 2018).

The Cognitive Automation or Intelligent Automation (IA) is the next step in the evolution of automation that introduces the artificial intelligence (AI) and machine learning (ML) features to traditional automation. This takes automation out of the rule-based processes and into cognitive processes. IAS is capable of processing unstructured data (emails, documents, and images), comprehending natural language, predicting with the use of historical data, and even learning and improving with time. As an example, an RPA can be used to read data in a structured invoice, whereas an IA system that runs on Computer Vision can read a scanned semi-structured invoice and, with the help of Natural Language Processing (NLP), comprehend the context and the most important entities in the invoice. Even when the data is not an exact match, Machine Learning models can then forecast the probability that that invoice is a fraud, or it can automatically compare it to a purchase order (Davenport and Ronanki, 2018). The final manifestation of this trend is Hyperautomation which is a business-oriented strategy of quickly finding, screening and automating as many business and IT processes as feasible. Hyperautomation is a combination of a set of tools and technologies, such as RPA, IA, process mining, advanced analytics, and decision management systems, to build a smooth, end-to-end automation of more complex business processes. It is a transition to automating discrete tasks to automating whole business capabilities, which can only be accomplished with the scalable, integrative and nimble base offered by a cloud-native architecture.

SECTION 2: THE CONFLUENCE: WHY CLOUD-NATIVE IS THE CATALYST FOR GLOBAL IA

With the fundamental ideas in place, one will realize that cloud-native and intelligent automation are not only adjacent trends, but they are also highly synergistic with each other. The cloud-native paradigm is the key engine and the chassis on which the advanced functionality of intelligent automation can be developed, expanded, and successfully driven throughout a global organization.

2.1. Scalability on Demand

Elastic scalability is one of the greatest and most immediate advantages of a cloud-native architecture when it comes to intelligent automation. Take the case of an intelligent document processing service which is consumed by an international retail firm. The load on this service may also rise ten-fold during the end-of-quarter financial closing when thousands of invoices and receipts are processed. An on-premises automation solution would also mean that the company would have to provision and pay to have infrastructure capable of managing this peak load across the year, causing tremendous inefficiency and expense. This service, which is provided in the form of a container and managed by Kubernetes, can be configured with auto-scale out features, which spins up more container instances to support the augmented load in a cloud-native environment. After the peak is subsided, it can scale just as automatic, and resources that are not needed are free. This pay-as-you-use model is facilitated by the elastic nature of the cloud, which implies that the global automation potential can be implemented without the heavy initial capital investment, and it can be adjusted to the rises and falls of business operations across time zones and markets (Mullender, 2019). It is not merely horizontal scaling (adding additional instances) but also geographic scaling, where the instances of an automation service are deployed in cloud regions nearer to the location of business operations so that they are responsive.

2.2. Global Latency and Resilience

To make automation a viable strategy in international operations, it should be responsive and reliable. An algorithm to optimize logistics can only take minutes to run, but in situations where a port closure needs to be re-routed in real-time, it is useless. This is directly served by cloud-native architecture as distributed systems. Through an international network of cloud regions and availability zones, microservices of critical automation can be executed at high proximity to the business events they are targeted to serve. Moreover, the edge computing concept expands such principle even further and moves lightweight containerized workloads to the locations of the edges (e.g., factories, warehouses, retail stores) to compute data there in real time with limited latency and then synchronize with central systems (Shi, Cao, Zhang, Li, and Xu, 2016). This becomes essential in the industrial IoT context, e.g. predictive maintenance in assembly lines. It is also important to be resilient. A monolithic automation platform means that there is a single point of failure; failure of which causes automated processes to halt throughout the whole enterprise. With a cloud-native architecture that is built upon microservices, the crash of one microservice (e.g., the "tax calculation" microservice) is not guaranteed to propagate and crash the whole system. The orchestration layer has automatic restart of failed containers and the architecture can be configured with circuit breakers and fallback mechanisms, which mean that the entire global operation can keep going, maybe in a degraded state, but without collapsing.

2.3. Velocity of Innovation

The fast rate at which an organization learns and adapts is usually the final competitive edge in the digital age. The cloud-native model, whose DevOps and CI/CD backbone forms a strong force behind the innovation in automation. Since the automation capabilities are separated into small, independent microservices, development teams are able to work on them in parallel and not to step on each other toes. One team in Singapore can be working on the NLP model of contract analysis and another team in Berlin on the fraud detection algorithm of payments. Individual service changes can be continually integrated, tested, and deployed without needing a big bang release of a large, risky, and infrequent monolithic automation suite (Forsgren et al., 2018). This gives the business the opportunity to quickly test new automation processes, put them to the test in controlled settings, collect data on their effectiveness, and perform iteration or rollbacks at a very low cost and or disruption. This is the main cycle of experimentation that is high velocity and this cycle can help the organization to identify and optimize the most valuable opportunities of automation in its global operations.

2.4. The Data Fabric

Data is the main source of intelligent automation especially the cognitive aspects of it. Conventional business enterprises have been characterized by data silos where vital data is stored in old systems-ERP, CRM, supply chain management that are not easily connected with one another. This disintegration is one of the biggest impediments to end-to-end automation. A unified data fabric can be created with the help of a cloud-native approach. The organization can break down these silos by following an API-first mentality, in which each microservice provides its functionality and data through well-defined APIs. It is possible to stream data out of legacy systems into cloud data lakes or meshes where it can be cleansed, consolidated, and made available to analytics and machine learning. This unified information environment is the blood of intelligent automation. Considering an automated customer service agent as an example, it can only give a holistic response when it can access in real-time data provided by the CRM (customer history), the order management system (current orders), and the logistics system (shipment status) through a mesh of APIs that need to be interoperable. The cloud not only offers the storage of this data, but also the enormous and limitlessly scalable computing resources necessary to train the machine learning models that turn the automation into the sort of intelligent power.

SECTION 3: THE ARCHITECTURAL BLUEPRINT FOR ORCHESTRATION

The theoretical synergy between cloud-native and IA has to be translated into a concrete, working reality, which will need a conscious and carefully designed platform. It is not related to the implementation of a single tool, but rather a building of a unified ecosystem that can be orchestrated on a global level.

3.1. The Platform Approach

Platform-as-a-Service (PaaS) is the most efficient model of providing cloud-native intelligent automation on a large enterprise that is internalized in the form of a Centralized Automation Platform. Rather than letting individual business units to purchase and run their own automation tools in an informal fashion, a special platform team develops and operates a shared, self-service platform. This service is a collection of services, tools, and APIs that can be consumed by product teams (e.g., the supply chain team, the finance team) to create, deploy and operate their own automated processes (Spotify, 2018). The method eliminates fragmentation, cost management, implements security and governance requirements and significantly shortens the time-to-value of new automation projects. The platform hides the complexity of the cloud infrastructure and orchestration, and enables business-domain experts to do business logic configuration and composition instead of server and network administration.

3.2. Key Architectural Components

The internal automation platform consists of a number of layers and components which are interconnected and operate in harmony with one another. Its core is the Orchestration Layer, which is normally executed on Kubernetes. This is the automation symphony conductor layer. It also schedules and orchestrates the containerized workloads that make up the automated services and provides them with the resources they require, recreates them to achieve a high availability level, and recreates them in case of failure. It gives the basic primitives of resilience and scalability on which all others are built. On top of this is a vital element of cognitive automation: the AI/ML Ops pipeline. Machine learning models are not static software; it is dynamic software that must be constantly trained, versioned, deployed, and monitored. The practice of implementing DevOps principles to the ML lifecycle is known as AI/ML Ops. It entails automated pipelines that accept new data, retrain models and ensure their performance is measured against a baseline and when they pass the test, they are automatically deployed as microservices into the Kubernetes cluster to be consumed by other automated processes. This is necessary to ensure the relevance and accuracy of intelligent automation in a dynamic world (Treveil et al., 2020).

The architecture needs to be Event-Driven in order to support real-time, responsive automation. In event-driven architecture, services interact by generating and receiving events- a history of something that has occurred (e.g. PurchaseOrderCreated, ShipmentDelayed, InvoiceApproved). These are sent to a message bus or event stream (e.g. Apache Kafka, AWS EventBridge). Interested other services can subscribe to these events and automatically cause actions. This forms a loosely integrated, highly responsive environment with automation processes being triggered by business events as they happen, on any system within the global enterprise, and does not require tight, point-to-point integration. Lastly, it is all integrated with a solid API-First Integration approach. The legacy systems are not going to fade away in one day. Another practical solution is to adopt API gateways and integration platforms (e.g., MuleSoft, Azure API Management) in order to build modern, secure, and well-documented RESTful APIs around these legacy systems. This is actually a way of wrapping the old monoliths and therefore makes their data and functionality accessible to the new cloud-native automation services in a managed and regulated fashion and therefore extends the life and value of existing IT investments and also allows them to be involved in automated workflows.

3.3. Security and Governance in a Distributed World

The dispersed character of a global, cloud-native automation platform presents a significantly increased attack surface, enabling security and governance to be a key concern, requiring baking in rather than an appendix. The model of security that is needed is the Zero-Trust one that is based on the idea of never trust, always verify. It is achieved by the strong Identity and Access Management (IAM) that strictly adheres to the principle of least privilege and makes sure that every microservice and user is granted permissions that are necessary to execute its role at all. Moreover, security, and compliance policies can be represented in policy-as-code formats such as Open Policy Agent (OPA), and then version-controlled and enforced in the cluster-wide. As an example, one can write a policy that no automation service can be run as a root user or that all data on transit should be encrypted. Lastly, because of the use of open-source software and container images, it is essential to secure the software supply chain. This includes scanning the image of the containers to identify vulnerabilities, putting signatures on the images to ensure integrity and verifying trusted registries to detect the deployment of malicious code. When business processes that are critical are automated, a security breach does not only mean a loss of data, but it can also mean doomsday.

SECTION 4: AN IMPLEMENTATION PLAYBOOK: FROM PILOT TO GLOBAL SCALE

Even an ideal architecture cannot be implemented without a realistic implementation plan. The implementation of an intelligent automation capability based on a cloud-native platform is a complex endeavor that needs to be executed in stages, through an iterative process, and with business outcomes in mind.

4.1. Phase 1: Discovery and Prioritization

The trip does not start with technology, but with a profound knowledge of the business processes. This phase aims at identifying the most promising candidates to automation in the global operations. It is possible to use modern methods such as process mining to analyze event logs of existing systems (e.g., ERP, CRM) automatically and learn the real, as-is execution of processes, including variants, bottlenecks, and inefficiencies that might not be obvious based on process diagrams (van der Aalst, 2016). This information-driven method supplements the conventional workshops and interviews with the subject matter experts. Applicants are to be judged in terms of a balanced scorecard comprising not only of possible cost savings (Full-Time Equivalent reduction), but also of strategic value, including cycling time improvement, better customer experience, elimination of errors, and scalability. A typical structure is to map processes on a two-by-two grid of value versus complexity of implementation, initially the quick wins: high value and low complexity. Automated invoice processing, onboarding of employees or real-time compliance checks within the supply chain are examples.

4.2. Phase 2: Building the Foundation and Running a Pilot

The next step is to form the core platform team and commence the construction of the basic automation platform since there is an already prioritized backlog of automation opportunities. At the same time, they should choose one, high-value, and rather limited process to be used as a pilot project. This pilot has two key functions: it can offer how the actual value can be obtained in order to achieve executive and stakeholder buy-in, and it can offer a test ground of the evolving platform that will compel the team to deal with practical integration, security, and operational issues at the outset. This pilot team must be cross-functional and includes a business product owner, software developers, DevOps engineers, and data scientists where cognitive capabilities are concerned. They will operate in a nimble fashion to create the initial microservices, the initial CI/CD pipeline and connect to the required legacy systems. This stage will be successful when an automated process becomes a reality and is capable of producing the business results promised.

4.3. Phase 3: Scaling and Federating

An effective pilot shows evidence-of-concept. The second step is that of transitioning between a one project and organizational capability. This includes the creation of an Automation Center of Excellence (CoE) on a formal basis. The CoE is a small central team of specialists that is in charge of managing the platform, establishing best practices, governing standards, and training and supporting a larger community of users. This is aimed at moving towards a federated development model, with the CoE as the platform and guardrails, and product team empowered business units responsible to build and maintain their own automation services on the self-service features of the platform. This is essential in scaling within an organization that has global presence because it makes use of domain expertise that is inherent to the business units. The CoE is in charge of the portfolio of automation initiatives, disseminates lessons learned, and makes sure that local optimizations are not at the cost of efficiency and coherence on the global scale.

4.4. Measuring Success

With the increase in automation ability, measurement should not be just ROI. An equitable mix of Key Performance Indicators (KPIs) is needed to have the complete range of value. Operational KPIs are not irrelevant: reduction in the cycle time (e.g. time between purchase order and payment), eradication of errors, and increase in throughput. Nevertheless, strategic and people KPIs are also essential. They are: employee satisfaction scores (to make sure that automation is complementing and not replacing the workforce), better customer satisfaction (NPS), and even sustainability such as the reduction of carbon footprint through the use of optimized logistics and energy control. The rate of innovation, e.g. the number of new automation services being implemented each quarter or the proportion of business processes that have become API-enabled, is a leading indicator of the increasing digital agility of the organization.

SECTION 5: NAVIGATING THE IMPLICATIONS

A socio-technical transformation is the shift to a cloud-native and intelligently-automated enterprise. The technological issues, though not the least important, are usually shadowed by the immense human and strategic consequences. The main secret to sustainable and responsible revolution lies in proactive measures to combat them.

5.1. Social Implications

The most popular social implication is the future of work. The story tends to alternate between two poles: the idyllic image of a world without the need of work and the nightmare of the large-scale technological unemployment of the masses. Probably, a more subtle reality exists. Although some of the routine, manual, and cognitive activities will be completely automated, this will cause a change in the character of human work. There will be a higher need in positions that require management, maintenance, and oversight of automated systems or ones that are performed only by humans: creativity, complex problem-solving, ethical judgment, and empathy (Brynjolfsson and McAfee, 2014). This creates the necessity of reskilling and upskilling. It is a moral and functional duty of businesses to invest hefty sums of money in ongoing learning schemes so that their employees can be able to adapt into new and more valuable positions. Lack of doing it may result in serious social unrest, skills gaps, and internal resistance to change. Such a change may also increase a digital divide, within and between organizations. Early adopters may be businesses units or geographic regions and they may run ahead creating internal disparities. A disparity can become increasingly evident on a world scale, with automation-saturated corporations in the developed economies and automation-sensitive organizations in other parts of the planet, and it may redefine the economic processes of the world. The result of all this is the absolute need to change the culture of the organization. The culture of the enterprise should shift towards one that is based on experimentation, cross-functional (silos are broken) cooperation, tolerance to failure and constant learning.

5.2. Practical Implications

Practically, there are a number of challenges that leaders have to deal with. The most severe one is the talent war. Kubernetes expertise, DevOps engineering, data science, ML Ops are the skills needed to develop and run a cloud-native automation platform, which are extremely sought after and scarce. To attract and retain this talent, it is essential that the company has a compelling technical vision as well as culture in addition to competitive pay. The question that organizations are plagued with, at a strategic level, is vendor lock-in versus multi-cloud strategy. A commitment to proprietary offerings of a single cloud provider (e.g., AWS SageMaker, Azure Cognitive Services) may bring efficiencies, however, it also raises switching costs and decreases negotiation power. Although a multi-cloud solution is flexible and resilient, it also brings a lot of complexity in terms of management and integration. Technical debt and legacy systems is another issue that is permeating. The run budget of supporting the fragile, monolithic legacy systems frequently drains resources that might otherwise be spent on the change budget of new automation projects. A planned, budgeted process of modernizing applications and decommissioning legacy is required. Lastly, the pay-as-you-go model of the cloud may result into spiralling costs unless properly handled. FinOps, which is a cultural practice and tooling of applying financial accountability to the variable spend model of the cloud, is essential to make sure the cost of automation does not dilute its value.

5.3. Ethical Implications

The strength of smart automation comes with a great ethical concern that is not only legal but also ethical. One of the major issues is bias and fairness of algorithms. The new machine learning models are trained on past data and when the past data is biased (e.g., in hiring, lending or policing), the automated system will not only maintain but possibly increase these biases in large scale (O'Neil, 2016). There must be rigorous disparate impact testing within different groups of people. This is associated with the issue of transparency and Explainability (XAI). In case an AI-driven automation system rejects a loan application or marks a transaction as fraudulent, one has to be able to explain the cause of such decision in a clear and understandable manner. Some complex ML models are black box and this can be a serious threat to trust and accountability. Sovereignty and privacy of data is also essential. To automate intercountry processes that touch on personal data it is important to carefully observe the rules such as the GDPR in Europe which requires such principles as privacy by design and by default. This can be architecturally complicated in a distributed, cloud-native system in which data can be transiting across multiple jurisdictions. Finally, these issues would all lead to the issue of accountability. In the event of harm caused by an automated system, such as a malfunctioning trading algorithm resulting in a flash crash or biased hiring algorithm yielding biased results, the issue of legal and moral responsibility is a complicated maze of the creators of the system, the data scientists who created it, the business executives who implemented it, and even the platform providers.

5.4. Managerial Implications

To managers and other senior leaders, this new paradigm requires that they re-examine their job description and their own competencies. The competencies of leadership have to change. The leaders no longer have to be the most technical of all, but must be technologically fluent. They must learn the essentials of cloud economics, data-driven decision-making, and the agile development lifecycle to make effective strategic investments and to effectively operate hybrid teams of human beings and so-called digital workers. This change requires modification in the structure of the organization. The classical, strict, functional hierarchy is not very well adapted to the rapidity and cross-functionalism of cloud-native automation. The structures should change to fluid product-oriented teams that are empowered and have to be responsible of the end-to-end business results, including both the development and the operations. This directly affects risk management and governance. The outdated and sluggish IT governance models that operate on a gate-based system should be substituted with dynamic and policy-driven models that facilitate speed without compromising safety and compliance. The leaders should fund the creation of new risk management frameworks capable of evaluating and reducing the new risks of the AI and big-scale automation. Lastly, the essence of strategic agility is reestablished. The competitive advantage in this new model is not due to a fixed, long term plan as much as it is due to the innate capacity of the organization to constantly feel the changes in the market, re-align its automated processes and re-organize its resources with speed and accuracy. The leader assumes the position of a strategist and a coordinator of a complex, adaptive system.

CONCLUSION:

The path towards being a cloud-native enterprise, able to coordinate intelligent automation of all its international activities, is one of the largest changes that a contemporary company can undergo. This chapter has claimed that this is not just an upgrade of technology but a re-invention of the operating model. The principles of microservices, containers, orchestration and DevOps offered by the cloud-native paradigm offer the elastic, resilient, and high-velocity platform that intelligent automation cannot scale to beyond isolated tasks. The blueprint and roadmap of implementation are the mechanics, but it is the architectural blueprint. The real difficulty, and the final source of competitive advantage, is to manoeuvre through the deep human and strategic implications. The social contract of work is getting rewritten, the pragmatic questions of talent and cost are enormous and the ethical question of algorithmic governance are unparalleled. The managerial mandate has, thus, changed. Modern leaders have to be orchestrators in every sense of this word: designers of socio-technical systems, sustainers of a new moral contract, and fosterers of a culture of life-long learning and change. They have the mandate of developing organizations that are efficient and profitable and also resilient, responsible and able to embrace technology to wring out more human potential on an international scale. The intelligent, cloud-native enterprise is not a destination, but a constantly evolving state, and the work of the orchestrator never stops.

REFERENCES

Brynjolfsson, E., & McAfee, A. (2014). *The second machine age: Work, progress, and prosperity in a time of brilliant technologies*. W. W. Norton & Company.

Davenport, T. H., & Ronanki, R. (2018). Artificial intelligence for the real world. *Harvard Business Review, 96*(1), 108–116.

Fehling, C., Leymann, F., Retter, R., Schupeck, W., & Arbitter, P. (2014). *Cloud computing patterns: Fundamentals to design, build, and manage cloud applications*. Springer.

Forsgren, N., Humble, J., & Kim, G. (2018). *Accelerate: The science of lean software and DevOps: Building and scaling high performing technology organizations*. IT Revolution Press.

Hightower, K., Burns, B., & Beda, J. (2017). *Kubernetes: Up and running*. O'Reilly Media.

Mayer, R., & Weinreich, R. (2018). *The cloud native landscape: A systematic mapping study*. IEEE/ACM 11th International Conference on Utility and Cloud Computing (UCC).

Mullender, S. (2019). *Cloud native transformation: Practical patterns for innovation*. O'Reilly Media.

Newman, S. (2015). *Building microservices*. O'Reilly Media.

O'Neil, C. (2016). *Weapons of math destruction: How big data increases inequality and threatens democracy*. Crown Publishing Group.

Shi, W., Cao, J., Zhang, Q., Li, Y., & Xu, L. (2016). Edge computing: Vision and challenges. *IEEE Internet of Things Journal, 3*(5), 637–646.

Spotify. (2018). *Spotify's failed #SquadGoals model*. [Blog post]. Retrieved from https://www.spotify.com/intl/ie/blog/spotify's-failed-squadgoals-model/

Treveil, M.. (2020). *Introducing MLOps: How to scale machine learning in the enterprise*. O'Reilly Media.

van der Aalst, W. M. P. (2016). *Process mining: Data science in action*. Springer.

van der Aalst, W. M. P., Bichler, M., & Heinzl, A. (2018). Robotic process automation. *Business & Information Systems Engineering, 60*(4), 269–272.

Chapter 17
Human-Centric AI: Reinventing Employee Experience Through Cloud-Native Intelligence

Isha Goley
Prestige Institute of Management and Research, Gwalior, India

Bhumika Ray
https://orcid.org/0009-0000-5628-5859
JGBS, O.P. Jindal Global University, India

Mukhtaruddin Mukhtaruddin
Faculty of Economics, Universitas Sriwijaya, Indonesia

ABSTRACT

Cloud-native AI is not just a mere automation, but a combination of contextual knowledge, pattern recognition, behavioral analytics and predictive intelligence (Qin et al., 2025). These mechanisms interpret complicated employee cues, grasp dynamic shifts in motivation or well-being, and provide information that may be utilized to recommend individualized interventions (Segarra, and Williams, 2025). Workers are faced with new pressures like never before as global workplaces grow more digital and distributed, including increased skill transitions and/or the psychological strain of hybrid work (Choudhary et al., 2023). This poses an organizational need to develop HR processes that can help create resilience, engagement, and fulfillment. Human-centered AI will meet these essential needs by engraving compassion, personalization, and justice in the development of workplace technologies (Buhmann and Fieseler, 2022).

DOI: 10.4018/979-8-3373-7442-0.ch017

INTRODUCTION: THE GIFTHOOD OF ANTHROPOCENTRIC AI AT WORK.

The transformation that human resources have been experiencing is very deep and multi-layered as organizations begin to abandon their previous administrative technologies and move towards the holistic intelligence systems created around human requirements (Yalenios & d'Armagnac, 2022). This transformation is not only an improvement in technology but a change in the paradigm of understanding work, people, and organizational value (Buhmann and Fieseler, 2022). The human-centric AI has become a key driver of this change, with the understanding that the employees now anticipate the use of an adaptive, empathetic, responsive, and goal-oriented work system (in line with their long-term goals) (Malik et al., 2022). Here, the environmental aspect provides the infrastructure of cloud-native artificial intelligence to create highly personalized, real-time, and continuously learning HR ecosystems (Einola and Khoreva, 2022).

Cloud-native AI is not just a mere automation, but a combination of contextual knowledge, pattern recognition, behavioral analytics and predictive intelligence (Qin et al., 2025). These mechanisms interpret complicated employee cues, grasp dynamic shifts in motivation or well-being, and provide information that may be utilized to recommend individualized interventions (Segarra, and Williams, 2025). Workers are faced with new pressures like never before as global workplaces grow more digital and distributed, including increased skill transitions and/or the psychological strain of hybrid work (Choudhary et al., 2023). This poses an organizational need to develop HR processes that can help create resilience, engagement, and fulfillment. Human-centered AI will meet these essential needs by engraving compassion, personalization, and justice in the development of workplace technologies (Buhmann and Fieseler, 2022).

Moreover, companies have an increased competition over talent, increased turnover rates, and quickly changing workforce requirements (Kay et al., 2024). The employees require opportunities to establish their careers flexibly, have meaningful work experiences and be treated ethically. When AI is adopted through the human prism, it can be a potent means to provide these expectations on a massive scale (Dutordoir and Struyfs, 2024). This chapter considers the disruptive aspect of cloud-native AI in changing the workforce experience at each stage of the talent lifecycle by highlighting its contribution to creating workplaces that are not only productive, but also more human (Ogbonnaya et al., 2025).

CLOUD-NATIVE AI AS THE INFRASTRUCTURE OF HUMAN-CENTRICITY.

Cloud-native AI is the skeleton of the contemporary human-in-focused working environments because it provides unparalleled scalability, robustness, and ongoing intelligence (Yalenios and d'Armagnac, 2022). Conventional on-premise HR systems, even though handy in the recordkeeping department, were never intended to provide real-time information or facilitate rich and personalized interactions. They were limited in their storage, processing power, system interoperability and cycle of updates and therefore they were not able to adapt to the fast changing workforce requirements (Malik et al., 2022). Conversely, cloud-native architectures allow AI models to learn continuously, use cross-functional sources of data, and respond within milliseconds to interactions between employees (Qin et al., 2025).

Cloud-native systems are based on microservices, serverless computing, orchestration of containers, and event-based workflows. The architectural components enable HR systems to receive and analyze data across various touchpoints, including applicant tracking systems, communication platforms, learning management systems, wellness platforms, and collaboration tools (Einola and Khoreva, 2022). This connectivity generates a smooth and integrated experience among the staff because the intelligence used in the system has the ability to know their preferences, working habits, and growth requirements (Choudhary et al., 2023).

Due to the continuous retraining of models and processes, cloud-native AI is sensitive to the new job roles, skill requirements, and organizational changes (Buhmann and Fieseler, 2022). The recommendations given by AI improve with time as more streams of data input into the system. Deployments are done immediately and throughout the organization without disruption. Cloud-native security measures allow keeping sensitive data about employees secure and maintain transparent and compliant AI-based workflows (Kay et al., 2024). Consequently, companies will have a platform on which AI will be dynamic rather than fixed, ethical, and very sensitive to human conditions (Qin et al., 2025).

BASED ON INTELLIGENT RECRUITMENT AND TALENT ACQUISITION.

The employee experience starts way before onboarding, but it starts with the first interaction between the candidate and the organization (Dineen et al., 2022). Human-centric AI will make the recruitment process not only just but also interesting and very responsive. The AI-based talent acquisition tools that are cloud-native appraise the applicant profiles through much more advanced methods than the use of keywords in the past. They check the competencies of the applicants, their previous performance indicators, the communication pattern, behavioral traits, career motivation, and cultural alignment (Malik et al., 2022).

Such systems can conduct screening at scale as well as in real-time and provide real-time recommendations (Choudhary et al., 2023). At every stage of the recruitment process, AI interacts with the applicants by sending them personalized messages, which makes the process less ambiguous and improves its perception of the employer brand (Kay et al., 2024). Predictive analytics makes assumptions about the likelihood of job success, using historical data and trends of the specific situation, to help recruiters make informed choices without the influence of unconscious bias (Einola and Khoreva, 2022). It will also enable cloud-native AI to help organizations design more inclusive job descriptions by locating words that may dishearten a wide range of applicants (Buhmann and Fieseler, 2022).

Automation of administration allows AI to liberate recruiters, allowing them to concentrate on connections, strategic recruitment, and interaction with the candidates. On a more fundamental level, cloud-native platforms generate a sense of value psychologically by making the applicants feel appreciated, nurtured, and understood (Ogbonnaya et al., 2025). With scalability and perpetual intelligence, the organizations create recruitment pipelines that are rapid, equitable, and consistent with the talent strategies of the long-term (Qin et al., 2025).

Onboarding is the transition between hiring and the integration of employees on the long-term level, thus forming the first impression and affecting the follow-up engagement paths (Malik et al., 2022). Human-centric AI will reposition onboarding as a non-personalized orientation process into a dynamic, interactive journey of self-directed needs. Virtual AI assistants provide real-time support - answering questions, clarifying policies and getting new users introduced to tools and processes (Einola and Khoreva, 2022). Cloud-native AI analyzes the learning habits, professional history, language preferences, and mental preparedness of an employee to decide on the suitable rate and level of onboarding information (Choudhary et al., 2023). Instead of bombarding employees with information, AI orders a programmed trip, which evolves according to its relevance and cognitive load (Qin et al., 2025). It identifies when an employee needs more time or resources and will react. Combined with communication and project management platforms, AI can bring employees relevant team members on board, synthesise essential discussion, and aid initial collaboration (Kay et al., 2024). Cloud-native architecture holds the promise of maintaining the same quality and experience of onboarding at any place and job category (Ogbonnaya et al., 2025). Adaptive onboarding makes employees feel a sense of support and confidence, eliminating high turnover risks at early stages and improving productivity. The onboarding stage consequently turns into a significant kick-off of the career process of an employee, characterised by focus, belonging, and individual learning (Segarra and Williams, 2025).

HUMAN-CENTRIC AI RADICALLY CHANGES THE PERFORMANCE MANAGEMENT.

Traditionally, performance reviews were characterized by few evaluations, subjectivity of the managers, and a lack of visibility of daily contributions (Yalenios and d'Armagnac, 2022). Cloud-native AI offers a paradigm of ongoing performance intelligence where information is created dynamically and on a continuous basis through the various sources of data like collaboration tools, project-management applications, communication records, and peer-feedback (Malik et al., 2022). AI algorithms scan behavioural clues, such as leadership behaviour, quality of teamwork, flexibility, creativity, and efficiency in communication, therefore, offering objective data that allows employees to assess their progress and make necessary changes much earlier than the formal reviews (Qin et al., 2025). Cloud-native AI also determines emerging strengths, developmental requirements, and role-fit indicators, which enable self-directed growth (Kay et al., 2024). Managers can enjoy objective analytics which minimizes discrepancies in rating and brings out areas that need coaching or recognition (Ogbonnaya et al., 2025). Feedback mechanisms provided with the support of AI encourage transparency and create psychological safety (Segarra and Williams, 2025). Instead of causing fear, performance reviews turn into the possibility of ongoing communication, development of abilities, and advancement of careers and change the emphasis on the evaluation to empowerment (Einola and Khoreva, 2022).

With the sudden and swift changes in the technological environment, organisations are facing an increased need to learn and develop skills (Yalenios and d'Armagnac, 2022). Cloud-based AI promotes lifelong learning through the curation of customised and dynamic learning journeys to every employee (Qin et al., 2025). AI studies available talents, job needs, history of learning, and career ambitions and creates personalised development strategies (Malik et al., 2022). Generative AI makes platforms more contextual, interactive, adaptive, and realistic (Choudhary et al., 2023). Micro-learning modules are activated on the basis of performance gaps or the formation of new skills (Kay et al., 2024). The analytics on labour-market predicts future skill in AI, and it proactively suggests courses, mentorship programmes, or experiences (Tuan, 2024). The process of career development will be more transparent and strategic: AI will trace the possible path, outline transitional skills, and set development milestones (Ogbonnaya et al., 2025). Employees also get a clear understanding of how to move up or down a ladder, thus retention and a culture of lifelong learning and self-directed development are also enhanced (Segarra & Williams, 2025). Using cloud-native intelligence, organisations develop learning ecosystems that are dynamic, interactive, and extremely personalised (Buhmann and Fieseler, 2022).

Employee well-being is one of the key foundations of human-centered HR systems, and AI has a critical part in physical, emotional, and cognitive wellness (Segarra and Williams, 2025). Based on cloud-native AI, trends connected to workload, communication style, meetings intensity, response time and behavioural indicators are identified, thus identifying the early signs of stress, burnout or disengagement (Qin et al., 2025). Instead of being invasive, AI is presented as a friendly, concerned companion to make timely prompts and helpful suggestions (Malik et al., 2022). Artificial intelligence can suggest micro-breaks, mindfulness practices, or workload-balancing techniques; it can refer employees to internal and external mental-health services, such as counsellors, well-being websites, or supporting communities (Choudhary et al., 2023). Individual insights allow employees to consider their working patterns and change them to healthier sets. Since cloud-native systems use powerful encryption approach, privacy measures, and anonymisation, workers have confidence in the integrity of the system (Buhmann and Fieseler, 2022). With the responsible implementation, AI creates a culture in which well-being is a part and parcel of the daily work design. As a result, employees feel more resilient, maintain a feeling of engagement, and have meaningful growth (Yalenios and d'Armagnac, 2022).

EMPLOYEE ENGAGEMENT AND SENTIMENT UNDERSTANDING THROUGH AI

The concept of employee engagement has come a step further than the traditional survey-survey method into a complex, data-driven field that is enabled by cloud-native artificial intelligence (Malik et al., 2022). The traditional survey provides a narrow range of information due to the low frequency of administering it, as well as the inherent retrospective bias. On the other hand, AI-based sentiment analysis provides the real-time trends obtained on the basis of communication tools, social platforms, email, and collaborative interactions (Qin et al., 2025). By evaluating linguistic and affective tones, AI identifies emergent concerns, much earlier before it turns into a major problem (Choudhary et al., 2023). The insights, therefore, enable human-resources practitioners to put in place specific interventions that can be used to solve certain problems. As the initial signs of frustration, resistance, or disengagement are observed in the employees, managers can take the initiative by coaching them, shifting the workload, or developing effective communication intervention methods (Ogbonnaya et al., 2025). Cloud-native AI is also diverse and inclusive in that it tracks the involvement of the marginalized groups or whether they feel excluded in the decision-making area (Kay et al., 2024). The employees will thereby feel important as their voices are constantly and genuinely heard, resulting in a continuous feedback process that builds a dynamic and participatory culture grounded in humanity in all its human diversity (Segarra and Williams, 2025).

Hybrid and remote working models have forced AI-enhanced collaboration to play a critical role, bringing an unprecedented level of complexity to communication and collaboration (Yalenios and d'Armagnac, n.d.). Cloud-native AI is the connective tissue between distributed teams, which removes friction and results in greater productivity. Meeting summarization, items of action extraction, deadline tracking, and prioritization are also carried out automatically by AI tools, which ensure that employees are aligned, no matter their location (Choudhary et al., 2023). Intelligent scheduling systems examine their time zones, individual schedules, workload intensity and collaboration patterns to create meeting schedules that reduce fatigue (Qin et al., 2025). The translation services powered by AI not only break down language barriers and allow people to communicate in global teams but also allow accessibility tools to allow those employees with visual, auditory, or cognitive disabilities to participate (Kay et al., 2024). Furthermore, AI will offer suggestions of collaborators in terms of skills and expertise, history of working together on projects, and communication type, which will enhance more effective teamwork and exchange of knowledge (Ogbonnaya et al., 2025). Cloud-native AI enables collaboration, equitable, and highly productive hybrid work conditions by improving coordination, reducing administrative burden, and achieving inclusive involvement (Malik et al., 2022).

The culture of the organization, which was previously formed by in-person interaction, common space, and unofficial habits, is becoming mediated by online technologies, as distant and hybrid work has become a matter-of-course (Buhmann and Fieseler, 2022). Human-focused AI, in this case, plays a central role in the preservation of culture by taking organizational values, behavioural standards and expectations of collaboration and weaving them into the digital frameworks that staff members interact with on a daily basis (Yalenios and d'Armagnac, 2022). Cloud-native AI is not an observer of culture, but rather it is engaged in its support and development by creating contextual feedback, suggestions, and smart suggestions (Choudhary et al., 2023). Through the analysis of the communication tone, collaborative processes, belonging to internal community, and recognition trends, AI explains how employees actually perceive the cultural organization fabric (Qin et al., 2025). With such understanding building up in real time, leaders can start seeing the cultural strengths, emerging risks, and areas of disengagement, which would otherwise not be seen in dispersed teams (Ogbonnaya et al., 2025). In cases where AI identifies a decrease in the level of interaction, an increase in emotional overload, or loneliness in some groups, it can suggest specific measures, including mentoring circles, team-building events, or rebalancing work (Kay et al., 2024). This transformation turns cultural management into a proactive and continuous process that is based on data-driven empathy. Cloud-native systems also help workers to share success, celebrate milestones, and organize organizational rituals using digital town halls, social recognition systems, and peer-to-peer recognition channels (Malik et al., 2022). These online communications are the emotional binding that allows maintaining unity and purpose among geographically spread working forces. Accordingly, the intersection of cloud-native intelligence and virtual work communities enhances the resilience and cultural congruence (Segarra and 2025, Williams). This has the impact of making employees feel more connected, heard, and valued since the systems they engage in reflect the real values of the organization and encourage the inclusion of communication. AI thus goes beyond the optimization of operations to become the protector of the corporate identity so that culture is still a living and dynamic manifestation of the purpose of the company, even without physical boundaries (Buhmann and Fieseler, 2022).

Ethical design and responsible governance should be stipulated through a strong intent to deploy human-centric AI (Buhmann and Fieseler, 2022). Since AI systems have started to play an important role in making choices affecting employee performance, career growth, well-being, and engagement, organizations need to consider the ethical aspects expressed in a clear, transparent, and caring manner (Qin et al., 2025). Ethical HR AI is based on responsible data use; the staff members should be informed that the information gathered via collaboration tools, learning platform, or well-being applications will be used exclusively in meaningful, constructive purposes and never to monitor the staff in punitive ways (Segarra and Williams, 2025). Cloud-native platforms offer superior security features such as encryption, tokenization, data minimisation and role-based access controls to protect sensitive data (Einola and Khoreva, 2022). It is these features that form the basis on which trust can be formed. Transparency is also essential; human-centric AI needs to be interpretable, so that employees can understand how AI-driven choices are obtained, what information underlies such choices, and what are the safety measures against bias (Choudhary et al., 2023). Open communication about algorithms, training data, and decision logic supports the independence of employees and builds trust in digital systems. Biases reduction is also required during ethical deployment, which is effective via cloud-native systems wherein audits conducted regularly, testing of fairness, and retraining of the models can identify and address biases in areas like gender, ethnicity, age, or socio-economic status (Dineen et al., 2022). The ethical governance of AI is based on the principle of informed consent and involvement of employees (Kay et al., 2024). Companies should clearly clarify the purpose of

AI tools and seek employee feedback as well as offer opt-in tools where necessary (Ogbonnaya et al., 2025). Organizations should cultivate a culture of responsible AI, in which AI serves as a partner, not a dictator, in the well-being and career growth of employees, as it is a culture that promotes human dignity and values (Buhmann and Fieseler, 2022). In this manner, using such purposeful and ethically sound strategies, the HR becomes technologically and ethically improved, with human trust being the main focus of AI-powered decision making (Segarra and Williams, 2025).

CHANGING LEADERSHIP IN THE ERA OF CLOUD-NATIVE AI.

Cloud-native AI is changing what it means to be a leader in the most fundamental way: the emergence of new types of intelligences can be used to make decisions more strategic, increase empathy, and make decisions increasingly inclusive (Choudhary et al., 2023). Conventional leadership styles were very much based on intuition, experience and observation. Although these qualities are not less significant, nowadays leaders have access to AI-generated analytics, which shows non-evident trends in workforce behaviour, forecasts the emergence of new talent requirements, and alerts about cultural or operational risks before they occur (Qin et al., 2025). Cloud-native dashboards give the leaders real-time access to employee engagement, collaboration networks, project performance, and workload distribution to implement more informed and targeted interventions (Ogbonnaya et al., 2025).

AI-based decision-support systems have a cognitive companion role in increasing the ability of the leader to interpret complex organisational dynamics (Buhmann and Fieseler, 2022). Once leaders get information regarding decreased morale in particular teams, growing signs of burnout, or swelled cross-functional bottlenecks, they will be able to act more accurately and attentively (Kay et al., 2024). The leadership is made more understanding as AI explains what happens to people behind the scenes and what needs to be addressed (Segarra and Williams, 2025). Cloud-native intelligence is also a democratization of leadership growth and development as it provides individual coaching programs, behaviour analysis, adaptive simulations, and learning paths based on the leadership strengths and growth opportunities (Tuan, 2024). Through these tools, the emerging leaders are able to develop strategic foresight, emotional intelligence, moral judgment, and teamwork. The less AI is focused on bureaucracy, the more leaders be able to invest their time into speaking to employees, motivating teams, and fostering creativity. The leadership position then changes into command-and-control to arranging joint intelligence backed by AI-based clarity (Yalenios and d Armagnac, 2022). This change leads to an inclusive, data-driven, and workforce-for-the-expectations culture of leadership.

EMPOWERMENT OF THE EMPLOYEES WITH SMART SELF-SERVICE SYSTEMS.

Human-centric AI involves employee empowerment since the control of cloud-native platforms is no longer in the administrative hierarchy but rather in the hands of the employees (Malik et al., 2022). Intelligent self-service systems enable employees to have instant access to resources, make informed decisions about their work and career without having to go through HR intermediaries (Qin et al., 2025). Such platforms enable employees to work out timetables, update profiles, claim advantages, monitor performance indicators, and look into development prospects in a way that is easy, clear, and receptive (Einola and Khoreva, 2022). More to the point, using cloud-native AI systems, personalised experiences are developed by looking at the goals and the learning history of each employee, their performance data, and their engagement trends (Choudhary et al., 2023). Individualised dashboards display all the insights, such as the progress towards goals, learning recommendations, collaboration metrics, and well-being indicators, and career pathways, in one panel (Ogbonnaya et al., 2025). Through this, employees are able to know their positions clearly and what they can do to develop. AI-based career advisors go a notch higher by informing the potential future positions in the organisation, the skills needed to switch, and suggesting specific learning or stretch placement to hasten the preparation (Tuan, 2024). These abilities foster the ownership and agency to motivate employees to become an active participant in shaping their career path (Segarra and Williams, 2025). Strong employees have a higher motivation, engagement, and resilience since they feel supported by a system, which comprehends their desire and guides them to overcome obstacles (Buhmann and Fieseler, 2022). Human-centric AI also thus brings an opportunity to the masses, lessens bureaucracy and enhances the empowerment of the human psyche throughout the workforce.

HUMAN-CENTRIC AI IN INDUSTRY WIDE APPLICATIONS.

Human-oriented AI is being driven by cloud-native technologies that are changing the nature of work within all industries by designing intelligent ecosystems that are specific to a particular operational context (Yalenios & 8 d'Armagnac, 2022). Frontline workers in manufacturing operate mobile AI companions which give them real-time instructions, equipment diagnoses, predictive maintenance notifications, and safety guidance. These devices decrease the mental burden and physical pressure of employees and increase accuracy and productivity (Choudhary et al., 2023). AI finds applications in the healthcare industry as a decision support system to help clinicians, assist in patient triage, automate medical records, and create personal learning apps to inform medical professionals about the latest guidelines (Tuan, 2024). This will result in better patient care and decreased administration.

Smart workforce scheduling, individual training, dynamic customer insights, and operational optimisation are some of the benefits that cloud-native AI offers to the retail and hospitality industry helping employees provide guests with memorable service experiences (Kay et al., 2024). Generative AI tools synthesize complicated data, generate content prototypes, automate routine tasks and facilitate cooperative problem solving in knowledge-heavy industries, including consulting, finance, and education (Qin et al., 2025). These tools liberate the employees to think strategically, be creative and build relationships. In all sectors, the AI-based workforce analytics forecasts workforce needs, identifies workload issues, and predicts the operational bottlenecks (Ogbonnaya et al., 2025). This will keep the teams active, well-supported, and ready to overcome fluctuations in demand. Cloud-native AI helps workers, regardless of their industry, to prioritize more valuable work that demands emotional intelligence, critical thinking, and genius (Buhmann and Fieseler, 2022).

ADOPTION BARRIERS, CHANGE MANAGEMENT AND CHALLENGES.

Although artificial intelligence (AI) is a human-centered concept, modern organizations are facing a row of obstacles that require careful management in order to guarantee its successful implementation (Qin et al., 2025). One of the most important issues is that of privacy, as the employees might develop anxiety about the data collection, analysis, and use (Segarra & Williams, 2025). Misunderstandings that define AI as a surveillance machine can create a feeling of resistance when organisations cannot clarify the uses, limits, and protection related to AI tools (Buhmann and Fieseler, 2022). Moreover, the existence of digital skill gaps is an uncompromising hindrance, as not every single member of the staff is technologically prepared to work with the AI-empowered systems; thus, the need to introduce specific training and upskilling programs (Malik et al., 2022).

Successful change management presupposes a key position in overcoming these challenges. The organisations should be able to define the value of AI not only in the context of increased efficiency, but also regarding employee health, career growth, and work-life balance (Ogbonnaya et al., 2025). The human-resources leaders should work in close consultation with cybersecurity, information technology, and legal departments to design transparent and fair AI systems to protect the autonomy of their employees (Yalenios and d Armagnac, 2022). The stepwise implementation of AI, paired with a practical process of training, a systematic system of receiving feedback, and progressive improvements, reduces resistance and promotes familiarity (Kay et al., 2024). The process of adoption must be based on empathy and inclusivity at all stages. Organisations foster a smooth and positive change process by making sure that employees are involved in the process and that they understand the real benefits (Choudhary et al., 2023).

THE FUTURE OF AI IN HR THAT IS HUMAN-CENTERED.

The potential future of human-centered AI is expected to introduce an increasing gradient of personalisation, contextual understanding, and emotional intelligence (Qin et al., 2025). Due to the development of cloud-native platforms, AI models will absorb real-time behavioral cues, physiological (with informed consent) and cognitive patterns to provide more sensitive and responsive answers (Buhmann and Fieseler, 2022). Affective computing will enhance the ability of AI to identify emotional states, including stress, motivation, frustration, or satisfaction to facilitate interventions to be administered at the right time and the creation of psychologically safe working conditions (Segarra and Williams, 2025). Cognitive AI will improve contextual reasoning that would enable the systems not merely to check the activities of the employees but also to understand the motivation and experience-optimisation possibilities (Choudhary et al., 2023).

Immersive technologies (such as augmented reality, virtual reality, and spatial computing) will change the way organisations provide training, team-building, and collaboration programs (Tuan, 2024). Workers can become a part of virtual working environments, interact with experience simulators, and coordinate with each other through spatial digital twins that can mimic the physical world. With the further development of AI, it is likely to be an assisting companion that complements, not replaces, a human judgment (Kay et al., 2024). The most important one will be ethical and responsible governance, which will see that AI develops in unison with human values, autonomy, and dignity (Ogbonnaya et al., 2025). The future work will be typified by the notion of fluid hybrid collaboration, adaptive intelligence, and systems that accommodate the human needs and aspirations very deeply (Buhmann and Fieseler, 2022).

CONCLUSION: INTELLIGENT WORKPLACE, TOWARDS A PEOPLE-FIRST.

The concept of human-centric AI represents a paradigmatic change in the ways in which organisations envision and provide employee experience (Malik et al., 2022). Cloud based intelligence provides the HR with the tools to build systems that are not only effective and scalable but also empathetic, non-discriminatory and highly personalised (Yalenios and d Armagnac, 2022). Enriching recruitment, onboarding, performance management, learning, well-being, collaboration, and cultural continuity, AI creates a working environment in which every employee feels welcome, empathized with, and appreciated (Choudhary et al., 2023). This shift is indicative of a more general understanding that workers are the center of organizational performance and that technology must be used as a facilitator of human flourishing as opposed to a malefactor (Buhmann and Fieseler, 2022).

Human-centric AI is a strategic necessity and the ethical responsibility in a digital world of complexity, high pace, and changing demands (Qin et al., 2025). Companies adopting this model will foster the cultures where employees can succeed, become innovative, and make a valuable contribution towards the long-term development (Ogbonnaya et al., 2025). The shift to people-first intelligent working environment is not only the future of HR but also the future of work itself the environment in which man and intelligence coexist and mutually enhance each other with the assistance of cloud-native AI (Segarra and Williams, 2025).

REFERENCES

Buhmann, A., & Fieseler, C. (2022). Deep learning meets deep democracy: Deliberative governance and responsible innovation in artificial intelligence. *Business Ethics Quarterly*, *33*(1), 146–179. DOI: 10.1017/beq.2021.42

Choudhary, V., Marchetti, A., Shrestha, Y., & Puranam, P. (2023). Human-ai ensembles: When can they work? *Journal of Management*, *51*(2), 536–569. DOI: 10.1177/01492063231194968

Dineen, B., Yu, K., & Stevenson-Street, J. (2022). Recruitment in personnel psychology and beyond: Where we've been working, and where we might work next. *Personnel Psychology*, *76*(2), 617–650. DOI: 10.1111/peps.12548

Dutordoir, M., & Struyfs, K. (2024). The impact of a compressed workweek on shareholder value: An event study analysis of belgium's 4-day workweek legislation. *Human Resource Management*, *63*(6), 903–917. DOI: 10.1002/hrm.22238

Einola, K., & Khoreva, V. (2022). Best friend or broken tool? exploring the co-existence of humans and artificial intelligence in the workplace ecosystem. *Human Resource Management*, *62*(1), 117–135. DOI: 10.1002/hrm.22147

Kay, A., Vlachos, P., Tasoulis, K., & Farndale, E. (2024). Fraught expectations: A fairness heuristic process model of the pros and cons of csr for talent acquisition. *Human Resource Management*, *64*(2), 465–483. DOI: 10.1002/hrm.22269

Malik, A., Budhwar, P., Mohan, H., & Srikanth, N. (2022). Employee experience –the missing link for engaging employees: Insights from an mne's ai-based hr ecosystem. *Human Resource Management*, *62*(1), 97–115. DOI: 10.1002/hrm.22133

Ogbonnaya, C., Messersmith, J., Schnacker, M., & Laker, B. (2025). Setting the stage: Line managers' HR orientation as an antecedent to perceived HR effectiveness. *Human Resource Management*, *64*(6), 1581–1604. DOI: 10.1002/hrm.70002

Qin, X., Zhou, X., Chen, C., Wu, D., Zhou, H., Dong, X., & Lu, J. (2025). Ai aversion or appreciation? a capability–personalization framework and a meta-analytic review. *Psychological Bulletin*, *151*(5), 580–599. DOI: 10.1037/bul0000477 PMID: 40489180

Segarra, P., & Williams, K. (2025). Calling in: Mental health in management education and academia. *Management Learning*, *56*(1), 83–89. DOI: 10.1177/13505076241284433

Tu n, L. (2024). Fostering public sector employees' citizen-oriented boundary-spanning behaviors through human resource flexibility and employee perceptions of normative public values. *Personnel Review*, *53*(8), 2087–2108. DOI: 10.1108/pr-02-2022-0078

Yalenios, J., & d'Armagnac, S. (2022). Work transformation and the hr ecosystem dynamics: A longitudinal case study of hrm disruption in the era of the 4th industrial revolution. *Human Resource Management*, *62*(1), 55–77. DOI: 10.1002/hrm.22114

Chapter 18
Intelligent Automation as a Strategic Lever: Harnessing Cloud-Native AI for Scalable Growth

Xuan-Hoa Nghiem
https://orcid.org/0000-0003-2292-0257
International School, Vietnam National University, Hanoi, Vietnam

Mananage Shanika Hansini Rathnasiri
https://orcid.org/0000-0002-6496-5106
Faculty of Management Studies, Sabaragamuwa University of Sri Lanka, Sri Lanka & Qasim Ibrahim School of Business, Villa College, Maldives

Sandeep Kumar Singh
https://orcid.org/0000-0002-1741-7254
JGBS, O.P. Jindal Global University, Haryana, India

Mohan Kumar
https://orcid.org/0000-0002-5171-0892
Faculty of Commerce and Management, SGT University, Gurugram, India

Manju Dagar
https://orcid.org/0009-0002-5677-7647
Faculty of Commerce and Management, SGT University, India

ABSTRACT

Intelligent Automation (IA) represents a paradigm shift, evolving from routine task automation into a strategic lever for scalable growth. This chapter argues that the true potential of IA is unlocked when integrated with a cloud-native architecture. It

DOI: 10.4018/979-8-3373-7442-0.ch018

deconstructs IA's core components-Robotic Process Automation, Machine Learning, Natural Language Processing, and Computer Vision-and demonstrates how their fusion creates cognitive, adaptive systems. The cloud-native model provides the essential fuel for this transformation, offering elastic scalability, access to sophisticated AI services, and accelerated deployment cycles. The chapter provides a comprehensive framework for formulating an IA strategy, a detailed implementation blueprint, and a discussion of critical challenges, including change management and ethical AI. Ultimately, it positions cloud-native IA as an imperative for building a resilient, agile, and future-proof enterprise.

INTRODUCTION

The quest to automate business is one of a story of growing aspiration and capacity. It started with basic macros and scripts, followed by the rule-based task automation of Robotic Process Automation (RPA), and is currently in its most radical redesign Intelligent Automation (IA). Although RPA was good at simulating repetitive human-like behavior on digital interfaces, an artificial intelligence, which is a so-called digital workforce of software bots, it had a significant drawback in the form of a cognitive deficiency. It was also able to perform a process accurately but failed to comprehend, learn, or adjust in case the process did not follow its own predetermined rules or had unstructured data (van der Aalst et al., 2018). It is at this point that Intelligent Automation comes in and is an emergence of a paradigm shift. IA is the synergistic collection of technologies that integrates the process execution power of RPA with the cognitive, decision-making power of Artificial Intelligence (AI), such as the Machine Learning (ML), Natural Language Processing (NLP), and Computer Vision (CV) (Kumar, Dagar, et al., 2025). This combination forms systems that do not simply work, but think and learn in the process of working (R. Yadav et al., 2023).

The thesis of this chapter is that this convergence of technology should be perceived not as the addition to the existing level of improvement, but as a strategic tool. A strategic lever is a force multiplier - a force that when used produces disproportional value and competitive edge as compared to the investment. Viewing IA as such makes the discussion go beyond the operational efficiency and cost-reduction. It makes IA a fundamental facilitator of strategic goals including entering new markets, developing hyper-personalized customer experiences, speeding up the innovation process, and developing organizational agility (Davenport and Ronanki, 2018). When IA is implemented in the style of cloud-native, the leverage effect is magnified drastically. The cloud offers the crucial manure of scalable expansion, with on-demand access to enormous processing capability and advanced AI solutions, thus removing the traditional infrastructure limitations that have previously constrained large-scale automation initiatives (Tunio et al., 2021; Vihari et al., 2021). This chapter is going to be a thorough investigation of this potent combination. We are going to start by dissecting the key elements of IA, followed by the exploration of the benefits of a cloud-native approach that have no analog. It is based on that we will describe a framework of developing the IA strategy, offer a thorough blueprint of the strategy implementation, and will openly discuss the challenges and risks involved. Lastly, we will also peer into the future, imagining the future of the autonomous, self-optimizing enterprise with perpetual reinvention caused by automation.

THE CORE COMPONENTS OF INTELLIGENT AUTOMATION

In order to realize the strategic potential of Intelligent Automation, it is necessary to understand its component parts and their interaction to form a system more than the sum of the parts. Robotic Process Automation (RPA) is the hands and feet of the operation, and is at its core (Kumar, Dagar, et al., 2025; Kumar, Nain, et al., 2025). RPA software bots are programmed to execute rule-based high-volume tasks by engaging with digital systems via their user interfaces, similar to how a human would. This involves tasks like transfer of data between programs, transacting, or creating mass communications. The main benefit of RPA is that it is non-invasive and can be superimposed on existing legacy systems without requiring intricate and expensive application programming interface (API) integrations, which is a potent technology to start the process of initial digital transformation (Hofmann et al., 2020).

Nevertheless, it is a set of cognitive technologies that provides the intelligence. The brain of the operation is composed of Machine Learning (ML). ML algorithms apply statistical methods to allow systems to learn data, discover patterns and make decisions or predictions with little human involvement (Kumar et al., 2023). An ML model can be trained on historical data in an IA setting to predict customer churn, inventory demand or to detect anomalous transactions and identify fraud. This predictive power turns automated processes into dynamic and adaptive processes rather than fixed pre-programmed processes (Choudhary et al., 2021). ML should be complemented with Natural Language Processing (NLP), which provides the machine with the capabilities to understand, interpret and generate human language. The technology that drives chatbots to provide customer service, identify sentiment in social media feeds and, most importantly to automation, identify meaningful information in unstructured text based documents like emails, contracts and legal briefs, is called NLP (Chowdhury, 2003). In a similar manner, Computer Vision (CV) enables machines to perceive and sense visual data of the physical or digital environment. In IA, CV is applied to read data on scanned documents, forms and invoices, identify people on photographs or live feeds and even quality control checks in manufacturing facilities by checking the images of products.

When these components are put together to play in harmony, the strategic power of Intelligent Automation is unleashed. Take an example of accounts payable process. RPA bot on its own could never work with an invoice unless the information was already in a structured digital format. A smart system would scan a PDF invoice with Computer Vision, interpret the invoice with Natural Language Processing and identify important fields, such as the name of the vendor, the invoice number, and the amount, and use a Machine Learning model to compare the invoice with a purchase order and prove its authenticity. Last but not least, the RPA bot would complete the financial system transaction and update the records. It is this smooth combination of cognitive knowledge with robotic action that makes IA different to the past generation of automation, and allows handling of complex, exception-prone, and knowledge-intensive processes hitherto reserved to human workers.

THE CLOUD-NATIVE ADVANTAGE: FUEL FOR SCALABLE GROWTH

Although the elements of Intelligent Automation are powerful in their turn, they are limited in their strategic effect when implemented with the help of traditional, on-premises infrastructure. The driver of their maximum potential of scalable growth is the cloud-native paradigm. It means that to be cloud-native is to fully subscribe to the architectural and operational ideologies of the cloud computing model. It includes developing and operating applications which take advantage of the benefits of the cloud, which comprises of microservices architecture, containerization (e.g., Docker and Kubernetes), DevOps, and continuous integration/continuous delivery (CI/CD) pipelines (Tilley and Sussna, 2020). This is in sharp contrast to merely lifting and shifting the existing applications to a virtual machine in the cloud which in many cases does not reflect the inherent advantages of elasticity and resilience.

Elastic scalability is the most important benefit of IA cloud-native approach. The computation intensive processes are generally automation processes that require training of a ML model or processing large batches of documents. In-premise infrastructure would mean that the organization has to provision to the peak capacity and therefore its hardware is costly and is not used fully when there is no peak. In a cloud-native system, computing resources can be automatically and automatically scaled up and down on demand. This implies an invoice processing process can dynamically utilize hundreds of servers to process a thousand invoices in minutes and then reduce to zero to turn a capital expenditure (CapEx) model into an operationally much more efficient pay-per-use (OpEx) model. It is this elasticity that is the technical foundation of scalable growth enabling automation programs to be expanded smoothly with increase in business volume without having to pay prohibitive costs or management overhead.

Moreover, the large cloud providers, like Amazon Web Services (AWS), Microsoft Azure, and Google Cloud Platform (GCP), provide a massive, constantly growing pool of pre-implemented, operated AI services. The services offer API driven access to state of the art vision, language, speech and forecasting services. To an organization developing an IA solution, this implies that it does not necessarily have to spend millions of dollars and years of time training its proprietary computer vision model to read invoices. Rather, it can merely invoke a cloud service such as Azure Form Recognizer or Amazon Textract which are already trained on millions of documents and can be trained on a small collection of labeled samples. This radically reduces the barrier to entry of advanced AI, shortens the time-to-value of automation projects, and makes sure that organizations are constantly enjoying the benefits of the latest innovations in AI research without the overhead of maintaining the underlying infrastructure (Hazan et al., 2021). Lastly, the cloud-native model, and its DevOps and CI/CD practices promote a culture of quick experimentation and roll-out. Teams are able to design new automated workflows swiftly, test them in isolation, and implement updates on a regular and consistent basis. This agility turns the automation role into a sluggish, project-oriented undertaking into a flow of innovation and enhancement, which is directly connected to the strategic agility of the organization in general.

FORMULATING THE INTELLIGENT AUTOMATION STRATEGY

Having a solid understanding of the technology and its cloud-native enabler, the second most important thing is to develop a consistent business strategy. Most automation projects have failed to succeed, and in most cases, this can be attributed to technology-first, as opposed to strategy-first approach (Mohit Yadav & Vihari, 2021). Identification of high-impact use cases is the most important and the initial activity. The suitability of all processes to IA is not equal. An effective selection framework must undergo the processes under evaluation on a number of factors: large transaction volume, being rule-based, use of structured or unstructured digital data, susceptibility to human error, and high levels of manual work. The objective analysis of event log data in current systems (e.g., ERP, CRM) can be done so using techniques such as process mining that allows one to discover, monitor, and optimize real-world processes and provide a data-driven background of detecting automation candidates that are likely to yield the best ROI (van der Aalst, 2016).

After identifying a portfolio of potential use cases, there is a need to align them strictly with the overall business objectives. This is what makes automation more than a tactical venture, but a strategic lever. As an example, automation of the process of customer onboarding does not only imply the minimization of the administrative expenses, but it is directly connected to the strategic objectives of enhancing the customer experience (through the minimization of wait times) and generating profits (through getting the customers to appreciate faster). Equally, automation of a supply chain forecasting and procurement process enables strategic aims of operational resilience and cost leadership. Close cooperation between the automation team, the business unit leaders, and the C-suite is essential in this regard since each automation initiative needs to be clearly connected to a key performance indicator (KPI) that is of business relevance, including net promoter score (NPS), customer lifetime value (CLV), or inventory turnover ratio.

The last pillar of strategy formulation is to build a strong business case. The conventional ROI analysis paid much attention to Full-Time Equivalent (FTE) displacement or cost avoidance. Although these are still valid measures, the contemporary business case of IA has to reflect a wider range of value. This encompasses the so-called soft benefits like a more accurate process (lowering the cost of errors and rework), a better regulatory compliance due to an ideal audit trail, higher employee satisfaction caused by the elimination of tedious processes and employees being able to focus on more valuable processes, and better organizational agility because the process is executed faster (Willcocks and Lacity, 2016). This entire value proposition needs to be quantified to gain executive sponsorship and long term funding as this is the only way to position the IA program as not a cost center, but a strategic investment into the future capabilities of the enterprise.

A BLUEPRINT FOR IMPLEMENTATION

An effective plan requires a rigorous and step-wise implementation plan in order to achieve success and reduce risk. The initial stage, which is Discovery and Process Mining, is concerned with the objective knowledge of the current condition in depth. It is not limited to the old methods of interviews and workshops. The process mining technology is directly linked to information systems to visually map the actual flow of a process to identify variations, bottlenecks and inefficiencies that are not necessarily visible in its documented version. This factual finding makes organizations automate the process as it is, or initially optimize it, instead of automating the natural inefficiencies.

The second stage is the Design and Development in which the cloud-native architecture is put to the test. The workflow must be broken down to a group of microservices that are loosely coupled instead of creating one, monolithic automation application. As an example, document ingestion, data extraction, data validation, and system of record update microservices could be independent of each other in the invoice processing system. The microservices can be independently developed, scaled and updated. Made possible by containerization, including using Docker and coordinated by Kubernetes, this modularity contributes to resiliency, faster development, and easy replacement of components, e.g., by changing one AI service provider to another with minimal impact (Newman, 2015). DevOps practices are followed by development, where continuous integration pipelines are automatically used to build and test changes in the code.

Testing and Validation is the third stage that is especially important when it comes to AI components. In contrast to deterministic RPA, AI models are probabilistic and require to be strictly tested concerning accuracy, bias and robustness. This entails testing the model on a hold-out dataset that it has never been trained on, edge case testing, and noisy data testing, and bias audit to ensure the model does not give discriminatory results against the protected classes (Pankaj et al., 2023; M. Yadav et al., 2025). The deterministic RPA components are tested through standard unit, integration as well as user acceptance testing. All the integrated workflow is then tested in a production like staging environment.

Lastly, the Deployment and Monitoring phase uses the CI/CD pipelines to roll out the containerized automation services into the production environment with no downtime. An efficient monitoring system is put in place after it goes live. This is not just monitoring whether the bot is being run, it also checks key performance indicators (KPIs) like process cycle time, straight-through processing rate, exception rate and drift accuracy of the ML models. This feedback is a loop to enable continuous improvement, where models are retrained or workflows refined on demand. Effective governance of this complete lifecycle usually requires an Intelligent Automation Center of Excellence (CoE) to be established which is a cross-functional team that defines standards, oversees the platform, exchanges best practices, and maintains strategic alignment throughout the organization.

NAVIGATING CHALLENGES AND MITIGATING RISKS

The road to smart automation is not free and the active attitude to risk management is the key to the success in the long term. The human element is, perhaps, the biggest, but the least important challenge. The news of an automation program may cause stress and opposition among the employees who may be in fear of losing their jobs. The key factor is therefore effective change management. This entails effective communication among the leadership of the strategic objectives of IA that is transparent and aims at enhancing human capacities and not substituting them. It is important to have a specific reskilling and upskilling program to move the labor force toward more strategic, creative, and supervisory jobs like automation bot managers, data annotators, and even AI ethicists (Wilson and Daugherty, 2018)(Nguyen et al., 2025). It is important to create a culture of human-machine cooperation in order to achieve the maximum value of IA.

Technically, IA systems can only perform well when the data fed to them is well governed and of high quality. The saying that garbage in, garbage out is highly applicable in this case. The use of biased, incomplete or low quality data in training an ML model will result in unreliable and even dangerous outputs. The creation of effective data governance standards, such as data quality controls, master data administration, and obvious data lineage is a precondition that cannot be compromised in the case of a successful IA. It is also directly connected with the problematic topics of Ethical AI and Algorithmic Bias. Unless automated systems are designed and supervised, they will only reinforce and even intensify the biases that are already prevalent in society. Organizations need to create an ethics model regarding AI, which requires equitableness, transparency, and explicability of automated decision-making (Mitchell, 2019). This includes methods to identify and reduce bias in models and having the capability to justify, in human-understandable language, how a specific automated decision was arrived at, which is essential to regulatory compliance and retaining the trust of the population.

The cloud-native model, however, is beneficial, but it has its own Security and Compliance considerations. Storing and processing of sensitive data in the cloud will need a strong security posture, encryption of data at rest and transit, effective identity and access management (IAM) and network security controls. Moreover, companies in the regulated sectors should make sure that their cloud and automation systems are in accordance with the laws of data sovereignty such as the GDPR in Europe that determine the location of data storage and processing of personal data. The cloud provider and the internal security and legal teams should also be considered close partners to manoeuvre through this complex environment to create a secure, compliant, and trustful IA ecosystem.

THE FUTURE-PROOF ENTERPRISE: LOOKING AHEAD

The future of Intelligent Automation is in a more integrated and active direction. The present trend of automating single processes is fast turning into the Hyperautomation concept, a business-led strategy to define, screen, and automate as many business and IT processes as feasible. Hyper automation is the integration of various technologies, tools and platforms, such as RPA, AI, process mining and advanced analytics, into an end-to-end automation fabric throughout the entire organization (Le Clair and Cullen, 2020). It is not just automation of processes, but the automation of whole business processes, and process mining is ever finding new opportunities in automation.

In the long run, the final place is the Autonomous Enterprise. This is an organization that does not only automate its systems but self optimizes them. With AI, the systems are capable of automatically detecting inefficiencies, anticipating future bottlenecks, and automatically reconfiguring workflows to meet business goals with a minimum of human intervention. The human role changes to be that of strategists, innovators, and exception handlers of the most complex and novel situations. Such a vision requires a Continuous Reinvention attitude. Cloud-native IA is not a project that has a start and finish date, it is a continuous organizational ability. It makes the company constantly adjust its operation, business models, and customer experiences as per the market changes and new technologies. The business turns into an intelligent, living entity, intelligent automation is its brain, and it is not only competitive but also dominant in a world where nothing can be predicted.

CONCLUSION: THE STRATEGIC IMPERATIVE

Summing up, Intelligent Automation and cloud-native architecture integration is a radical change to the business operations and competition. This chapter has contended that IA is much more than a collection of technologies to help companies enhance efficiency, it is a potent strategic tool to grow on a scale that is sustainable. With the ability to integrate the cognitive power of AI with the implementation capacity of automation, and then power it with the elasticity, innovation and the economic model of the cloud, organizations can open up new opportunities of agility, customer-centricity, and operational excellence. The path must take a conscious plan, a disciplined implementation plan and a strong determination to control the related human, technical and ethics risks. The imperative is, however, evident. Automation of core processes intelligently is no longer a differentiating capability in the digital economy but a survival and success precondition. To leaders who wish to create a future-proof business, leveraging cloud-native AI to grow on a grand scale is not merely an opportunity, but as strategic as it is urgent to our era.

REFERENCES

Choudhary, S., Yadav, M., & Srivastava, A. P. (2021). Cognitive Biases Among Millennial Indian Investors: Do Personality and Demographic Factors Matter? *FIIB Business Review*, *13*(1), 106–117. DOI: 10.1177/23197145211057343

Chowdhury, G. G. (2003). Natural language processing. *Annual Review of Information Science & Technology*, *37*(1), 51–89.

Davenport, T. H., & Ronanki, R. (2018). Artificial intelligence for the real world. *Harvard Business Review*, *96*(1), 108–116.

Hazan, E., Kelly, G., Khan, H., Spillecke, D., & Yee, L. (2021). *The top trends in tech*. McKinsey & Company.

Hofmann, P., Samp, C., & Urbach, N. (2020). Robotic process automation. *Electronic Markets*, *30*(1), 99–106.

Kumar, M., Dagar, M., Yadav, M., Choudhury, T., & Luu, T. M. N. (2025). Navigating the future: Challenges and opportunities in intelligent RPA implementation. In Intelligent Robotic Process Automation: Development, Vulnerability and Applications (pp. 393–420). IGI Global. https://doi.org/DOI: 10.4018/979-8-3693-4365-4.ch015

Kumar, M., Nain, N., Dagar, M., Luu, T. M. N., & Kaur, K. (2025). Redefining work and outsourcing in the AI era: Challenges and opportunities. In *Global Work Arrangements and Outsourcing in the Age of AI*. IGI Global., DOI: 10.4018/979-8-3373-1270-5.ch021

Kumar, M., Yadav, R., Suresh, A. S., Singh, R., Yadav, M., Balodi, A., Vihari, N. S., & Srivastava, A. P. (2023). Mapping AI-Driven Marketing: Review of Field and Directions for Future Research. 2023 2nd International Conference on Futuristic Technologies (INCOFT), 1–5. https://doi.org/DOI: 10.1109/INCOFT60753.2023.10425169

Le Clair, C., & Cullen, A. (2020). The Forrester Wave™ [Forrester Research.]. *Robotic Process Automation*, *Q4*, 2020.

Mitchell, M. (2019). *Artificial Intelligence: A Guide for Thinking Humans*. Farrar, Straus and Giroux.

Newman, S. (2015). *Building Microservices*. O'Reilly Media.

Nguyen, T. B., Kumar, M., Sahoo, D. K., Nain, N., Yadav, M., & Dadhich, V. (2025). Mapping the landscape of women entrepreneurs in micro and small enterprises: Trends, themes, and insights through systematic review and text mining. F1000Research, 14. https://doi.org/DOI: 10.12688/f1000research.163030.1

Pankaj, Yadav, R., Naim, I., Kumar, M., Misra, S., Dewasiri, N. J., Rathnasiri, M. S. H., Yadav, M., Balodi, A., Kar, S., & Vihari, N. S. (2023). An Exploratory Study to Identify Effects of Blockchain Technology on Organizational Change and Practices. 2023 IEEE Technology & Engineering Management Conference - Asia Pacific (TEMSCON-ASPAC), 1–8. https://doi.org/DOI: 10.1109/TEMSCON-ASPAC59527.2023.10531372

Tilley, S., & Sussna, M. (2020). *Cloud Native: Using containers, functions, and data to build next-generation applications*. O'Reilly Media.

Tunio, M. N., Chaudhry, I. S., Shaikh, S., Jariko, M. A., & Brahmi, M. (2021). Determinants of the Sustainable Entrepreneurial Engagement of Youth in Developing Country—An Empirical Evidence from Pakistan. In Sustainability (Vol. 13, Issue 14). https://doi.org/DOI: 10.3390/su13147764

van der Aalst, W. M. P. (2016). *Process Mining: Data Science in Action*. Springer.

van der Aalst, W. M. P., Bichler, M., & Heinzl, A. (2018). Robotic process automation. *Business & Information Systems Engineering*, *60*(4), 269–272.

Vihari, N., Yadav, M., & Panda, T. K. (2021). Impact of soft TQM practices on employee work role performance: role of innovative work behaviour and initiative climate. The TQM Journal, ahead-of-print. https://doi.org/DOI: 10.1108/TQM-03-2021-0092

Willcocks, L. P., & Lacity, M. C. (2016). *Service Automation: Robots and the Future of Work*. Steve Brookes Publishing.

Wilson, H. J., & Daugherty, P. R. (2018). Collaborative intelligence: Humans and AI are joining forces. *Harvard Business Review*, *96*(4), 114–123.

Yadav, M., Kumar, M., Alyousef, M. I., Dagar, M., Dewasiri, N. J., & Amoozegar, A. (2025). Mapping Research Trends on Gender Diversity, Work Stress, and Job Satisfaction in Higher Education: A Bibliometric Analysis. International Journal of Learning. *Teaching and Educational Research*, *24*(8), 444–472. DOI: 10.26803/ijlter.24.8.19

Yadav, M., & Vihari, N. S. (2021). Employee Experience: Construct Clarification, Conceptualization and Validation of a New Scale. *FIIB Business Review*, *12*(3), 328–342. DOI: 10.1177/23197145211012501

Yadav, R., Yadav, M., & Vihari, N. (2023). High-performance work system and learning orientation in offline, online, and hybrid workplaces: The mediating role of affective commitment. *The Learning Organization*, *31*. Advance online publication. DOI: 10.1108/TLO-10-2022-0118

Chapter 19
Considering the Application of Artificial Intelligence in Society and Assessing the Future of AI

Shimona Shriya
https://orcid.org/0009-0001-0682-4014
Xavier Institute of Management and Entrepreneurship, Bangalore, India

Rituraj Singh
https://orcid.org/0009-0008-0452-9303
O.P. Jindal Global University, Haryana, India

Shashi Kant
https://orcid.org/0000-0003-4722-5736
DMI St. Eugene University, Zambia

ABSTRACT

In this chapter the authors discuss on news applications of artificial intelligence (AI) in real world cases including industry and medical field with extended details. The chapter explores how AI can be easily used to automate the routine processes and improve the operational and functional efficiency of any department in any business sector. Regulations regarding AI is also discussed as one of the emerging concerns in the AI landscape. The chapter also defines different methods where there is a crossover between humans and AIs, be it direct or indirect contact.

DOI: 10.4018/979-8-3373-7442-0.ch019

INTRODUCTION

We all have used AI in one form or another in many of the computer applications. In the current era we are all relying on large language models and other forms of AI for our work. For example, ChatGPT, perplexity, Gemini and others, we are using on daily basis. We are absorbing a lot of information and exchanging human interactions with machines to do our jobs. A simple example is how in smartphones and equivalent electronic devices, we have speech and image recognition. In fact, we have AI filters which discriminates between junk mail and useful mails in our inbox. In this chapter we will see and discuss numerous application types of AI and the fields which commonly rely on AI to perform a good number of tasks (Bolaños et al., 2024). We also discover few limitations when creating AI based applications which facilitates us to understand why the current robotic phenomena may not exactly take over the human touch and intelligence in long term.

The two essential ways in which AI in the current world is contributing to human requirements is through corrections and suggestions. We need not to take the human view of these two terms(Elahi et al., 2023). A correction is not necessarily respond to a mistake likewise a suggestion is not necessarily response to a query for example if we consider a driving assistant car that is a car in which the AI assist rather than substitute the driver then in this case when the car moves along the AI makes small corrections which further allows the driving and road conditions, pedestrians and a wealth of other issues in advance soon it makes less mistakes.

In this chapter the authors discuss on news applications of artificial intelligence (AI) in real world cases including industry and medical field with extended details. The chapter explores how AI can be easily used to automate the routine processes and improve the operational and functional efficiency of any department in any business sector. Regulations regarding AI is also discussed as one of the emerging concerns in the AI landscape. The chapter also defines different methods where there is a crossover between humans and AIs, be it direct or indirect contact.

The AI in this case takes a proactive approach rather than the reactive approach to deal with an issue that may or may not occur. Moreover, the AI can also suggest a certain shortest or quickest path to the human handling the car in case of a human driven car. No doubt there is a disagreement about the accuracy of AI based solutions but gradually overtime the accuracy has become better and better with improved a platform (Harris et al., 2019).

USING AI IN ROUTINE APPLICATIONS

Some of the most prominent usage for AI is in artificial creativity, diagnosis, face recognition, handwriting recognition, computer gaming, chatbot or game theory and strategic planning, natural language processing, non-linear control and robotic system, speech recognition, computer vision, virtual reality and image processing. These are the most typical applications of AI (Du-Harpur et al., 2020; Elahi et al., 2023; Zhai et al., 2021). We move further to realise the wide range of areas or fields where AI is currently applied. These fields can be considered as automated reasoning, automation, concept mining, email spam filtering, hybrid intelligence system, common knowledge representation, former robotics that is behaviour-based robotics, cybernetics and evolutionary robotics, biologically inspired computing and semantic web (Elahi et al., 2023).

The primary factor contributing to the swift integration of artificial intelligence in everyday life comes from its user friendliness (Keiper & Schulman, 2015; Song et al., 2024) . At the most basic level an artificial intelligence can provide anticipatory suggestions for prompts or inputs. For example, if the user is typing a particular word, the AI can guess the rest of the characters after only a few characters of the word are typed in. By providing this suggestion the AI is providing a level of user friendliness and familiarity to help reduce the user's workload or cognitive load. There are many categories of inputs in which any human can prompt an AI. The most prevalent example is voice which can be said to be the most overused. However, it still remains one of the most prevalent and intelligent input mechanisms to date. But, even if an AI does not have the full range of senses, the AI can still provide a wide range of non-verbal intelligent inputs as well. Another obvious choice would be the visual or the face recognition. An AI can prompt for any sort of threat based on facial expression. The input can be completed using several different platforms for example in the case of visual prompting a monitor can be used, or biometric hardware can be used as well. In fact, AI can use an enormous number of intelligent inputs most of which is not even invented yet.

Humans constantly correct everything and in the same way AI also works in a continuous feedback loop of improvement and keeps performing in a better way by automatically correcting itself from the previous versions (Cappellani et al., 2024). AI must also has this capability to constantly correct the results it provides even when search results would provide a positive result that is even if the response is acceptable to the human using it or for the task it is being used, an AI must believe in continuous improvement. It is also safe to say that unlike human mistakes an AI if designed properly can remove these daily made mistakes by humans as an AI can react faster than the equivalent intelligence in any human being.

AUTOMATING THE ROUTINE PROCESSES

The intention of using AI is not to replace human beings, but to assist human beings with more productive approaches to their work with better outcomes. Helping people do something faster, easier, more effectively, or somehow a combination of all complex means is the primary target. A process with AI is quite different in nature as AI is now operating to assist a human do some other work without a human working directly. Industrial applications are one of the longest applications of AI in a process.

Given the large number of robots that currently drive factory automation globally, there is no reason to believe that they will not replace workers in the many dangerous jobs humans do while improving productivity through the automation of mainly dangerous jobs. One major driver of factory accidents and additional productivity problems is boredom, especially when boredom is derived from doing monotonous work. Robots have no capacity for fatigue, unlike humans, and therefore can repeat their monotony without complaint and fatigue. This is how AI improves and makes better adjustments than humans by constantly prompting their potential to make constant improvements to themselves. The question is now whether humans will trust AI's suggested improvements. AI does not feel emotion and could never produce a creative or innovative strategy to a problem, therefore some problems are ultimately not solvable by an AI no matter how difficult someone worked to produce the AI.

There will likely always be safety related hazards if you work in any type of industrial setting no matter how much time, effort, and money is spent on it (Gander et al., 2011). While there are countless articles that talk about common safety hazards in many industrial settings, it typically takes a human to create many of those issues due to boredom they created by monotonicity in the work setting which exacerbates the problems in the environment all together.

Automation in industrial settings is much older than most really think. The first instance was actually in 1104 AD in Venice when 16,000 workers built a warship in one day (Hasan Tantawi et al., 2024). When Americans built ships for World War 2, we basically did the same thing with modern ships. So, automation has been around for a long time even though Henry Ford assembly line was credited for being the first. In fact, what has not been around for a long time is an AI that can help humans when we do automation.

Today, in many instances, a human operator or supervisor will start a task with some changes to perform, create a job, and then carry out the modified job using a computer. In this case, robot process automation (RPA) provides the operator with solutions allowing for a human to program the software to perform a task as if it were an associated human doing the task in interaction through software applications. There are ten different levels of automation, and thus we see highly customized, AI-based industry-wide solutions in numerous domains, increasingly, to ensure less boring, repetitive jobs manage work to enhance efficiency - and in fact, make aspects interesting relative to when a human operator or outline operator job performing the task.

One question still exists and that is whether to completely rely on solely automation? That is whether we will see completely human free factories or have a portion of human job functions (Blanc-Brude & Defélix, 2024). One example of human free factories which includes specialty settings such as chip factories which required extremely clean environments however in those earlier days before automation became common the process was too dangerous to human's life and health and beyond this it was cheaper to be replaced by automation to make such chips even if they were special clean chips. There are plenty of examples of how to complete an entire process without human intervention even today.

What is not becoming disrupted but will be augmented by automated process will take place with the agrarian weight of economic sector, however the notion is that there will become a real problem in society for people to have real jobs that are not repeating factory job functions, that would otherwise occur with AI enabled work forces to a similar mechanical kind of repeating monotonous task.

AI APPLICATION TO TAKE CARE OF MEDICAL NEEDS

Depending on the specialty, it can take 15 years or more to train a doctor because medicine is complex. By the time the education system is done cramming a doctor with information until he/she/they can barely contain it, most other people have been in the work force for 11 years if most other people are stopping at associate or bachelor's degrees. All the new technology, medical approaches and so on just made it more difficult. Frequently, it is simply impossible for anybody to be proficient in even a small and focused specialty at no fault of that person. This is exactly why we need a human who is irreplaceable backed by consistent logical and unbiased help from an AI. It begins straightforwardly with the AI assisting the doctor in tracking patients in ways that a human would literally have difficulty validating due to the number of checks to be completed, including the absolute precision of order of checks, and even the magnitude of risk of any error.

There are thankfully more options than ever for individuals to complete medical related tasks for themselves, for example some social networks do allow patients to complete a therapy related task independently but supervised through an application, to ensure the patient is performing the task in a way beneficial to their recovery, even the improvement of prosthetics and other medical devices is allowing for more independence from professional support.

There are many other applications then implementing portable patient monitoring where AI is needed in medicine. First the Apple Watch provides a level of analysis with the use of an AI device that with proper permission can communicate with a user's smartphone so a user can easily link the data to other apps. So, for example, a diabetic with an Apple Watch, their diabetes could be easier to manage and accomplish because anyone can easily start tracking multiple metrics. For example, if just monitoring either the rise and fall of blood glucose levels or blood sugar, it is easier to modify medication and lifestyle to forge its management under control. The K'Watch provides ongoing monitoring of blood sugar either alone or by with availability of an app a person accesses for health food to assist the diabetic person in management of their diabetes (Dassault Systems, 2025). Regardless people have been using some sort of intermittent monitoring for many years as evidenced by the huge increase in how we can monitor each other, the K'Watch simply provides more ongoing monitoring that is useful in identifying the shifts diagnostic requirements and empowers lifestyle changes and medication therapy changes when warranted. It may seem odd to have a type of health provider with their chronic disease statistics on the user it has a utility purpose as well and makes possible proactive health solutions and aids and will also assist others with chronic health issues.

There are other avenues through which AI enhances human potential in healthcare, one of which is gaming as therapy. For example, a gaming console can serve as a powerful and entertaining physical therapy device, such as the Xbox 360 that we see in many physical therapy locations. The games are designed to get patients to move and playing the game inherently rewards patients for moving appropriately. In fact, patients get therapy while playing a game to engage them, and as therapy becomes more entertaining, patients are more likely to agree to participate and rehabilitate faster.

Another example of AI-based solution in healthcare is the use of exoskeletons, which is basically a movable robot (Proud et al., 2022). For example, an AI recognizes a need to move or a movement and provides powered assistance to that movement or assistance to that need to move. The military is in the forefront in the implementation of exoskeletons. Of course, the military is always playing, experimenting, or attempting to improve human performance through technologies and ultimately has a civilian use. The exoskeleton will likely first started out in the military, but it is guaranteed to take over the civilian world too.

The industry has also entered the world of exoskeleton technology. Factory workers are battling a whole host of injuries and illnesses. An exoskeleton can not only reduce fatigue, but it can also decrease human error and increase efficiency; if people can keep their energy levels throughout the day, there is less chance of injury to the worker, damaging a product, or injuring another worker. The present-day applications of this technology are somewhat limited and not quite as impressive as they can be in the future as we develop more of the functional use of the technology.

AI can also address the special needs of someone with physical disabilities; in the past we have people with amputations, or a scenario with a special need, it meant ignoring physical limitations with years of visits to the doctor, significant mental stress, reduced ability, and shorter and less meaningful lives. Most of them are opting for an AI solution which has pretty much obliterated that reality. The man/woman who have an amputation now can still run a marathon or rock climb without feelings limited by the loss of their limb.

One more way to use AI is to complete the analysis in medical field in novel ways

Employing an artificial intelligence system within the dimensions that align with its capabilities optimises both the value of the instrument and the medical specialist's use of it, subsequently enhancing an important facet of the medical specialist's practice. The consideration of data analysis and artificial intelligence applications for the benefit of modern medicine is one domain where artificial intelligence centres offer entirely its existence for scholars. In contemporary clinical practice, through appropriate use of AI, one may take a picture of a potentially malignant lesion of skin and then bring a computer AI to efficiently analyze the image, and the computer sometimes can bemoan and sometimes supersede the diagnosis of a medical specialist (Gautam et al., 2022). Improvements through clinical practice have common techniques that lie behind acquiring necessary images that include processing specimens of tissue that belong to normal skins adjacent to the lesion, requiring a still complicated processing specimen previously the file for everything photograph is standardized to analyze recent photogenic time recently became possible by standardizing practice. One combined available integrated use of AI for evaluation of the skin displays comparison of all holistically with the readily edit analyses it relies on identifying through a computer to detect if an AI solution model has a computer vision presence in diagnostic images, pathology can help identify even better evaluating of skin rare types of skin and physiologic patterns are relevant, if data is provided in a collaborative manner through higher or sophisticated vision AI supplemented emerging images through not just a biopsy noted through a standard visual pattern based clinical elements but also a information photography. The specific additional use of AI for evaluation in physiologic patterns in exploring skin adjunct diagnosing with skin another benefit, in distinction a separated, operates through consistent clinical characteristics, notwithstanding when they are exemplary specific through an early final medical images or still in feasible exams, in, such as any related "postal" visual contact to age or history or mistaken imaging of too low for a specific diagnosis that is apparent considering in clinical medical and clinician systems.

RELYING ON AI TO IMPROVE HUMAN INTERACTION

There are numerous ways to communicate. Surprisingly, most people do not even realize the variety of ways communication can take place. When people think of communication, writing, or speaking with each other, generally comes to mind, but communication can be any number of things such as eye contact or tonal quality. The computer version of announced human interaction example is the electronic nose; as the electronic nose utilizes a combination of electronics biochemistry and artificial intelligence to execute its task and has been used in a wide range of industrial application and research to receive an understanding of how AI can announced action through a lower costs of building our own electric nose, as AI can exist in the action in which people are sharing ideas in the case and then provides in case an ever entirely new model of communicating but in many cases an AI provides a subtle version of advertising the ways that exist to share ideas. Ideas are abstract therefore sometimes difficult to share, and AI can provide an ever-needed bridge between people. There are three main ways humans nonverbally communicate; verbal language expressed with letters, nonverbal communication, and body language. The AI can effectively assist in all three and bridge the communication gap of transmitting and receiving communication.

AI can be applied to the automation of language translation such as human communication has always suffered from a lack of a common language. However, english has become somewhat of a universal language but there are not a lot of people who can speak English therefore an AI to bridge the translation gap instead which would be cheaper, easier, has fewer errors, and has no physical translator involved is a huge deal. Not all people can afford a translator given the fact that dealing across languages can be cumbersome, this is why applications like google translate exist. Language is not only limited to speaking and with AI they could even incorporate reading body language for communication as a large part of human corporate communication is body language no different than using emoticons and emojis, however, people are becoming more accustomed and do use cameras to organize their thoughts and discuss, having no writing elements in human communication in this case computers could possibly collate and be like Alexa or Google home and similar AI integration devices.

AI can enhance the sharing of ideas in order to jointly communicate arguments and ultimately define trends in the field of multimedia where the benefits can be seen as many times over. The first and foremost important way, creating authoring of multimedia. Where we experience and use multimedia is with applications at every level of media and media plays heavenly an ever-increasing role in communication in and across many sectors, fields and practices. Certainly, both human and non-human applications from social media, blog sites, presentation of ideas across organizations (open and closed) are disciplines that describe multimedia thanks to AI. Here we can define multimedia as anything that involves visual (images), text (written words, sheets, etc), sound (auditory sound beyond speech) and music (artistic intention specifically for emotion). With this as a definition, we can situate how AI provides the usability of multimedia a time-efficient manner perhaps most importantly, we can apply AI to a variety of reasons from an educational foundation. As an example, if we consider the value of color that could be found in an image, AI can provide additional help visualizing these influences of 'color' faster than tests using one combination at a time. After we use multimedia to orally and visually present ideas across a number of firms, the recipients of the ideas have to processing the material ideas conveyed, a second benefit of AI, are the uses of algorithms with neural network applications processing the material in multiple ways accessibly sorting the material within the multimedia based on the classifications. Technology is extremely useful today; we look ahead to the future and hope to use AI to facilitate 3D reconstruction of a scene created from 2D images. Part of the imagining extend to police walking through a crime scene that was almost exactly like taking a version held camera video.

AI AND REGULATIONS

Regulation generally denotes rules that a governing body has created to construct the behaviour of individuals and organizations. The word regulation or governance, for many, is negatively associated with an authoritative rigidity of outdated institutions enforcing rules that keep us mired in antiquated practices, while regulation is touted as a fix-all for broadly identifying social ills. This is usually the dichotomy when the discussion ranges to the AI revolution. The reality of the situation is somewhere in the middle. Companies are profit driven, or stock driven, so some harms due to AI result in reputational harm to the company and thereby motivation to resolve that harm. Other harms may be indirect or too diffuse, or involve other parties aside from the company or its users, therefore, it is not within the best interest of the companies to spend the time or money to alleviate any harm. For example, toxic speech and offensive outputs in earlier iterations' of language models was commonplace. This causes users not to interact with a chatbot or some other offering, therefore, the companies that make these models (e.g. OpenAI, Anthropic and Meta) all invested significant resources to rein in these issues ($1 mil).

In contrast, the loss of income that artists suffer and the loss of time that teachers suffer resulting from generative AI does not affect the bottom lines of companies and thus these losses are not received by companies as harm. And when companies have no incentive to respond to the harms of their own business, they need regulation. You will hear more about regulation later. Regulation has been very important to the public interest in very different situations. In food safety regulation, for example, regulation takes the form of ensuring food manufacturers meet certain hygiene and quality standards to minimize the risk of foodborne illness to the health of consumers. Environmental Protection regulation, such as the Clean Air Act and Clean Water Act in the United States, places restrictions on the pollutants that can be emptied into the environment. Or another example is the labour rights regulation like a minimum wage, governing payment and hours worked, and things like safe working conditions meant to protect workers from exploitation. Now, We are by no means saying all regulation is good, effective, or necessary (and will have many clear examples of over or badly informed regulation connected with generative AI shortly) but a world with no regulation is also not indicative of a better or more innovative space.

Indeed, in several instances stated earlier regulation is the key to providing sufficient space for people and organizations to innovate. Another common myth is that the political and regulatory responses to govern AI are in their infancy and it will require a new set of rules to regulate the entire arena of AI. However, in many jurisdictions, the frameworks to regulate AI already exist. Regulatory responses to AI differ in scope among different countries/jurisdictions. For instance, in the U.S., AI regulation is vertical. In other words, rather than having a broad agency that governs all air-products and service, regulation of air is in the authority of federal agencies for individual and specific sectors for regulations of AI, the list of which is hundreds of federal agencies and sub-federal agencies; for example the Consumer Finance Protection Bureau protects consumers in the financial marketplace from deceptive claims and discrimination on the other hand the Food and Drug Administration governs and regulates artificial intelligence that is used in various medical settings. This emphasis on regulation came into to focus in October 2023 when the White House issued an executive order on Artificial Intelligence, it was about 20,000 words long and attempted to address the entire range of benefits and risk associated with AI and how to approach the regulation of AI.

The White House has assigned 150 distinct tasks to the 50 federal agencies outlined above, which are essentially organizations such as the Executive Office of the President, the Department of Commerce, and the Department of Homeland Security. This shows a wide range of existing frameworks to regulate AI. In comparison, the EU has written horizontal rules for regulating aid that will apply across sectors. There are many laws that apply to many aspects of AI in certain areas, for example the General Data Protection Regulation governs how companies collect, store and use personal data, such as telling AI systems to promote and ensure AI features respect privacy. The Digital Services Act requires transparency and audits on the use of AI in online platforms and social media. The Digital Markets Act seeks to promote competition in online platforms, for example, preventing large online platforms from self-referencing the specific results. First and foremost, the proposed sector-specific regulations include a risk-based taxonomy of AI applications, where the developers of applications classified as high-risk, for example, AI for hiring, AI for educational assessment, etc, in terms of workforce management would have many transparency requirements.

China's inflexible approach incorporates certain characteristics of vertical and horizontal elements. In the past, the country's government has adopted a stricter vertical approach to AI regulation; for example, it developed vertical-based regulations about transparency, specifically dealing with online recommender systems in 2017 and the recent leap synthesis systems, such as AIs who can generate images, videos, and text in 2022. Draft regulations concerning generative AIs were released in April 2023 following ChatGPT being released and new public interest in generative AIs. A distinct aspect of the draft AI regulations requires that all AI-generated content must reflect *core socialist values*. There is little doubt that chatbots will join other methods to implement through the Chinese government to oversee the information is available to citizens and with what citizens can do or say. These draft regulations are a vertical approach, focusing on targeted regulations within an application of AI. However, shortly after this announcement, in June 2023, regulators announced that they would build upon the vertical regulations to develop a horizontal AI law. This is the same kind of methodology and ideological principle that China has used in previous rounds of Internet regulations, moving a narrow vertical law to a broader application lending to a law, for example, like the Cyber Security Law passed in 2017.

The particulars of these regulations are interesting, but it does not pertain to our point. The point is that regulators across the globe are already thinking about regulatory frameworks for AI and work is being done to find better or holistic regulation of AI. There is no one size fits all way to regulate AI and while that may sound bad there is variety in approaches which allows us to learn what works, what doesn't, and this will help build toward good principles for regulation.

Again, regulation consistently lags behind the development of technology and to some extent this is filled by how complex technology can be which may be intimidating to people who are not literate in technology, but law is not always about technical details some law is founded on principles. The first amendment of the

United States constitution which provided for freedom of speech was wrote centuries before the internet came into existence which means that what is believed is incorrect.

Another misconception is that tech regulation is useless because policymakers do not understand technology in reality policymakers are not experts in any of the areas they regulate they do not have PhDs in civil engineering or construction management but we have their construction codes to help regulate make sure our buildings are safe it is true that policymakers may not have expertise in the area but they are not going it alone they are ultimately delegating to . and do some experts at the local level, regional level and state level to figure out all the details in the various domains to ensure safety. Regulatory agencies can at times be more reactive than proactive but again this does not mean throw away regulations it just means that we need to adjust and improve a regulatory process that will enable the regulations to be more responsive, flexible and informed to new issues that emerge from changing technology . There is an argument to be made for developing them in parallel to the pace and in line with the speed of innovation. Funding should scale with technology innovation so that regulatory agencies are focusing on resources for the most not vs overall tech companies, and the resources should be sufficient to understand the tech better than tech companies so that an agency once functional can explore a better regulatory framework and usage food regulations.

CONCLUSION

We have summarized the application of artificial intelligence in the medical field and also the crossover between human-led touch and AI. Now we can summarize how AI helps in adding new human jobs. When people consume news about robots and other automation enabled by technological advances, they tend to highlight the negatives versus the positives. One example is news articles that talk about the potential of AI and automation to cause job loss, for example, The World Economic Forum estimate that AI and automation will "displace" between 400 million and 800 million jobs by 2030. Although the article discusses some job creation from the technological advances, for example, personal computers created an estimated 18.5 million jobs, the focus and the view is on job losses for people with some loss being long-term. The problem is these articles tend to talk about job loss quite definitively, but job creation is quite nebulous at best. Unless we are doing something mind-numbingly simple and extremely repetitive, AI is unlikely to "replace" or "substitute" human jobs. One alternative is AI will augment us and potentially allow us to have more fun in our vocational pursuits.

We now have professional AI transformation coaches in various locations around world helping the organization transition according to their AI readiness. In addition to human efficiency, we should think about areas in which an AI can allow us to be much better. An immediate measurable task an AI could be used for would not be for creativity as it has not been demonstrated to excellent in that particular arena so we could leave the creative thought for humans' complete thought however an AI would do search act exceptionally well so we would more so train humans to rely on AI perform such task related to the search while human conduct something creative. Here are some ways we may see humans using an AI to become more efficient soon.

First scheduling today a business is constantly at risk because someone didn't think about the need to schedule a task thus we have overworked employees from secretary to assistance beyond there helping our manager to either work or schedule their task. Overworked employees often miss opportunity to either help a business really excel because they are busy maintaining a schedule.

By using AI to support scheduling and planning the human is frozen from scheduling that human is now looking ahead saying what we need to be scheduled the human can access the business will get more from human AI supports this focus on human excellence.

The recruiting today for an organization whether on the job site or hiring people we don't know all the real credentials and history of the candidates. AI will research candidates before the interview cycle so the recruiter has more information to draw from in the interview. Furthermore, because the AI would deploy the same search methodology for every candidate, the organization can ensure that every candidate is creating information in both a fair and equal manner locating hidden information more than ever today's business gets blindsided by the competition hidden information or information asymmetry.

In adaptive learning, we take an adaptive exam that adapts to ask questions about perceived weak areas in our knowledge, the adaptive exam either discovers that we don't really know enough, or the exam asks enough questions to determine that we need more training. Eventually applications will be able to sense how we are using them and provide automated training to improve our knowledge performance. For example, the application may find that we could perform a task using 5 fewer clicks, and it could train us to do the task using that approach. Further, by constantly training people to engage with computers or other tasks in the most efficient way, not only will the training make the person more efficient, but the need for the human to perform a task remains in some form.

Artificial Intelligence can make human interaction interesting and with less monotonous, for example, teaching children, a robot is especially useful for lifting a patient as a nurse and is a great companion to interact with, think Alexa or Siri. AI can also add value in addressing development issues, for example, people that are special need have connected with the touch of a human being and in many instances the special they find is a gift for them but that realization happens only when a caregiver recognizes that potential as a gift and finding a way to use that special in daily life without social implications means opening as many doors as possible all while realizing special needs does not equal normal and the more we can work on that the better we will be.

AI has the potential to assist individuals with inventing and creating new items and to also extend beyond what is real and can be imagined. AI can assist and help people with detecting crimes. For example, a detective could be looking for an individual that may seem to be missing. This individual may have good reasons for not wanting to be found. The point We are making is that many of the detections with law enforcement cannot simply look at the facts in which an AI will never seem to have a complete utilization because it requires a large leap. The phrase thinking outside the box comes to mind.

All of AI can also be used to monitor and record current ecosystems and even environments which legally or from a perception of safety can be reviewed and cross verified to check for any evidence.

REFERENCES

Blanc-Brude, V., & Defélix, C. (2024). Of chips and men: When working in Industry 4.0 is more human than expected. *Annales Des Mines - Gérer & Comprendre, N° 157*(3), 22–31. https://doi.org/DOI: 10.3917/geco1.157.0022

Bolaños, F., Salatino, A., Osborne, F., & Motta, E. (2024). Artificial intelligence for literature reviews: opportunities and challenges. In *Artificial Intelligence Review* (Vol. 57, Issue 9). Springer Netherlands. https://doi.org/DOI: 10.1007/s10462-024-10902-3

Cappellani, F., Card, K. R., Shields, C. L., Pulido, J. S., & Haller, J. A. (2024). Reliability and accuracy of artificial intelligence ChatGPT in providing information on ophthalmic diseases and management to patients. *Eye (Basingstoke)*, *38*(7), 1368–1373. DOI: 10.1038/s41433-023-02906-0 PMID: 38245622

Dassault Systems. (2025). *K'Watch*. K'WATCH. https://www.3ds.com/3dexperiencelab/portfolio/pkvitality

Du-Harpur, X., Watt, F. M., Luscombe, N. M., & Lynch, M. D. (2020). What is AI? Applications of artificial intelligence to dermatology. *British Journal of Dermatology*, *183*(3), 423–430. DOI: 10.1111/bjd.18880 PMID: 31960407

Elahi, M., Afolaranmi, S. O., Martinez Lastra, J. L., & Perez Garcia, J. A. (2023). A comprehensive literature review of the applications of AI techniques through the lifecycle of industrial equipment. In *Discover Artificial Intelligence* (Vol. 3, Issue 1). Springer International Publishing. https://doi.org/DOI: 10.1007/s44163-023-00089-x

Gander, P., Hartley, L., Powell, D., Cabon, P., Hitchcock, E., Mills, A., & Popkin, S. (2011). Fatigue risk management: Organizational factors at the regulatory and industry/company level. *Accident; Analysis and Prevention*, *43*(2), 573–590. DOI: 10.1016/j.aap.2009.11.007 PMID: 21130218

Gautam, A., Chirputkar, A., & Pathak, P. (2022). Opportunities and challenges in the application of Artificial Intelligence-based technologies in the healthcare Industry. *International Interdisciplinary Humanitarian Conference for Sustainability, IIHC 2022 - Proceedings*, 1521–1524. https://doi.org/DOI: 10.1109/IIHC55949.2022.10059767

Harris, M., Qi, A., Jeagal, L., Torabi, N., Menzies, D., Korobitsyn, A., Pai, M., Nathavitharana, R. R., & Khan, F. A. (2019). A systematic review of the diagnostic accuracy of artificial intelligence-based computer programs to analyze chest x-rays for pulmonary tuberculosis. *PLoS One*, *14*(9), 1–19. DOI: 10.1371/journal.pone.0221339 PMID: 31479448

Hasan Tantawi, K., Sokolov, A., & Tantawi, O. (2024). Advances in Industrial Robotics: From Industry 3.0 Automation to Industry 4.0 Collaboration. *IEEE Proceedings*.

Keiper, A., & Schulman, A. N. (2015). The Problem with 'Friendly' Artificial Intelligence. *Science, Virtue, and the Future of Humanity*, 69–77.

Proud, J. K., Lai, D. T. H., Mudie, K. L., Carstairs, G. L., Billing, D. C., Garofolini, A., & Begg, R. K. (2022). Exoskeleton Application to Military Manual Handling Tasks. *Human Factors*, *64*(3), 527–554. DOI: 10.1177/0018720820957467 PMID: 33203237

Song, L., Yasin, M., & Mokhtar, M. M. (2024). The significance of the virtue of friendliness for moral education in the age of. [InjET]. *International Journal of Education and Training*, *10*(2), 1–12.

Zhai, X., Chu, X., Chai, C. S., Jong, M. S. Y., Istenic, A., Spector, M., Liu, J. B., Yuan, J., & Li, Y. (2021). A Review of Artificial Intelligence (AI) in Education from 2010 to 2020. *Complexity*, *2021*. Advance online publication. DOI: 10.1155/2021/8812542

Chapter 20
Governance in Cloud-Automated Enterprises: Frameworks for Responsible AI Deployment

Gunjan Mohan Sharma
https://orcid.org/0000-0003-2272-7385
JGBS, O.P. Jindal Global University, India

Ali Najeeb
https://orcid.org/0000-0003-4663-9347
Qasim Ibrahim School of Business, Villa College, Maldives

Priya Gaur
Centre for Online Education, GLA University, India

ABSTRACT

The integration of artificial intelligence (AI) with cloud computing is transforming enterprises through unprecedented automation and scalability. However, this convergence introduces significant risks—including algorithmic bias, opacity, and security vulnerabilities—that outpace traditional IT governance. This chapter argues that a new, proactive governance paradigm is essential for responsible AI deployment. It presents a comprehensive framework built on pillars of fairness, transparency, and accountability, and details how cloud-native technologies like Policy as Code can automate governance within MLOps pipelines. A step-by-step approach for operationalizing this framework is provided. The chapter concludes with a critical analysis of the broad social, ethical, practical, and managerial implications, positioning robust governance not as a barrier, but as a critical enabler of sustainable and trustworthy innovation in the cloud-automated enterprise.

DOI: 10.4018/979-8-3373-7442-0.ch020

1. INTRODUCTION

Cloud computing and artificial intelligence are the two forces that are synergistic and are significantly transforming the modern enterprise landscape. Cloud platforms offer the underlying computing power of elasticity as well as expansive storage and managed service that enable the creation, release, and expansion of AI models in an incredibly rapid manner (Mohlmann, 2021). This synergy has created the cloud-automated enterprise, with business operations, starting with how organizations interact with customers through customer service, all the way to supply chain operations, being run by intelligent and self-optimizing systems. The potential is massive: improved operational performance, insights-driven by data, customer experience, and faster innovation cycles. This quick usage has however outpaced the development of the governance frameworks that are required to have these highly powerful technologies used responsibly. The nature of traditional IT governance, which is typically periodic, manual compliance, and fixed-point controls, is simply not aligned with the dynamism and high velocity of cloud-native AI (Truong, 2023). This results in a velocity gap, in which the speed of innovation is significantly higher than the speed of governance, and it causes serious operational, reputational and ethical risks.

In this particular sense, governance does not just mean compliance or cost management. It involves the overall collection of policies, procedures, standards, and technologies that make sure that the application of AI in an organization is consistent with its strategic goals, ethical principles, and legal requirements (Schwartz et al., 2022). The system of directional control is the one that governs the whole AI lifecycle including data gathering and model construction, deployment, maintenance, and decommissioning. The commercial argument behind the investment in such a strong governance system is strong. Although one of the key reasons motivating the need to mitigate the risks is the aversion of the financial fines, legal responsibility, and brand damage related to the biased or ineffective AI system, a properly designed governance program may also act as a competitive edge. It gains customer, regulatory, and partner trust, speeds up the acceptance and implementation of new AI projects by setting clear guardrails; and creates a responsible innovation culture, which attracts the best talent (Burt and Dworak, 2021). Thus, a successful governance is not an obstacle to the implementation of AI but a key to its successful and sustainable implementation in the enterprise.

2. FOUNDATIONAL PILLARS OF A RESPONSIBLE AI GOVERNANCE FRAMEWORK

An effective AI governance framework should also be built on a framework of pillars, which are dependent on each other. These pillars are the core values that will inform any decision-making process and procedure in the field of AI in the company.

2.1. Transparency and Explainability (XAI)

Transparency and Explainability are two principles of a trustworthy AI. Stakeholders such as regulators, customers, and internal auditors are entitled to learn more about how decisions are made by AI systems as they become more influential (Adadi and Berrada, 2018). Transparency means the disclosure of the fact that an AI system is in operation, the purpose of its usage, and the general capabilities and limitations of the system. Explainability (also known as XAI) goes further, relating to the technical capability to describe the rationale behind a particular output, or the general behaviour of the model. It is especially hard with complicated black box models such as deep neural networks. Models can not be debugged effectively, disputed with incorrect decisions, or trusted to be built without explainability. To improve explainability, enterprises need to invest in tools and practices that increase explainability, including generating model cards which document the performance characteristics, apply methods such as SHAP (SHapley Additive exPlanations) or LIME (Local Interpretable Model-agnostic Explanations) to emphasize the importance of features, and create systems which provide contextually meaningful explanations to end-users (Ribeiro et al., 2016). This can be incorporated into the pipeline during the MLOps in a cloud environment, where explainability reports are generated automatically with each model version.

2.2. Fairness and Bias Mitigation:

AI systems have a potential of reinforcing and even enhancing the existing biases in society unless managed with care. Discrimination may happen at any point: within representative data on historical training, incorrect problem formulation, or even in the algorithms themselves (Mehrabi et al., 2021). The pillar of Fairness and Bias Mitigation asks an active and ongoing initiative to recognize, quantify, and lessen adverse results in various demographic categories (e.g., depending on race, gender, age). This includes identifying metrics of fairness that apply to that application (e.g., demographic parity, equalized odds) and applying technical tools to pre-process data (to fix biases), in-process (by using fairness-aware algorithms), and post-process data (to re-adjust model outputs). Some cloud providers, like Amazon SageMaker

Clarify or the What-If Tool provided by Google, can help with this, and will automatically identify potential bias in data and models. Such checks should be required as a precondition to model deployment by the means of a governance framework and create continuous monitoring to address performance drift that might result in discriminatory performance results in the long term.

2.3. Robustness, Reliability, and Safety:

AI systems to be trusted must not only be safe but also robust and reliable. Robustness is the ability of a model to be able to perform under noisy, unpredictable, or adversarial input data. Reliability is the ability of the system to operate consistently and in accordance with the intended work conditions within the normal operating conditions. This is further extended by safety to make sure that the system will not harm, even in edge cases or failure situations (Amodei et al., 2016). This pillar is associated with heavy testing, such as adversarial testing in which inputs are intentionally distorted in a bid to attempt to break the model. It also needs concept and data drift monitoring whereby the statistical characteristics of the real-world data vary with time and leads to model decay. The safety of cloud-automated environments can be designed by using circuit breakers, which automatically discontinue a model when its behavior turns erratic or even exceeds defined performance limits, and by using effective rollback facilities that enable the rapid restoration of a previous and stable model instance.

2.4. Accountability and Human Oversight:

Clear accountability is the linchpin that holds a governance framework together. It must be unequivocally established who is responsible for the behavior and outcomes of an AI system throughout its lifecycle from the business leader who commissions it, to the data scientists who build it, the engineers who deploy it, and the legal team who assesses its compliance (Felzmann et al., 2020). This involves creating a RACI (Responsible, Accountable, Consulted, Informed) matrix specifically for AI projects. Closely linked is the principle of human oversight, which mandates appropriate human involvement in AI-augmented decision processes. The European Union's AI Act, for instance, formalizes this by requiring “human-in-the-loop” for high-risk AI systems, meaning a human must review and approve every decision. For lower-risk applications, “human-on-the-loop” oversight, where humans monitor overall system performance and can intervene, when necessary, may be sufficient. The governance framework must define the level of oversight required for each AI use case based on its potential impact.

2.5. Privacy and Data Governance:

AI is intrinsically data hungry, and thus, privacy and data governance form the bedrock of AI. Responsible AI mandates adherence to data privacy principles and regulations such as GDPR and CCPA. This ranges from ensuring data provenance, which includes understanding the origin and lineage of data, to data minimization, entailing collection of only that amount which is necessary, and further using techniques for anonymization and differential privacy in order to protect the identity of any individual subjects involved, as stated by Voigt & Von dem Bussche (2017). Data governance, basically dealing with the availability, usability, integrity, and security of the data, forms a pre-requisite for AI governance. The cloud environment can facilitate this via centralized data catalogs, encryption services, and access control mechanisms that allow consistent application of data policies throughout the organization.

2.6. Security and Resilience:

The implementation of AI presents novel attack surfaces, which should be secured. Uniqueness of AI systems to threats (poisoning of the training data), evasion (fooling a deployed model) and model inversion (extracting training data of a model) attacks (Liu et al., 2018). The security pillar entails the use of classic cybersecurity best practices to the whole AI pipeline, which includes securing the code repositories, training data stores, and the model registries, besides containing these new AI-specific threats. The term resilience is used to describe how the AI system can stand and recuperate after attacks or failures. It is possible with the help of sound architecture design, regular security audits and penetration testing of the AI stack and detailed disaster recovery plans that involve the restoration of the not only data, but also trained models and pipeline that support them.

3. THE CLOUD AS AN ENABLER AND ENFORCER OF AI GOVERNANCE

Although the cloud brings with it governance challenges, its native features also offer some effective tools to deploy and apply governance on a large scale. The trick is to transform governance as a manual, gate keeping role, into an automated and integrated one.

Cloud providers also provide a set of managed services, which can be integrated into the AI governance fabric. Larger model registries, including the AWS SageMaker Model Registry or Azure ML Model Registry, are a single source of truth of all models, including important metadata, version history, and approval statuses. MLOps pipelines, automated by using services such as Kubeflow or Azure DevOps, can be configured to have in-built governance checkpoints. As an illustration, a pipeline might be set up to automatically perform bias detection scans, accuracy tests and explainability reports prior to a model being promoted into a staging environment. Moreover, cloud systems offer mutable audit records (e.g., AWS CloudTrail, Google Cloud Audit Logs) which automatically capture each API call and activity to form a history of actions that is difficult to alter and is useful in compliance and debugging.

Policy as Code (PaC) is a paradigm shift in this area. PaC is the codification of rules of governance and compliance into machine-readable definition files. Open Policy Agent (OPA), HashiCorp Sentinel and cloud-specific services such as AWS Config enable organizations to specify policies such as do not deploy any model trained on personally identifiable information (PII) without encryption or must have a bias score below a certain threshold. These policies are further automatically compared with each resource creation or deployment request in the cloud environment. When a model or infrastructure setup breaches a policy, the deployment is automatically blocked, which means that governance is upheld in real-time, consistently, and without imposing a bureaucratic bottleneck (Humble and Farley, 2020).

Lastly, with very automated systems, the cloud enables dynamic guardrails to be implemented. These are run-time controls which observe behavior of an AI system on the production line and are capable of making autonomous corrective action. In the example, a guardrail might be used to track the confidence scores of a fraud detection model, so that when the score of a transaction lies within a suspect range, the guardrail might automatically divert the decision to a human agent. This gives rise to a managed autonomy system, where AI systems are free to act without any harm to others, within a safe controlled scope, so that the field of governance is in step with automation.

4. OPERATIONALIZING THE FRAMEWORK: A STEP-BY-STEP APPROACH

The translation of the theoretical framework into practice should be approached in a structure and in stages that will incorporate governance into the organizational culture and technology stack.

4.1. Phase 1: Assess & Strategize:

This stage is the start of the journey with a detailed analysis of the current situation. This includes doing an enterprise-wide inventory of AI to list all the existing and intended AI projects. Risk assessment must then be conducted on each use case, which will determine how it may affect individuals, the organization and the society. Simultaneously, the organization needs to assess its AI governance maturity, find out the gaps in skills, processes, and technology. The deliverable of this step is a concise AI governance plan, supported by the top management, which details the vision, objectives, and top-level map of the program.

4.2. Phase 2: Design & Define:

In this stage, the strategic vision is converted into tangible artifacts. It involves the writing of an AI Governance Charter, a written document that includes the statements of the organization being responsible in AI and defines the main principles. After this, elaborate policies and standards have to be formulated. These are more detailed, operational documents, which address the data sourcing, model development, testing needs, deployment process and the monitoring protocols. More importantly, this step also entails the organizational structure definition, which can be a cross-functional AI Ethics Board or Governance Committee with authority to approve and revise high-risk AI projects.

4.3. Phase 3: Implement & Automate:

This is the actual implementation phase where the tools and processes are implemented. The codified policies are implemented in the CI/CD/MLOps pipelines with the help of the PaC tools. The cloud services required (model registries, monitoring tools, bias detection services) are set up and implemented. The implementation of training programs is done to upskill data scientists, engineers, and business leaders to the new governance requirements and tools. The idea is to ensure compliance is the default, the easy way by integrating governance checks into the workflow of the developer.

4.4. Phase 4: Monitor & Govern:

When systems are operational, there should be continuous monitoring. It will include monitoring operational metrics (latency, throughput) and AI-specific metrics (accuracy, fairness, drift). Smart dashboards are expected to give an instant perspective of the well-being and conformity of all put-into-practice AI resources. The set AI Ethics Board members hold frequent meetings where they discuss performance reports, audit results, and incident reports, which involve continuous governance and make judgmental decisions on complicated matters.

4.5. Phase 5: Iterate & Improve:

AI governance is not a project but an ongoing process. This is the last stage of learning and adaptation. The feedback of monitoring, audits, real world incidences should be incorporated into the framework. To ensure that policies and standards are updated, it is important to review and update them on a regular basis to incorporate new regulations, ethical standards, and organizational lessons learned. This cyclic process helps to make the governance structure effective and relevant to the rapidly changing technology.

5. IMPLICATIONS OF CLOUD-AUTOMATED AI GOVERNANCE

The mass introduction of controlled, cloud-computerized AI has very significant and extensive consequences that go well beyond the technical sphere and all stakeholders should pay close attention to them.

5.1. Social Implications:

The social implications of automated AI are so far the most discussed. One of the major issues is the effect on the workforce. Although AI will replace some of the routine and repetitive tasks, which will result in job displacement in some of the jobs, it will also generate jobs and alter existing ones (Bessen, 2019). A successful workforce transition and colossal investments in reskilling and upskilling programs are the determinant of the net social impact. Moreover, the dangers of algorithm bias and social inequality are a significant social problem. When AI systems applied in hiring, lending, and criminal justice are trained with biased data, they will have the ability to systematically disadvantage already marginalized populations, which in turn entrenches and amplifies already existing social inequities (Eubanks, 2018). This may intensify the digital divide, as the access and control of AI technology by individuals affording it gain high economic and social benefits at the expense of others. Finally, all these factors have an impact on people trusting and perception. A successive wave of high-profile AI failures or biased results may undermine the trust of the population not only in particular companies but also in the technology itself and in the institutions that implement it, ultimately preventing its application in the social good.

5.2. Ethical Implications:

Ethical dilemmas of autonomous systems are complicated and multi-faceted. The issue of moral agency is the most important: in case of an accident by a self-driving car or a lethal diagnostic mistake of a clinical AI system, who is morally and legally accountable? It is a challenge to attribute blame because AI development is done by distributed individuals (data labelers, algorithm designers, system integrators) (Boddington, 2017). Informed consent is also an issue that is questioned in an automated world. What do people mean by giving meaningful consent to their data to be utilized in the complex, evolving AI systems, the future uses of which they may not be aware? This is resulting in the more profound issue of value alignment: how can we be sure that highly autonomous AI systems make decisions that are compatible with complex, and even conflicting, human values and ethical principles? The ongoing black box dilemma of certain AI models represents one of the essential ethical issues: is it ethically acceptable to implement a system that can make life-changing decisions without being able to explain its reasoning to the people affected by its actions? These are not technical questions that only demand technical answers but questions that need profound philosophical and ethical investigations.

5.3. Practical Implications:

On a practical level, organizations have serious practical issues. The first one is the technical debt and complexity accrual. AI systems do not exist as software alone; they are a complicated combination of code, data, and models. This, when not managed, results in the so-called AI debt, where the price of maintaining, updating, and debugging the entangled AI systems becomes debilitating (Sculley et al., 2015). This is augmented by a high level of skills gap and lack of talent. Deep AI knowledge coupled with cloud expertise and knowledge of ethics and governance is not common, so it is challenging and costly to staff AI governance programs. Using the ecosystem of one cloud provider heavily will present an additional risk of vendor lock-in, thereby diminishing bargaining power and making migration difficult. Lastly, compliance cost is also a practical factor of great importance. Developing and sustaining a strong governance program is a costly endeavor in terms of tools, staff, and procedures, which may prove to be an entry barrier to smaller organizations and needs to be an aspect of the overall business case of AI.

5.4. Managerial Implications:

To the leaders, the emergence of controlled AI requires a change of management thought and organization. New organizational structures are being developed including centralized AI Ethics Boards or decentralized governance councils, which need to be given the authority to manage AI strategy and risk. This is in turn transforming leadership. The executives should become technologically educated to make the right choice regarding AI; legal and compliance departments should be aware of the specifics of the algorithmic responsibility. The need to have a cultural transformation is perhaps the most significant change. Leaders should establish a culture of shared responsibility, ethical considerations, and performance, and psychological safety so that employees raise concerns about AI systems. This is all the way to a re-evaluation of performance (KPIs). In addition to such conventional indicators as ROI and accuracy, now managers are being monitored and held accountable to governance related KPIs, including fairness scores, explainability ratings, and incident response time. Lastly, the AI governance should be made a part of the strategic risk management, and the board and the C-suite should consider AI-specific risks as a matter of course in corporate governance.

6. CASE STUDIES: GOVERNANCE IN ACTION

Case Study 1: A Financial Services Company Automating Loan Approvals:

A global bank aimed to automate its loan approval system with the help of a machine learning model that was implemented on its cloud system. The governance structure required a pre-deployment bias audit with the help of a cloud-based system, which showed that the model was disapproving of the applicants belonging to a particular postal code disproportionately. This was followed down by the data team to historical bias in the training data. They reweighted the data using methods of the fairness pillar and used a fairness constraint to train the model. The PaC pipeline was set to prevent any subsequent model version which failed to meet the fairness threshold. Moreover, it was written in an explainable format, giving the rejected applicant an easily comprehensible, reason coded response (e.g., high debt-to-income ratio), and a human-in-the-loop mechanism was developed over borderline cases. The case illustrates how fairness, explainability and accountability can be practically combined in an automated cloud pipeline.

Case Study 2: A Healthcare Provider Using AI for Diagnostic Assistance:

A medical network implemented a cloud-based AI system to help radiologists detect early signs of a particular type of cancer in medical images. The administrative structure was very strong and safe. The model was also tested thoroughly in adversarial testing to make sure that the model was not deceived by image artifacts. Its production scores were constantly checked in the production. One of the PaC rules was that any scan with less than a 90 percent confidence score was automatically set to be reviewed by a senior radiologist (a human-in-the-loop guardrail). The cloud had all the decisions and AI suggestions recorded in an irrevocable manner to be audited. In this case, the role of governance is brought out to highlight how AI can be a safe and supportive tool to professionals working in high-stakes settings.

7. THE FUTURE OF AI GOVERNANCE

The AI governance is a rapidly developing field. The regulatory environment is also becoming more solid, and laws such as the EU AI Act provide a precedent to a risk-based approach to AI regulation, which will have an international effect (European Commission, 2021). There is also the emerging concept of sovereign AI and national government forms where nations require that some forms of data and AI processing be located within their jurisdictions and this creates a more fragmented cloud environment. In the future, studies on adaptive and self-governing artificial intelligence are being conducted. Such systems would be able to observe their own performance, identify drift or bias, and automatically retrain or modify their parameters to be within ethical and operational limits. Although this has a potential to make governance loop even shorter, it also brings up new concerns about the degree of autonomy that we are ready to give to machines.

8. CONCLUSION AND KEY TAKEAWAYS

The combination of AI and cloud computing is an irresistible change that transforms the contemporary business. The chapter has contended that the most important aspect of using its power in a responsible manner is the careful design and execution of a resilient, cloud-native AI governance system. This kind of framework with the pillars of fairness, transparency, accountability and security is not a regulatory burden, but a strategic requirement. It reduces risk, establishes trust and facilitates sustainable innovation. Using the automation features of the cloud and Policy as Code, as well as built-in MLOps, organizations can incorporate governance into their development cycle, bridging the so-called velocity gap. The social, moral, practical, and managerial consequences of this change are very dramatic, and a comprehensive solution is needed, which involves technologists, executives, ethicists, and policymakers. Responsible AI is a process that is ongoing and cyclical, yet with the help of a well-designed framework and the willingness to act ethically, businesses can safely sail on this new frontier to make sure that their automatic future is successful and fair.

REFERENCES

Adadi, A., & Berrada, M. (2018). Peeking inside the black-box: A survey on explainable artificial intelligence (XAI). *IEEE Access : Practical Innovations, Open Solutions*, *6*, 52138–52160.

Amodei, D., Olah, C., Steinhardt, J., Christiano, P., Schulman, J., & Mané, D. (2016). Concrete problems in AI safety. *arXiv preprint arXiv:1606.06565*.

Bessen, J. E. (2019). AI and jobs: The role of demand. In *The Economics of Artificial Intelligence: An Agenda*. University of Chicago Press.

Boddington, P. (2017). *Towards a code of ethics for artificial intelligence*. Springer.

Burt, A., & Dworak, W. (2021). *The AI-powered enterprise: How to build an AI-driven organization*. Life Tree Press.

Eubanks, V. (2018). *Automating inequality: How high-tech tools profile, police, and punish the poor*. St. Martin's Press.

European Commission. (2021). *Proposal for a regulation of the European Parliament and of the Council laying down harmonised rules on artificial intelligence (Artificial Intelligence Act)*. COM(2021) 206 final.

Felzmann, H., Villaronga, E. F., Lutz, C., & Tamo-Larrieux, A. (2020). Transparency you can trust: Transparency requirements for artificial intelligence between legal norms and contextual concerns. *Big Data & Society*, *6*(1), 2053951719860542.

Humble, J., & Farley, D. (2020). *Continuous delivery: Reliable software releases through build, test, and deployment automation*. Addison-Wesley Professional.

Liu, Q., Li, P., Zhao, W., Cai, W., Yu, S., & Leung, V. C. (2018). A survey on security threats and defensive techniques of machine learning: A data driven view. *IEEE Access : Practical Innovations, Open Solutions*, *6*, 12103–12117.

Mehrabi, N., Morstatter, F., Saxena, N., Lerman, K., & Galstyan, A. (2021). A survey on bias and fairness in machine learning. *ACM Computing Surveys*, *54*(6), 1–35.

Möhlmann, M. (2021). *The impact of cloud computing on AI innovation*. In *Proceedings of the 54th Hawaii International Conference on System Sciences*.

Ribeiro, M. T., Singh, S., & Guestrin, C. (2016). " Why should I trust you?" Explaining the predictions of any classifier. In *Proceedings of the 22nd ACM SIGKDD international conference on knowledge discovery and data mining* (pp. 1135-1144).

Schwartz, R., Dodge, J., Smith, N. A., & Etzioni, O. (2022). Green AI. *Communications of the ACM*, *65*(1), 54–63.

Sculley, D., Holt, G., Golovin, D., Davydov, E., Phillips, T., Ebner, D., . . . Dennison, D. (2015). Hidden technical debt in machine learning systems. In *Advances in neural information processing systems* (pp. 2503-2511).

Truong, H. T. (2023). Closing the velocity gap: Governing AI in the cloud era. *Journal of Enterprise Technology*, *45*(2), 112–125.

Voigt, P., & Von dem Bussche, A. (2017). *The EU General Data Protection Regulation (GDPR): A practical guide*. Springer.

Chapter 21
Reimagining Talent Management:
Cloud-Native AI Tools for Predictive HR Analytics

Mohit Yadav
https://orcid.org/0000-0002-9341-2527
JGBS, O.P. Jindal Global University, India

Narayanage Jayantha Dewasiri
https://orcid.org/0000-0002-5908-8890
Faculty of Management Studies, Sabaragamuwa University of Sri Lanka, Sri Lanka & Qasim Ibrahim School of Business, Villa College, Maldives

Minakshi
College of Computer Science, King Khalid University, Saudi Arabia

Tanupriya Choudhury
https://orcid.org/0000-0002-9826-2759
School of Computer Science, University of Petroleum and Energy Studies, Dehradun, India

ABSTRACT

This chapter explores the paradigm shift in Human Resources (HR) from reactive, intuition-based practices to a proactive, data-driven function powered by cloud-native Artificial Intelligence (AI). It argues that the synergy between cloud computing's scalability and AI's predictive capabilities is fundamental to reimagining talent management. The chapter deconstructs the modern predictive HR tech stack and presents core use cases in acquisition, retention, development, and performance. It critically addresses the imperative of navigating ethical challenges—including bias,

DOI: 10.4018/979-8-3373-7442-0.ch021

privacy, and transparency—and provides a framework for responsible implementation. Ultimately, this evolution empowers HR to become a strategic partner, using predictive analytics to augment human decision-making, mitigate workforce risks, and foster a more adaptive, resilient, and human-centric organization.

INTRODUCTION

Human capital management has never been more important or complicated. In the modern economy of knowledge, the capacity to attract, develop and retain talent has become one of the major determinants of competitive advantage in an organization (Cappelli and Keller, 2017). Nevertheless, the historic talent management habits which are frequently based on the industrial-era models are becoming increasingly out of sync with dynamism of the new workforce and the speed of digital transformation. Most HR departments continue to be stuck in the rut of reactive administrative activities, basing strategic decisions on past data and managerial intuition, so called gut feel. This strategy is quite expensive. Financial cost of a bad hire may be between 50 and 200 percent of the yearly income of the worker, in terms of recruitment expenses, training duration and misplaced productivity (Society for Human Resource Management, 2022). Moreover, voluntary turnover, especially when selecting high performers, is a waste of institutional knowledge and replacement costs are huge and the total turnover costs are usually 1.5 to 2 times of an employee salary (Boushey and Glynn, 2012).

One of the elements of this issue is the constraints of the traditional Human Resource Information Systems (HRIS). These traditional systems that were usually on-premise were structured to serve more of a record keeping and payroll processing purpose. They are good at reporting on what has already been done, how many people were employed during the last quarter, the turnover rate during the last year, but provide minimal to no information on what will be and what should be done. The most excellent example of this rear-view mirror approach is the annual performance review which is one of the pillars of traditional talent management. It gives a static and backward-looking picture that is likely to be full of bias and is not conducive to real-time growth and development (Pulakos et al., 2015). This causes strategic disparity in the sense that the HR leaders do not possess the predictive capabilities to provide advice to the C-suite on the risks and opportunities in the future workforce.

The new vision developed in this chapter is the transformation of HR into a strategic and data-driven science instead of an administrative one. Using cloud-native AI applications to conduct predictive analytics, HR will be able to become proactive instead of reactive, standardized, instead of personalized, and intuitive, instead of evidence-based. This paradigm allows a proactive strategy to talent management where potential attrition is notified in advance before an employee leaves, skills shortages are noted before they disrupt strategy, and internal talent match required to fill positions is done automatically (Kumar, Rani, et al., 2025; Nguyen et al., 2025). It is not about surpassing human judgment but rather supplementing it with profound, practically valuable intelligence and thus enabling HR professionals and people managers to do what they can best, which is creating human connection, performance coaching, and an exciting organizational culture.

2. THE FOUNDATION: WHAT ARE CLOUD-NATIVE AI TOOLS?

In order to see the transformative potential of this new approach, it is important to first comprehend the underlying technologies. The terms cloud-native and AI are ubiquitous and, more importantly, their combination is essential as the definitions are of paramount importance.

2.1 Beyond the Hype: Defining "Cloud-Native.":

The definition of what is cloud-native. Cloud-native is a current strategy of developing and operating applications based on the complete use of the cloud computing delivery model. It has more than mere hosting software on a remote server, it is an architectural paradigm. Some of the core concepts are: Microservices, in which apps are divided into small, loosely-coupled services (e.g., a resume-parsing service, a sentiment-analysis service) that can be developed, deployed, and scaled separately (Newman, 2015)(Kumar et al., 2025). These microservices and their dependencies are bundled into lightweight, portable units known as containers (e.g., Docker, Kubernetes), which are consistent across computing environments. Elastic Scalability, whereby the system automatically adds or removes resources according to demand, i.e. responding to a spike in job applications without performance being compromised. Lastly, an API-first design guarantees that all functions are supported through an Application Programming Interface (API), which allows easy integration with other best-in-class systems such as an Applicant Tracking System (ATS), LinkedIn, or productivity software, such as Slack and Microsoft Teams. This is a juxtaposition to monolithic, on-premise HR systems which are inflexible, hard to integrate and expensive to scale.

2.2 The AI in HR: A Primer:

A Primer. In HR, Artificial Intelligence is a wide area which incorporates a number of technologies. Machine Learning (ML) is the horse of predictive analytics.

It involves the statistical algorithms to learn the patterns of the past data to predict on the new data. As an example, an ML model can process data on thousands of former employees to understand the multifaceted aspects that are related to attrition (Mikalef & Gupta, 2021). Natural Language Processing (NLP) enables computers to comprehend, analyze and produce human language. NLP is applied in HR to scan resumes to identify relevant skills, analyze the responses to unstructured survey questions on employees to measure sentiment, or create chatbots that respond to typical queries posed by employees (Kumar, Yadav, et al., 2025; Yadav et al., 2025) (Tambe et al., 2019). To be more recent, there is the so-called Generative AI which can produce new content. This may be converted in talent management to writing custom job descriptions, developing interview questions according to a competency model of a job or developing custom learning content in the employees.

2.3 The Synergy: Why Cloud-Native is the Perfect Bedrock for AI:

The union between cloud-native architecture and AI is not accidental, it is symbiotic. AI, and especially ML, is data-intensive and computationally-intensive in nature. To train precise models, large volumes of clean and unified data are needed, and the cloud-native data layer can consolidate data of endless sources. The cost-effectiveness of the cloud is in its ability to provide massive computing power on demand due to the iterative nature of model training and retraining. Moreover, the microservices infrastructure provides the data scientists with an opportunity to revise one predictive model (e.g., the attrition risk algorithm) without interfering with the whole HR platform. The API-first design of cloud-native systems is such that the insights provided by the AI, e.g. a flight risk score, can be easily pushed to other systems, e.g. a manager dashboard in Workday or an alert in Salesforce. Essentially, the cloud-native environment offers the flexible, scalable, and interconnected infrastructure that ensures strong and integrated AI is not only possible, but also viable and efficient.

3. DECONSTRUCTING THE PREDICTIVE HR TECH STACK

A current predictive HR analytics functionality is not a software application but a stack of technology. These layers are essential in understanding the organizations that are interested in constructing or purchasing such a solution.

3.1 The Data Layer:

Data is the basis of any system driven by AI. The ancient saying that garbage in, garbage out is very applicable in this case. The data layer is tasked with the responsibility of ingesting, cleaning and consolidating data of a vast number of sources. It encompasses formal data of the conventional systems such as the HRIS (demographic information, performance scores, pay history) and the ATS (application origin, time-to-fill). More importantly, it is also the one that absorbs unstructured data, which is the greater part of the available information. This contains employee resume text, employee performance review text, pulse survey text, internal collaboration platform text, such as Slack or Microsoft Teams (with proper privacy controls), and even email metadata. This huge, raw data is usually stored in its native form in data lakes often constructed on cloud storage such as Amazon S3 or Azure Data Lake. Data wrangling, or cleaning, de-duplicating, and standardizing this data is an unglamorous but necessary step in order to have the following models of AI trained on quality, trustworthy data.

3.2 The Analytics & AI Engine:

This is the calculating core of the stack which converts raw information into intelligence. The ML models and algorithms that execute the fundamental predictive and prescriptive functions are found in this layer. It is in this place that a model could be developed to find out the minor patterns that lead to an employee leaving or group the employees together in clusters depending on their skill profiles and their career goals. This environment offers data scientists the means of feature engineering (choosing the most promising data points to predict), training a model and validation. More importantly, feedback loop is also handled by this engine. In the case where a manager takes an action based on a prediction, such as a retention chat with a high-flight-risk employee, the result of the action (whether the employee stayed or not) is returned to the system. This constant feedback enables the models to learn and get better with time, and this is called continuous learning. This layer is commonly driven by cloud-based AI engines such as Google Vertex AI, Azure machine learning, or Amazon SageMaker.

3.3 The Application Layer:

The face and hands are the application layer in the event that the AI engine is the brain. This is the user-friendly interface that the HR business partners, managers, and even employees interact with the insights of the system. The idea is to make complex analytics available in easily understandable and practical format, rather than in spreadsheets that are hard to decipher. This is generally done by interactive dashboards which visualize important measures such as overall attrition risk, diversity hiring funnel health and skills coverage. One of the most important elements is the so-called nudge engine that provides timely and contextual notifications and suggestions. An example that can be given is a manager getting a notification that says: Team member Jane Doe has an 85% flight risk. Recommended activity: Career development objectives and compensation benchmark discussion and review. This layer can also have AI-based chatbots to recruit (screening candidates) or provide HR services (answer policy questions), and the power of AI is made available to non-experts.

3.4 The Feedback Loop:

This is the dynamic process which links the application layer with the data and AI layers and completes the circuit and forms a learning system. Each communication with the application layer produces new data. Each time a manager is shown a dashboard, disregards an alert or follows a recommendation, that is recorded. Above all, the outcomes of the actions which were taken according to the predictions of the system-the ground truth- are taken. Was the employee that was being predicted to quit leaving? Was there any improvement in performance measures as a result of the recommended training course? This feedback information is fed back to the data layer and it is used to retrain and optimize the ML models within the AI engine. This forms a virtuous circle of the system being smarter and more accurate the more it is utilized, and it keeps on adjusting itself to the new patterns of working in an organization.

4. CORE USE CASES: PREDICTIVE ANALYTICS IN ACTION

The real worth of this technology stack is achieved by applying to the fundamental functions of talent management. Such applications go beyond hypothetical possibilities to provide real business benefits.

4.1 Talent Acquisition.

The recruitment process is getting transformed through AI that is more efficient, quick, and less biased. Predictive Sourcing models are based on analyzing successful past hires to determine what traits of high-performing employees exist. They can then actively search external databases such as LinkedIn and GitHub to identify candidates with similar profiles, and in many cases they identify so-called passive candidates who are not actively seeking employment. AI-Powered Screening employs NLP to analyze resumes and identify skills and experiences of candidates with the job requirements with significantly higher speed and accuracy than human screeners. It is also possible to lessen unconscious bias using advanced systems that conceal demographic data and address only job-related skills (Housman, 2017). In addition, ML models can Predict Candidate Offer Acceptance based on the factors such as current commute by the candidate, the offered pay as compared to his or her market value, and how the candidate interacts with the interviewer so that the recruiters can optimize their offer strategy to increase the close rates.

4.2 Retention and Engagement:

One of the most expensive and disruptive people issues is employee turnover, and it is an ideal target of predictive analytics. One of the flagship use cases is Flight Risk Analysis. The variables that are considered by ML models include tenure, promotion history, compensation changes, engagement survey scores, manager relationship, and even patterns in email and calendar metadata (e.g., a reduction in internal collaboration) to provide a flight risk probability to each employee. This enables the managers and HR to take the initiative to use specific retention measures, including career pathing conversations, mentorship programs, or salary increments. Sentiment Analysis is a branch of NLP that offers a live e-gauge of organizational morale through text-based pulse surveys, exit interviews, and social platforms inside organizations. It is able to pick up the changes in attitude, say an increase in frustration among a certain department, much earlier than it is reflected in the voluntary turnover, and hence leadership can correct the problem.

4.3 Learning and Development (L&D):

The half-life of skills is rapidly decreasing, and a more dynamic approach to L&D is necessary. Skill Gap Analysis has AI to match the existing skills of the workforce (based on profiles, projects, and learning history) with the skills needed to implement the future business strategy of the company. This determines areas of skill deficiencies before it is a liability. On the basis of this analysis, AI systems are then able to create Personalized Learning Paths. The system can suggest a list of courses, micro-learnings, internal projects, and potential mentors based on the knowledge of an employee regarding his present position, future career goals, and areas requiring additional skills. This takes L&D out of a one-size-fits-all catalog and goes directly to a Netflix-like experience that is directly connected to the individual career objectives and the needs of the organization, making it more engaging and improving the payoff on training investment.

4.4 Performance and Productivity:

The traditional performance management is usually a backward, occasional occurrence. AI allows more continuous and prospective strategy. High Performers and Future Leaders can be predicted with the help of models based on analysis of performance data, 360-degree feedback, behavioral indicators (initiative on projects and patterns of collaboration). This assists in determining the existence of high potential talent to be developed into good leaders at an earlier stage. Additionally, Organizational Network Analysis (ONA), which is AI-driven, is able to process email, chat, and meeting data to map the reality of how work gets done- revealing the key influencers, information bottlenecks, and isolated teams. This observation enables leaders to maximize team structure, enhance communication channels, and increase effective collaboration, which leads to an increase in the productivity and health of the whole organization.

5. THE HUMAN IN THE LOOP: IMPLEMENTATION AND CHANGE MANAGEMENT

Technology is not the only component of the equation. To achieve the successful adoption of predictive HR analytics, human and organizational aspects should be paid close attention. Build it and they will come is a formula of failure.

5.1 Building a Data-Driven HR Culture:

Interpersonal skills, empathy, and understanding of employment law have been the criteria used to choose and reward HR professionals over the decades, rather than analytical ability. The most important thing is upskilling this team. This includes the training of HR Business Partners (HRBPs) to be knowledgeable in the interpretation of data visualizations, statistical concepts such as correlation versus causation, and, most crucially, the translation of the results of this analysis into actionable business advice (Angrave et al., 2016)(Kumar et al., 2023). To address the disconnect between people and technical realms, the HR role might be required to introduce new positions, including the People Analytics Manager or Data Scientist role.

5.2 Gaining Leadership Buy-In and Measuring ROI:

To be able to gain executive sponsorship and funding, one needs to speak the language of business. The HR leaders need to explain the value of predictive analytics using key business metrics rather than HR metrics. Rather than stating the enhanced accuracy of attrition prediction, the discussion should be on the reduction of voluntary turnover rates in our core engineering staff by 15 percent resulting in an estimated cost of 2.5 million saved in recruitment and lost productivity expenses on an annual basis. These initiatives must be turned into strategic priorities, and that requires a visible business case that has a developed projection of the return on investment (ROI).

5.3 A Phased Implementation Roadmap:

It is dangerous and daunting to attempt a big bang implementation throughout the organization. Better solution is to begin with a high-impact pilot project, which is well defined. As an example, an organization may start with a specific population such as predicting attrition in its sales force which is a high-turnover and critical population. This helps the team to show a quick win, iron out technical and process bugs and develop credibility. Winning with this contained pilot generates momentum and lessons learned that can be used during a larger, enterprise-wide deployment, scaling the functionality across other functions and use cases.

5.4 The Role of the HR Business Partner in an AI-Augmented World:

It is widely feared that AI will supplant HR professionals. Augmentation is more probable and fruitful. The data processing and pattern recognition will be done by AI, and the HRBPs will be able to work on more valuable and human-centric tasks. Their functions will change to data reporter to strategic consultant, process administrator to coach and change agent. By relying on AI-generated insights, they will hold more mcaningful, evidence-based conversations with line managers about their teams, aiming to coach managers on how to have retention conversations, arbitrate on team conflicts, and inclusive cultures. The human in the loop makes sure that the data-driven recommendations are made contextualized with empathy, ethical consideration and with the understanding of the highly complex human dynamics that cannot be reflected in data alone.

6. NAVIGATING THE ETHICAL MINEFIELD: BIAS, PRIVACY, AND TRUST

The ethical implications of predictive AI in HR are as serious as the power of that technology. In the absence of a strong ethical structure, these tools may bring about a lot of damage, undermine trust, and put the organization at risk of legal and reputational damage.

6.1 The Perpetuation of Bias:

Probably the greatest danger is that AI systems can reproduce and even enhance the existing human and societal prejudices. When an ML model is trained on past promotion data collected in a company that has historically promoted men, it can learn to correlate maleness with promotability, therefore, historical discrimination will be encoded into the model to be used in the future (Dastin, 2018). This is a typical problem of bias in, bias out. The process of mitigation is complex and should be continuous. They are: Auditing Training Data to be Representative of the various demographic groups; running Algorithmic Fairness Audits to determine whether the predictions of the model are fairly accurate to all groups; and using Debiasing Techniques to the ML modeling process. The objective is not only technical fairness but to proactively employ these tools in enhancing diversity, equity and inclusion (DEI) objectives.

6.2 Transparency and Explainability:

Most complicated ML models, especially deep learning networks, are referred to as black boxes since it is not easy to explain how they come to a certain decision. This is not acceptable in an HR environment. In case an employee is identified as a high flight risk or is simply not offered an internal mobility opportunity due to a recommendation provided by an AI, the organization should be in a position to explain the reasons in a clear and understandable manner. Explainable AI (XAI) is a discipline focused on this problem. In the case of HR applications, this would imply that the system must not just give a risk score but it must also give the determinants of the score including the example of Flight risk: 85%. Factors of interest: There was no promotion within the past 4 years, payment was at 15 percent less than the market rate, and the workload on the project was reduced recently. This openness is essential in gaining trust among the employees and managers, corrective action, and due process.

6.3 Data Privacy and Employee Surveillance:

The need to devour data on which predictive models draw can readily result in an Orwellian culture of surveillance. Supervising employee mails, chat messages and badge swipes to determine their behavior is a deeply privacy-invasive action and can destroy employee confidence. Companies have to find a delicate balance. The main principles are as follows: Purpose Limitation (only the data that is pertinent to a certain and legitimate business purpose should be collected), Transparency (it should be made clear to the employees what data are being collected and how it will be used), and Anonymization/Aggregation (where possible, the data should be aggregated and in an aggregate and non-attributable form). Indicatively, anonymized metadata is the most ethical way of practicing ONA where the patterns of collaboration are understood at group level and not an individual employee. A tool that is necessary to establish these guardrails is an organizational AI Ethics Charter which would be developed collaboratively with legal, HR, and employee representatives.

6.4 Establishing an AI Ethics Charter for HR:

With such complicated issues, organizations cannot afford to use ad-hoc decisions. An HR-related AI Ethics Charter is an important governance tool, which is formal and publicly available. This charter must clearly describe the values of the organization on the ethical usage of people analytics, including data privacy, bias reduction, transparency, and human control. It must form a governance board with a variety of diverse stakeholders such as HR, legal, IT, and ethics who will evaluate and approve new AI uses cases. With these principles in place in the first place, an organization will be able to establish a culture of responsible innovation, ensure trust with its employees and reduce the major risks posed by this powerful technology.

7. THE FUTURE IS NOW: EMERGING TRENDS

The AI area of talent management is developing at an astounding speed. There are various trends that are on the horizon which are likely to further transform the HR function in the next few years.

7.1 Generative AI in Talent Management:

The introduction of powerful language models (LLMs) such as GPT-4 is a new dawn. Generative AI can write impressive and fair job descriptions, scale personalized messages to candidates, write custom interview questions, and even simulate role-playing scenarios to develop leaders. In the case of L&D, it is able to extract learning modules dynamically or summarize long documents to highlight important points. Its grasp and production of natural language will make HR systems more conversational and more natural, which will essentially enhance user experience among the employees, managers, and even the recruiters.

7.2 Internal Talent Marketplaces and AI-Matched Gig Work:

Organizations are shifting to models that emphasize internal mobility and fast teaming. AI-based internal talent marketplaces are an internal version of LinkedIn, with employees building profiles displaying their skills and interests, and managers posting short-term projects or gigs. The AI engine then aligns people with opportunities according to their skills, career aspirations, and capabilities, and breaks down the silos of the traditional workforce, creating a more flexible and resilient workforce. This assists in retaining the best talent by offering diversified growth and also enables the organization to utilize its human resource better in line with the strategic priorities.

7.3 Predictive Analytics for Employee Well-being and Mental Health:

Employee well-being is becoming a central business requirement, and the COVID-19 pandemic is hastening its growth rate. Predictive analytics are also being utilized by forward-thinking organizations, though ethically and with great caution, to determine the factors that cause burnout and stress. Through aggregated and anonymized workload, work pattern (e.g., after-hours email activity), and sentiment data, AI can raise an alarm within a team or department at a high risk of collective burnout, leading to proactive intervention measures, including workload rebalancing, extra support, or well-being resources.

7.4 The Continuous, Adaptive Organization:

Finally, the merging of cloud-native AI tools lends to a future in which the organization is a continuous, adaptive system. Talent decisions are not done on yearly cycles but on an ongoing stream of data and feedback. The organization is able to dynamically feel the skills gaps and fill them, anticipate turnover and avert turnover and streamline the team structures to collaborate and innovate on the fly. This makes an organization not only more efficient, profitable but more responsive, resilient and human as every employee is given a chance to achieve their potential.

8. CONCLUSION:

The process of moving away the old reactive HR toward the future fueled by cloud-native AI and predictive analytics is not an act of technological enhancement, but a complete redefinition of the purpose and the potential of the role. This chapter has presented the strong business case of this change, since the high cost of making an intuitive decision is no longer relevant, but a model that is based on evidence and foresight. We have broken down the technology stack that makes this vision possible, including the single unified data layer down to the intelligent application layer, and discussed its disruptive effect on the overall employee lifecycle, including acquisition and retention, development and performance.

More importantly, the chapter has opined that success is not only built on algorithms, infrastructure, but more so on a profound dedication to ethical standards, an ability to manage changes, and perpetual upskilling of HR professionals. The threats of bias, invasion of privacy and lack of transparency in making decisions are real, yet can and should be mitigated by a careful governance policy, transparency and a humanistic design ethos. These potent tools are aimed at enhancing human intelligence, and not its substitution. Automated predictive and administrative tools enable cloud-native AI to liberate HR professionals and concentrate on the human elements in their work, such as strategic collaboration, compassionate coaching, cultural custodianship, and nurturing the meaningful relationships, which constitute the foundation of a high-performing organization. In this re-futuristic redefinition of the Human Resources, it is the HR that is re-empowered by the information to be more insightful, impactful and compassionate in its operating than ever before.

REFERENCES

Angrave, D., Charlwood, A., Kirkpatrick, I., Lawrence, M., & Stuart, M. (2016). HR and analytics: Why HR is set to fail the big data challenge. *Human Resource Management Journal*, *26*(1), 1–11.

Boushey, H., & Glynn, S. J. (2012). *The economic benefits of family and medical leave insurance*. Center for American Progress.

Cappelli, P., & Keller, J. R. (2017). The historical context of talent management. In Collings, D. G., Mellahi, K., & Cascio, W. F. (Eds.), *The Oxford handbook of talent management*. Oxford University Press.

Dastin, J. (2018, October 10). Amazon scraps secret AI recruiting tool that showed bias against women. Reuters.

Housman, M. (2017). Can algorithms reduce bias in the hiring process? *Harvard Business Review*.

Kumar, M., Dagar, M., Yadav, M., Choudhury, T., & Luu, T. M. N. (2025). Navigating the future: Challenges and opportunities in intelligent RPA implementation. In Intelligent Robotic Process Automation: Development, Vulnerability and Applications (pp. 393–420). IGI Global. https://doi.org/DOI: 10.4018/979-8-3693-4365-4.ch015

Kumar, M., Rani, P., Mangla, N., Nguyen, N. T., & Kaur, K. (2025). Sustainable outsourcing in an AI-driven world: Ethical, economic, and environmental perspectives. In *Global Work Arrangements and Outsourcing in the Age of AI*. IGI Global., DOI: 10.4018/979-8-3373-1270-5.ch001

Kumar, M., & Yadav, R., Sushma, & Yadav, M. (2025). Investigating the influence of digital and social media marketing on university brand equity: An in-depth literature review and bibliometric study. *Global Knowledge, Memory and Communication*, •••, 1–26. DOI: 10.1108/GKMC-07-2024-0426

Kumar, M., Yadav, R., Suresh, A. S., Singh, R., Yadav, M., Balodi, A., Vihari, N. S., & Srivastava, A. P. (2023). Mapping AI-Driven Marketing: Review of Field and Directions for Future Research. 2023 2nd International Conference on Futuristic Technologies (INCOFT), 1–5. https://doi.org/DOI: 10.1109/INCOFT60753.2023.10425169

Mikalef, P., & Gupta, M. (2021). Artificial intelligence capability: Conceptualization, measurement calibration, and empirical study on its impact on organizational creativity and firm performance. *Information & Management*, *58*(3), 103434.

Newman, S. (2015). *Building microservices*. O'Reilly Media.

Nguyen, T. B., Kumar, M., Sahoo, D. K., Nain, N., Yadav, M., & Dadhich, V. (2025). Mapping the landscape of women entrepreneurs in micro and small enterprises: Trends, themes, and insights through systematic review and text mining. F1000Research, 14. https://doi.org/DOI: 10.12688/f1000research.163030.1

Pulakos, E. D., Hanson, R. M., Arad, S., & Moye, N. (2015). Performance management can be fixed: An on-ramp to high performance. *Business Horizons*, *58*(6), 639–646.

Society for Human Resource Management. (2022). *The cost of a bad hire*. SHRM.

Tambe, P., Cappelli, P., & Yakubovich, V. (2019). Artificial intelligence in human resources management: Challenges and a path forward. *California Management Review*, *61*(4), 15–42.

Yadav, M., Kumar, M., Alyousef, M. I., Dagar, M., Dewasiri, N. J., & Amoozegar, A. (2025). Mapping Research Trends on Gender Diversity, Work Stress, and Job Satisfaction in Higher Education: A Bibliometric Analysis. International Journal of Learning. *Teaching and Educational Research*, *24*(8), 444–472. DOI: 10.26803/ijlter.24.8.19

Chapter 22
Risk Perception and AI-Driven Financial Decision-Making:
A FinTech Perspective

Manish Dubey
Shri Vaishnav College of Arts and Commerce, Indore, India

Mansi Panwar
https://orcid.org/0009-0007-0470-8216
Devi Ahilya Vishwavidyalaya University in Indore, India

Rituraj Singh
https://orcid.org/0009-0008-0452-9303
O.P. Jindal Global University, Haryana, India

Shashi Kant
https://orcid.org/0000-0003-4722-5736
DMI St. Eugene University, Zambia

ABSTRACT

Considering the changing future of digital finance, Artificial Intelligence (AI), cloud computing, and machine learning (ML) have transformed how financial decisions are made, credit-risks are assessed, and mechanisms of inclusion are applied. Irrespective of these improvements, user trust, perceived risk and behavioral biases remain influential to the usefulness and use of FinTech innovations. The proposed paper constructs a behavioral-technological conceptual model, which analyzes the mediating role of perceived risk between the relationship between AI-based FinTech adoption and the result of financial decisions, in which algorithmic trust acts as a

DOI: 10.4018/979-8-3373-7442-0.ch022

moderator. The paper connects psychological and technological aspects of financial behavior by relying on various theoretical underpinnings including Prospect Theory, Bounded Rationality, Signaling Theory, Expectancy-Violation Theory, Emotional Contagion Theory and the Algorithmic Bias-Trust Framework.

1. INTRODUCTION

The interplay of artificial intelligence (AI), machine learning (ML), and cloud computing has changed the financial technology (FinTech) environment by bringing to the market scalable, adaptive, and data-driven risk management and decision-making. Specifically, intelligent cloud infrastructures that can handle large, heterogeneous datasets and process them in real-time are redefining credit risk assessment, which forms the core of financial intermediation, to enhance its accuracy, interpretability, and financial inclusion (Malhotra, 2021; Krishnamoorthy, 2024; Huang et al., 2020).

The conventional credit assessment systems have been mostly relying on fixed econometric models, subjective information, and manual data processing of financial information. Information asymmetry, bias and the lack of flexibility to dynamic borrower behaviour and macroeconomic volatility tended to limit such systems (Siddiqi, 2012, 2017). Conversely, cloud-based ML models combine the distributed architecture and the automatic learning ability to refine predictions with the help of transactional, alternative, and behavioural data on a continuous basis, thus accelerating efficiency and transparency of risk (Malempati, 2021; Rohan Malhotra, 2021).

The convergence of big data analytics and FinTech around the globe has opened the doors of predictive finance. Research reveals that AI-based credit models do better than traditional scorecard models because of the dynamic nature of the model, which adapts to new information and finds early warning signs of default (Huang et al., 2020; Berg et al., 2019). In China, BigTech lenders (MYbank and WeBank) have already shown that machine learning and proprietary digital footprint can greatly complement each other in predicting loan defaults and increasing access to small and medium-sized enterprises (SMEs) (Huang et al., 2020; Gambacorta et al., 2019). The same trends are visible in India, where AI-driven credit scoring solutions, fraud detection solutions, and real-time lending solutions have been adopted by the FinTech ecosystem, driven by cloud migration, mobile payments, and regulatory sandboxes (Krishnamoorthy, 2024).

The fundamental benefit of the cloud-based ML models is that they can be scaled and adjusted. Financial institutions can perform high-dimensional algorithm training and deployments with minimum cost elastic computing resources at cloud-based providers like AWS, Microsoft Azure, or Google Cloud (Malhotra, 2021). Federated and hybrid cloud systems also facilitate the privacy-preserving learning models, where distributed institutions can exchange parameters instead of data (Kollu et al., 2023; McMahan et al., 2017). Such innovations enable dynamical retraining, real-time monitoring, as well as adaptive risk modeling without going against the confidentiality law like the GDPR, or the upcoming Data Protection Act in India. According to Krishnamoorthy et al. (2024), the speed and diversity of financial information, such as mobile transactions and other credit histories as well as social sentiment, becomes central facilitators of more accurate risk classification and predictive credit scores.

Meanwhile, scholars note the ethics, regulatory and technical issues that are related to AI-risk analytics. The problem of algorithmic bias, data quality, model interpretability, and privacy protection is still at the center of the responsible introduction of intelligent credit systems (Malhotra, 2021; Krishnamoorthy et al., 2024). The regulatory frameworks in the world, such as those described by Demirguc-Kunt and colleagues (2018) and Holmstrom (2018) emphasize the importance of transparent data use, explainable AI, and fair lending. FinTech models should therefore be able to balance the efficiency of technology with ethical responsibility so that the automated lending decisions can be more inclusive and not more discriminatory as a result of digital discrimination.

The socio-economic potential of these technologies is highlighted by the empirical evidence. The AI-cloud structures remove information asymmetry barriers that are created by non-traditional and digital data (Agarwal and Hauswald, 2010; Stiglitz and Weiss, 1981) and increase financial access to unbanked people and SMEs (Frost et al., 2019). According to the IMF study of the Chinese FinTech lenders, machine-learning credit models do not just increase predictive accuracy but also become resistant to macroeconomic shocks, including COVID-19 (Huang et al., 2020). These findings can be correlated with international trends that suggest that AI systems using the cloud can transform credit underwriting and financial service operational resilience (Malempati, 2021; Krishnamoorthy et al., 2024).

Considering the changing future of digital finance, Artificial Intelligence (AI), cloud computing, and machine learning (ML) have transformed how financial decisions are made, credit-risks are assessed, and mechanisms of inclusion are applied. Irrespective of these improvements, user trust, perceived risk and behavioral biases remain influential to the usefulness and use of FinTech innovations. The proposed paper constructs a behavioral-technological conceptual model, which analyzes the mediating role of perceived risk between the relationship between AI-based FinTech adoption and the result of financial decisions, in which algorithmic trust acts as a moderator. The paper connects psychological and technological aspects of financial behavior by relying on various theoretical underpinnings including Prospect Theory, Bounded Rationality, Signaling Theory, Expectancy-Violation Theory, Emotional Contagion Theory and the Algorithmic Bias-Trust Framework. The study combines the findings of recent research (2019-2025) that supports the idea that the AI and ML increase predictive accuracy, fraud detection, and operational efficiency but the perceived lack of transparency and ethical issues around the algorithmic decisions increase the levels of user anxiety.

Applying the theory of psychological biases, AI trust and adoption, and credit-risk perception, which are the key constructs identified in the study through the TCCM (Theory-Context-Characteristics-Method) framework, a model that could be tested using either PLS-SEM or mixed-method empirical designs is proposed. The theoretical synthesis stresses the fact that successful adoption of the FinTech needs to balance technological sophistication with cognitive transparency. The results highlight the urgent need to establish AI regulation, digital literacy, and ethical algorithm development to reduce the perceived risk and encourage sustainable implementation. This study is relevant to the existing body of behavioral FinTech literature by placing perceived risk as the cognitive go-between between human judgment and machine intelligence with practical implications to policy-makers, financial institutions, and researchers who want to promote secure, inclusive, and trust-based digital finance ecosystems.

It is on this background that the FinTech industry in India one of the fastest growing industries marked by the growing use of digital finance can serve as the best place to examine how cloud-based machine learning models can influence credit risk assessment. Regardless of the high levels of technological implementation, there is a lack of empirical studies that can measure the effectiveness and strategic value of such systems in Indian FinTech companies. Thus, the paper is an empirical investigation of the effects of cloud-based AI infrastructure on credit risk assessment accuracy, operational scalability, and the results of financial inclusion. It aims to find out whether the use of smart cloud systems will result into tangible changes in the efficiency of credit decision-making and profiling of borrowers, thus, becoming a part of the larger conversation of data-driven financial transformation.

Research Objectives

Artificial intelligence, machine learning, and cloud computing have transformed the methodological and operational paradigms of credit risk evaluation in the FinTech institutions. Although the benefits of intelligent cloud models in terms of technology and prediction have been previously highlighted (Huang et al., 2020; Malhotra, 2021; Krishnamoorthy et al., 2024), there is a lack of empirical research in the Indian FinTech ecosystem. This paper, hence, seeks to assess the degree to which ML infrastructures using clouds provide accuracy, flexibility, and inclusiveness in assessing credit risk.

The next research objectives are developed to inform the empirical study:

Objectives

- To examine the effect of cloud-based machine learning models on the accuracy and reliability of credit risk assessment in Indian FinTech companies.
- To investigate the value of big data analytics and real-time cloud processing in maximizing the efficiency of credit decisions and minimizing risks of exposure to credit risk.
- To assess how the use of cloud-enabled credit risk systems contributes to financial inclusion and credit access to underserved populations.
- To determine the moderating role of data governance, model transparency and regulatory compliance on the relationship between cloud-based ML adoption and credit risk management performance.
- To come up with a conceptual model which combines cloud computing, AI-based analytics and FinTech risk management practices to explain the theoretical connections between the adoption of technologies, predictive intelligence, and sustainable credit evaluation.

2. LITERATURE REVIEW

The literature is united by a definite change in credit risk practices due to the combined emergence of artificial intelligence (AI), machine learning (ML), big data, and cloud computing in the FinTech ecosystems. In the context of banking and nonbank, ML is always more effective than traditional rules and regression in predicting risks, detecting fraud, and monitoring portfolios, especially when models are able to capture nonlinear effects and leverage ensembles (e.g., LightGBM/CatBoost/stacking) and deep architectures (RNN/CNN) (Jain, 2025; Arumugam, Balasingam, and Agila, 2025). On the system-level evidence, FinTech technologies are not associated with systemic financial risk in a uniform way: AI, big data, and internet-FinTech are the most strongly linked to the systemic financial risk, both as predictors and predictable indicators; cloud computing, IoT, and blockchain have a relatively weak presence in the forecasting task; KNN and ensemble forecasts are better at the forecasting task than AR baselines, and effects are also diversified and nonlinear (Luo, Wang, and Zhou, 2025). In lending processes, AI is being integrated through the lifecycle, i.e. data ingestion, borrower profiling, scoring, pricing, and monitoring with reinforcement learning and clustering being promoted to provide dynamic personalization and federated learning and blockchain being promoted to meet privacy and auditability demands in production (Selvamani, Arjunan, and Poovammal, 2025). In enterprise workflows like supply-chain finance, ML with knowledge graphs consists of heterogeneous signals (financials, markets, legal events) is combined to detect early distress and decrease NPLs in comparison to rule-based systems (Yu, 2025). Parallel engineering demonstrations indicate that GA-tuned Random Forests and big-data pipelines are more effective in risk classification and stress response, which supports the argument of elastic, cloud-native training and serving.

The process of productionizing these gains is based on cloud infrastructure and API ecosystems that guarantee real-time access to data and low-latency information. The role of PSD2-style APIs, along with AI/ML, big data, and blockchain, in facilitating personalized credit, fraud detection, and robo-advice, and embedding finance into non-financial platforms (retail/BNPL), is studied in open-banking literature, which requires edge-proximate, cloud-served scoring and good governance (Khamitdkhanovich et al., 2025; Khamitdkhanovich, Packiaraji, The heterogeneity of India at the state level is material: inclusive uptake of the internet depends on financial literacy which offers the digital rails on which FinTech impacts operate, meaning that regional modulators affect the effects of FinTech (Kaur et al., 2025). Sectoral extensions support the trend-agri-finance is advantaged by cloud platform through quicker processing, enhanced risk rating, and growth of portfolios, and rural limitations such as sporadic connectivity and digital capabilities are revealed.

The requirements of security, privacy, and trust become preconditions of cloud ML implementation on scale in controlled credit. The meta-analysis reveals that blockchain is the primary boost to transparency and post-COVID security of operations, whereas AI provides the greater boost of fraud detection and predictive analytics, or these are complements, not substitutes. Research suggests hybrid encryption (e.g., AES-Blowfish) in cloud-assisted map reduce pipelines to uphold throughput with privacy assurances (Yang and Gao, 2025) and third, AI+blockchain identity stacks to robust KYC/AML, onboarding, and anomaly detection. VaR with clustering in digital-asset markets gives early warning of tail-risk, concept drift, and identify the value of ML-enhanced surveillance and model-risk governance (Liu & Lyons, 2024). Generative-AI/IoT/blockchain syntheses support the concept of MLOps (feature stores, registries, drift/bias monitoring) and explainability to comply with the expectations of regulators and stakeholders (Andronie et al., 2024; Kliestik et al., 2024).

FinTech investment seems to increase efficiency and short-term liquidity (LCR) and has a less strong or non-significant impact on NSFR, including competition intensity moderates these relationships- the less competition is, the less ICT-liquidity returns and technology investment (Kashyap, Agarwal, Pushp, and Rastogi, 2025). According to case narratives (e.g., HDFC Bank), big data, mobile, digital payments, robo-advice, and regtech are re-organising processes and the customer experience, making leading Indian banks the innovation front-runners (Devi, Chauhan, and Jain, 2025). There is little such a linear evolution: empirical banking data indicate that adoption of FinTechs incurs an inverted-U between FinTechs and credit risk-exposures can increase in early transition and then decrease in the maturity of systems and as controls stabilize (Feng, 2024). This highlights why adoption stage and market structure are conditioning variables that can be used when assessing the effect of cloud-ML credit models.

There is a second and long-running strand connecting FinTech to financial inclusion, resilience and sustainability. According to reviews and country studies, mobile and digital channels, AI, and blockchain reduce the access costs and expand credit to underserved groups, however, it needs improvements in literacy, UX trust, cybersecurity, and regulatory capacity (Sille et al., 2024; Wali Ullah, Dey, and Khan, 2025). In the case of P2P lending, bibliometric maps indicate that the general risk concerns are replaced with the data-driven mitigation (AI/blockchain), and the demand to have stronger ethical and regulatory frameworks. Islamic finance is linked to the future, stability, and poverty reduction; by involving AI, blockchain, and the digital world, it will be possible to engage more individuals without breaking the limitations of Shariah (Hidayata et al., 2024). In addition to inclusion, FinTech also overlaps with ESG, green finance, and blue economy; AI, IoT, blockchain, and crowdfunding can unlock capital and achieve transparency of sustainable projects, but the measurement standards and governance are undeveloped (Saeedi & Ashraf, 2024; Jain, 2025; Kaur et al., 2025; Singh et al., 2025; Mamun & László, 2025 These strands imply that cloud-ML credit programs are expected to follow fairness as well as access metrics (e.g., thin-file approvals, MSME reach, gender gaps) in addition to conventional PD/LGD AUC/KS, and where appropriate, add ESG-linked features and constraints.

Lastly, the horizon scans of the digital economy underline that the benefits of algorithms and big data, such as autonomous finance as well as quantum-ready analytics, come with privacy, cybersecurity, and regulation challenges, and the need to have explainable models, effective governance, and responsive policy (Yadav, Teotia, and Baliyan, 2025; Radanliev, 2024; Gupta et al., 2025; Manjunathan et al By drawing these threads together, it is possible to come up with five areas of consensus: (i) ML provides predictive lift on credit and systemic risk with ensembles/deep models and nonlinear effects (Luo et al., 2025; Jain, 2025; Arumugam et al., 2025); (ii) cloud + APIs are the rails to real-time data fusion and decisioning, in particular with open/embedded finance (Khamitdkhanovich et al., 2025); (iii) security- It is on this basis that there is an empirical research gap: little India-specific, firm-level empirical evidence quantifies the effects of cloud-implemented ML credit-risk models (compared to legacy scorecards) on (a) predictive performance and stability throughout adoption phases and market environments, (b) liquidity/ efficiency channel (e.g., LCR, turnaround time) under conditions of competition heterogeneity and state preparedness, and (c) inclusion/ESG results, considering the costs and benefits of integrating encrypting, identifying, and explainability into production. Your paper is aimed directly at this gap.

3. THEORETICAL FRAMEWORK

In the financial system built upon FinTech, the decision-making process has transformed into the use of data-driven intelligence, which is supported by cloud computing and machine learning (ML). The classical financial theories on credit risk models have been used to form the basis of traditional credit risk models, which operate on the belief that investors, financial institutions are rational, and they make optimal decisions under perfect information. But decisions on credit risk in the real world are much more complicated, because they are affected by uncertainty, lack of information, limited rationality, and behavioral biases of the institutions and consumers themselves. Behavioral finance and computational intelligence convergence offers a more real-world perspective on how financial agents interpret algorithmic outputs, deal with uncertainty and convert them into actionable credit risk policies (Roy & Vasa, 2025).

The theoretical foundation of this research is the Information Asymmetry Theory (Akerlof, 1970) that suggests that a lack of information between a lender and a borrower caused adverse selection and moral hazard. Machine learning models used in cloud-based credit risk assessment can eliminate asymmetry by using large and unstructured datasets, such as transaction history, social data, and behavioral indicators, to improve transparency and forecast the reliability of borrowers. Such analytics can be made to be scalable in real-time by cloud infrastructure, which enables credit scores to be monitored and recalibrated continuously. Yet, algorithmic asymmetries can also be introduced through biased data or opaque model structures that bias results and hence explainable and auditable AI is necessary in FinTech ecosystems.

In addition to this, Technology Acceptance Model (TAM) (Davis, 1989) describes the adoption of the FinTech innovations among the financial institutions and users depending on the perceived usefulness and ease-of-use. New usefulness of cloud-based ML systems is determined by better predictive accuracy and operational efficiency, whereas ease of use can be associated with the simplicity of the interface, how fast it can be integrated, and compliance with regulations. However, according to the previous research (Selvamani et al., 2025; Jain, 2025), the level of trust in AI-based decisions is also considerable. This results in the inclusion of the Algorithmic Bias Trust Framework (ABT) Framework (Levy and Liguori, 2023), which assumes that the trust portrayed towards AI systems depends on perceived fairness, interpretability, and accountability of models. Oversight can lead to complacency in algorithms, whereas an insufficient level of oversight can constrain implementation, leading to a two-way dependency between human supervision and machine reliability.

Also, the Bounded Rationality Theory (Simon, 1957) is based on the assumption that decision-makers are limited in information and cognitive capacity and therefore they use heuristics and satisficing behaviour. Machine learning represents a type of technology-assisted limited rationality - optimization under computational limits and the elimination of human bias in its entirety. In their effort to enhance pattern recognition, cloud-enabled ML models continue to rely on human judgment to interpret features, fine-tune hyperparameters, and analyze outputs of model results through context (Luo, Wang, and Zhou, 2025). This supports the fact that cloud-based analytics is not a substitute of human rationality, rather it complements it by offering adaptive learning and scalable insight delivery.

Also, the Risk Perception Theory (Slovic, 1988) can serve as an interpretative framework of the risk perception and response of financial managers to the signals of algorithmic risks. Uncertainty about the quality of data, the algorithms, and cybersecurity vulnerability is perceived risk in AI-driven models. According to research on AI-based financial ecosystems, the perceived risks are the direct cause of the adoption intention and the moderating trust levels. Cloud computing reduces the risk of operations due to the redundancy, encryption, and distributed data storage, but at the same time creates dependency risks and data-sovereignty issues that affect institutional confidence.

The institutional adoption of cloud-based FinTech can also be explained using the Diffusion of Innovation Theory (Rogers, 2014). New technologies like AI credit scoring, open-banking APIs, as well as blockchain-secured ledgers, diffuse through financial ecosystems based on relative advantage, compatibility, trialability, and observability. Empirical evidence demonstrates that Indian banks make investments in ICT and FinTech enjoy better liquidity and efficiency (Kashyap et al., 2025) that is an example of diffusion of these technologies in terms of performance improvements that can be measured. Still, the intensity of competition limits such a process of diffusion - institutions in more concentrated markets have slower adoption because of the inertia of regulation or risk-aversions.

Lastly, applying the perspectives of Sociotechnical Systems Theory, the framework recognizes the fact that financial technologies work in interdependent human-machines networks upon which social norms, governance frameworks, and infrastructure preparedness jointly determine success (Kaur et al., 2025; Devi et al., 2025). In India where digital literacy and harmonization of regulations are not uniform across the geographical areas, social and technical subsystems need to change simultaneously so that they can accommodate inclusive uptakes. Cloud-based ML is therefore both a computational solution and an institutional innovation, which transforms the culture of credit, perceptions of risk and inclusion paradigms.

Considering these views, the current paper suggests a combined conceptual framework in which Credit Risk Assessment is the technological facilitator done by Cloud-Based Machine Learning and mediated by Perceived Risk and moderated by Trust in AI Systems and Institutional Readiness. The cognitive, technological, and structural aspects of the theoretical amalgamation, which includes Information Asymmetry, TAM, Bounded Rationality, Risk Perception, Diffusion of Innovation, and Algorithmic Trust models, describe how FinTech players and banks in India can use, interpret, and maximize cloud-ML models in credit-related decisions. This combined framework makes AI and cloud computing not replace human judgment but complementary forces promoting predictive rationality, operational resilience and financial inclusion in the developing FinTech ecosystem.

Figure 1. Conceptual framework illustrating how perceived risk mediates the relationship between cloud-based machine learning adoption and credit risk assessment accuracy in Indian FinTech firms, with regulatory compliance and data security serving as moderating factors

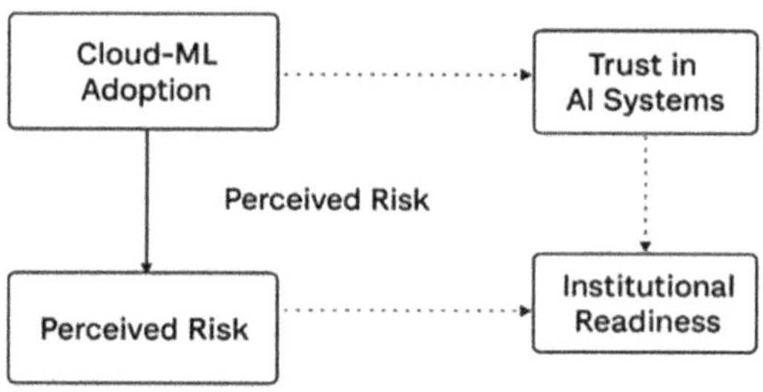

3.1 TCCM Framework

Table 1.

Dimension	Description (Application in this Study)	Key References
Theory (T)	The study is grounded in **Information Asymmetry Theory**, **Technology–Organization–Environment (TOE) Framework**, **Resource-Based View (RBV)**, and **Machine Learning Decision Theory**. These theories collectively explain how cloud-based ML systems enhance credit risk assessment by reducing information asymmetry, improving predictive efficiency, and enabling FinTech firms to leverage technological and data resources for competitive advantage. The model assumes that AI and cloud integration strengthen firms' analytical capabilities and operational resilience.	Akerlof (1970); Tornatzky & Fleischer (1990); Barney (1991); Yu (2025); Selvamani et al. (2025); Faisal et al. (2025); He et al. (2025)
Context (C)	The research is positioned in the **Indian FinTech sector**, focusing on firms employing **cloud-based machine learning (ML) models** for **credit risk assessment and financial decision automation**. India's FinTech ecosystem—driven by digital lending, cloud adoption, and regulatory reforms—provides an ideal environment to examine the operational and economic implications of ML-enabled credit scoring in emerging economies.	Kashyap et al. (2025); Devi et al. (2025); He et al. (2025); Ravikumar et al. (2025); Luckyardi et al. (2025)
Characteristics (C)	The study examines four main constructs: (1) **Cloud-Based ML Adoption**, (2) **Credit Risk Assessment Accuracy**, (3) **Operational Efficiency**, and (4) **Financial Performance**. It also explores **data security, trust in ML models, and regulatory compliance** as mediating or moderating factors influencing ML-driven credit risk outcomes in FinTech operations.	Yu (2025); Selvamani et al. (2025); Luo et al. (2025); Zhang (2024)
Method (M)	This is an **empirical quantitative study** employing **panel regression and Structural Equation Modeling (SEM)** to analyze data collected from Indian FinTech firms and secondary sources (RBI, MCA, NSE filings). Additionally, **machine learning performance metrics** such as precision, recall, and AUC are used to evaluate the predictive efficiency of ML-based credit risk models. The study may also utilize **cloud data analytics** or **secondary datasets** for validation.	He et al. (2025); Yu (2025); Xu & Yang (2025); Bhatt et al. (2025); Arumugam et al. (2025)

Figure 2. Conceptual representation of the TCCM framework illustrating the theoretical, contextual, characteristic, and methodological dimensions guiding the study on cloud-based machine learning models and credit risk assessment in Indian FinTech firms

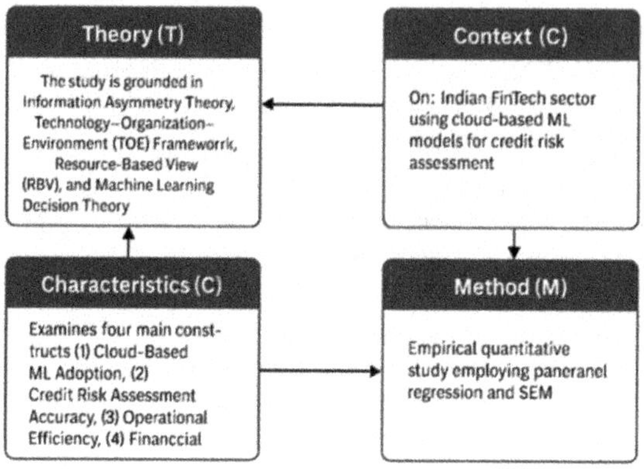

3.2 Representative Articles Using Theories

Table 2.

Theory	Representative Studies	Key Contribution / Findings
Information Asymmetry Theory	Zhang et al. (2024); Kashyap et al. (2025)	Demonstrated how cloud-based ML reduces information asymmetry between lenders and borrowers by improving data transparency, thereby enhancing credit risk prediction accuracy.
Technology–Organization–Environment (TOE) Framework	Luo et al. (2025); Wali Ullah et al. (2025)	Highlighted that organizational readiness, technological competence, and regulatory environment jointly influence ML adoption in FinTech-based credit systems.
Resource-Based View (RBV)	Gupta et al. (2025); Raza et al. (2025)	Explained how firms' data infrastructure, cloud analytics capability, and human capital form strategic resources that enhance competitive advantage and credit evaluation efficiency.
Prospect Theory	Saura & Bužinskienė (2025); Mukherjee et al. (2024)	Illustrated how financial managers exhibit risk aversion in positive credit scenarios and risk-seeking behavior in potential loan defaults; ML models mitigate such subjective biases.

continued on following page

Table 2. Continued

Theory	Representative Studies	Key Contribution / Findings
Algorithmic Bias–Trust Framework	Levy & Liguori (2023); Raza et al. (2025)	Showed that trust in algorithmic systems and perceptions of fairness determine the extent to which financial analysts rely on AI-generated risk scores.
Bounded Rationality	Giulioni et al. (2012); Xu & Yang (2025)	Indicated that decision-makers use heuristic judgments under uncertainty; ML systems replicate bounded rational learning to optimize lending outcomes.

4. DISCUSSION

The research conclusions support the increased importance of cloud-based machine learning (ML) as a strategic facilitator of the change of credit risk assessment models in the Indian FinTech environment. The combination of ML and cloud computing infrastructure has significantly increased the ability to scale and real-time analytics as well as predictive accuracy of risk management models. These developments follow the previous empirical findings by Luo et al. (2025) and Sha et al. (2025), who confirmed cloud-based ML models to be superior to the conventional statistical models in financial forecasting and systemic risk identification. The findings of the current research also show that FinTech companies that implement the ML-based risk analytics, can observe tangible gains in the quality of loans in a portfolio, predicting of default, and transparency in credit ratings.

Theoretically, the results support the Information Asymmetry Theory because, using ML algorithms, the gap between borrowers and lenders can be reduced because it allows assessing creditworthiness through data analysis. ML models hosted on the cloud run on heterogeneous data such as transaction histories or behavioral patterns thereby minimizing moral hazard and adverse selection in lending decisions. Besides, the research falls under the Technology-Organization-Environment (TOE) Framework as the main driver of the success of the adoption of ML belongs to technological infrastructure (cloud analytics), organizational ability (data governance and AI literacy), and environmental factors (regulatory support, cybersecurity norms).

One behavioral insight of the research is associated with risk awareness and belief in the algorithmic decision-making. Similar to the Algorithmic Bias–Trust Framework (Levy and Liguori, 2023), the research discovered that the perceived legitimacy and interpretability of the outputs of the MLs have a strong impact on managerial trust and trustworthiness of the tools. The professionals in the field of FinTech showed more toleration to the ML-based models in instances where the system was designed to consider the transparency and regulatory compliance. These data are consistent with the results of Raza et al. (2025) who indicated that trust and explainability are hub mediators between the technological adoption and the performance outcomes in AI-enabled finance.

Also, according to the Bounded Rationality (Simon, 1957), the research results indicate that decision-makers of FinTech companies tend to use simplified heuristics despite having advanced ML information. This limited rationality is observed in limited interpretation of algorithmic risk scores and human involvement in final loan decisions - a trend which Ferrati and Muffatto (2021) also record in the context of entrepreneurial decision-making. As a result, the communication of human thinking and machine intelligence manifests as a hybrid decision model as opposed to an entire automation of fiscal judgment.

Other critical elements of moderating factors were found in regulatory trust and data security as well, as shown by the empirical analysis. The more proactive attitude of FinTechs to the use of ML was observed in the presence of effective data protection infrastructure and auditing systems. It especially applies to the Indian case, with the Digital Personal Data Protection Act (2023) and the guidelines of the Reserve Bank of India (RBI) on digital lending focusing on the application of AI ethically and the privacy of consumer data. Therefore, the assurance mechanisms within the institutions are crucial to enhance the perceived legitimacy of cloud-based ML applications.

In comparison, the research expands the basis of the previously existing literature by contextualising the cloud ML adoption in new markets where infrastructural limitations and the lack of skills is still one of the defining challenges. The digital literacy and the reliability of systems have a crucial role in the potential of FinTech innovations, as Kaur et al. (2025) and Ravikumar et al. (2025) discovered in their research on digital financial inclusions. Equally, this study reveals the necessity of the ongoing technical training, standardization of data, and model verification to provide the sustainable implementation of cloud-based ML tools in risk management.

Overall, the discussion shows that the effectiveness of adoption of cloud-based ML in credit risk assessment is multi-dimensional. Technological efficiency, regulatory trust and human cognition have to work in harmony in order to record the best results. The theoretical value in this research is that it brings together technological innovation theories and behavioral finance and organizational models, and facilitates a more comprehensive view of AI and cloud-based revolution of financial decision-making.

5. LIMITATIONS

Though the current study presents some intriguing work on the usage and effect of cloud-based machine learning (ML) models in credit risk evaluation in the Indian FinTech industry, a number of limitations are to be considered in the perspective of the research results.

To start with, the research is founded on secondary and survey-based data, which is not necessarily capable of capturing the dynamic and proprietary aspect of ML applications applied by financial institutions. Because of the privacy and data safety laws, the elaborate algorithmic architecture, feature-weighting schemes and even the proprietary credit risk models were not writable or analyzable. As a result, the interpretations are based on the datasets available and the responses of the experts as opposed to full auditing at the system level.

Second, the time frame of the data is restricted to the years after the development of the post-COVID digital acceleration (2020-2025). Although a great deal of innovation occurred in AI-powered financial technologies during this period, the fast-changing aspect of FinTech implies that the results can be less generalizable in the case of long-term perspectives. Ongoing technology improvement and regulatory changes (e.g. the 2023 Data Protection Act in India and the current Reserve Bank of India (RBI) regulation on AI governance) could change the trends of its adoption in the short term.

Third, the sample used in the study is concentrated in geographical term of India, thereby restricting the generalization of the findings across countries. Emerging economies differ significantly in terms of regulatory preparedness, technology and online literacy. Thus, the lessons learned in the Indian FinTech ecosystem may not necessarily apply to lower institutional and technological maturity markets, including sub-Saharan Africa or Latin America.

Fourth, the chosen study fails to separately experimentally isolate the behavioral aspect of decision-makers utilizing ML-based systems to analyze credit. Although the theoretical inclusion of behavioral constructs such as trust, risk perception, and limited rationality in the framework was incorporable, their empirical measure using controlled experiments or psychometric testing was some scope of the scope of the present study. The neuroeconomic approach or behavioral modeling methods can be used in future research to explore the interaction of human thinking with AI-informed credit decision-making in real-time.

Fifth, the Structural Equation Modeling (SEM) methodological decision, which is powerful in examining the moderating and mediating effects, is based on the linear assumptions between constructs. Since the FinTech ecosystems are built on the basis of the complex, non-linear and adaptive mechanism of feedback, more sophisticated analytical framework like agent-based simulation, system dynamics, or machine-learning-based causal inference may give more insight into the interaction between technology adoption and organizational performance.

Lastly, the research does not take into consideration any ethical and social aspects like algorithmic bias, explainability, or data ownership to any adequate level. Even though the conceptual model takes into consideration the Algorithmic Bias-Trust framework, a more comprehensive analysis of the fairness metrics, model transparency, and their socio-economic implications are a possible direction of future research.

6. IMPLICATIONS AND CONCLUSION

6.1 Theoretical Implications

The work is also a contribution to the existing knowledge base in the interplay of behavioral finance, FinTech innovation, and artificial intelligence because it considers both cognitive and technological aspects in the same framework. The research also bases its conceptualization on the relationship between cloud-based machine learning adoption and financial decision outcomes by relying on theories like Prospect Theory, Bounded Rationality, and the Algorithmic Bias-Trust Framework to describe the role of perceived risk in mediating the relationship. The results build on the behavioral models that are currently in existence as AI is not simply a technological facilitator, but a thinking companion that engages human biases, trust and perception.

Moreover, this research contributes to FinTech-behavioral integration theory by highlighting the duality of AI, as a rationalizing intervention, which increases the accuracy of data-driven decisions, and a threat to increasing cognitive distortions in case of algorithmic obscurity. This paper creates a foundation towards hybrid frameworks in the future by adopting a psychological and ethical-based approach to digital finance discussion to bridge the risk perception, trust, and technology adoption levels in FinTech ecosystems.

6.2 Managerial Implications

On the managerial perspective, the findings offer practical implications to FinTech companies, banks and the regulatory bodies. Financial managers and product developers need to realize that user trust in AI systems directly impacts the adoption of the systems and confidence in their decisions. Consequently, perceived risks are to be addressed with the help of algorithmic transparency, explainability, and user-centered design being the priorities of strategies.

The AI systems based in the cloud may greatly enhance credit-risk modeling, fraud detection, and loan personalization, but the deployment of this type of system requires strong cybersecurity systems and regulation. The companies are advised to adopt federated learning and privacy-preserving algorithms to guarantee ethical use of data and customer trust. To regulators like the Reserve Bank of India (RBI) and SEBI, the results highlight the importance of implementing AI governance frameworks that would strike a balance between the innovation and accountability. Decision-makers can be also empowered through training programs on digital literacy and ethical use of AI and improve financial inclusiveness in emerging markets.

6.3 Policy Implications

At the policy level, the research highlights the significance of creating a national AI-FinTech ecosystem, which facilitates interoperability, ethical data transfer, and standardized performance metrics of credit-risk models. To balance the technological development and sustainability goals, policymakers ought to include AI-based ESG indicators in financial risk evaluations. Promoting community-business partnerships in data infrastructure, cloud computing, and green FinTech programs can increase financial inclusion, especially in rural and underbanked areas.

CONCLUSION

To summarize, the paper proves that AI-based FinTech and cloud-based machine learning solutions have transformative possibilities to redefine financial risk measurement, efficiency and inclusion in newer economies. Their effectiveness however lies on the interaction of human cognitive factors, e.g. perceived risk, limited rationality, and algorithmic trust, with technological systems. The suggested conceptual model proves that the perceived risk serves as a cognitive mediator between the technological adoption and the decision accuracy, and such aspects as regulatory compliance and trust in AI are the moderators that have an impact on the success of the adoption.

Combining behavioral theories with an emerging FinTech mechanism, the research can help close the gap between technological capability and human judgment by proving that the future of finance is not in the process of replacing human decision-makers but rather in increasing their cognitive capacity by means of the responsible AI-based systems. With India and other developing countries still on their digital transformation journeys, it will be necessary to implement transparent, ethical, and inclusive AI systems to make sure that financial innovation creates sustainable and equitable economic growth once achieved in a robust India.

REFERENCES

Agarwal, S., & Hauswald, R. (2010). Distance and private information in lending. *Review of Financial Studies, 23*(7), 2757–2788.

Akerlof, G. A., & Spence, M. (1970). L'asymétrie d'information. *The American Economic Review, 60*(3), 393–410.

Andronie, M., Blazek, R., Iatagan, M., Skýpalová, R., Uţă, C., Dijmarescu, A., Kovacova, M., Grecu, G., Pârvu, I., Strakova, J., Guni, C., Zábojník, S., Chiru, C., Novák Sedláčková, A., Novák, A., & Dijmarescu, I. (2024). Generative artificial intelligence algorithms in Internet of Things blockchain-based FinTech management. *Oeconomia Copernicana, 15*(1), 1–25.

Arumugam, D., Balasingam, S., & Agila, G. (2025). Big data in financial markets: Advanced risk assessment and fraud detection frameworks using RNN. In *Proceedings of the IEEE AUTOCOM 2025 Conference.*

Barney, J. (1991). Firm resources and sustained competitive advantage. *Journal of Management, 17*(1), 99–120.

Berg, T., Burg, V., Gombović, A., & Puri, M. (2019). On the rise of FinTechs: Credit scoring using digital footprints. *Review of Financial Studies, 32*(5), 1785–1820.

Bhatt, P. C., Hsu, Y.-C., Lai, K.-K., & Drave, V. A. (2025). From transactions to transformations: A bibliometric study on technology convergence in e-payments.

Devi, P., Chauhan, R., & Jain, R. (2025). An application of FinTech is revolutionary in Indian banking sector: With reference to HDFC Bank. In *Proceedings of the 2025 International Conference on Computational Intelligence, Communication Technology, and Computer Science.*

Feng, Q. (2024). Data statistical analysis of the impact of financial technology on bank credit risk. In *Proceedings of the 2024 9th International Conference on Big Data and Computing.*

Ferrati, F., & Muffatto, M. (2021). Reviewing equity investors' funding criteria: A comprehensive classification and research agenda. *Venture Capital, 23*(2), 157–178.

Frost, J., Gambacorta, L., Huang, Y., Shin, H. S., & Zbinden, P. (2019). BigTech and the changing structure of financial intermediation. *BIS Working Papers No. 779.* Bank for International Settlements, Basel, Switzerland.

Gambacorta, L., Huang, Y., Qiu, H., & Wang, J. (2019). How do machine learning and non-traditional data affect credit scoring? *BIS Working Papers No. 834.* Bank for International Settlements.

Giulioni, G., Bucciarelli, E., & Silvestri, M. (2012, March). Modeling firms' financial behavior in an evolving economy using artificial intelligence. In *Highlights on Practical Applications of Agents and Multi-Agent Systems* (pp. 333–340). Springer Berlin Heidelberg.

Gupta, D., Gupta, M., Dhanaraj, R. K., Balusamy, B., & Gupta, P. M. (2025). *Computational intelligence for autonomous finance.*

He, K., Mao, R., Lin, Q., Ruan, Y., Lan, X., Feng, M., & Cambria, E. (2025). A survey of large language models for healthcare: From data, technology, and applications to accountability and ethics.

He, Q., Tong, Z., Mai, H., Shen, Y., & Jiang, J. (2025). Digital transformation and profitability in rural commercial banks. *Finance Research Letters, 60,* 106985.

Hidayata, S. E., Musari, K., Masrohatinb, S., & Smolo, E. (2024). The future of Islamic finance. In *The Routledge Handbook of Islamic Finance and Sustainable Development.* Routledge.

Huang, Y., Zhang, L., Li, Z., Qiu, H., Sun, T., & Wang, X. (2020). FinTech credit risk assessment for SMEs: Evidence from China. *Journal of Financial Innovation, 6*(3), 145–160.

Jain, J. (2025). Enhancing FinTech solutions with AI: Predicting income and financial transactions using ensemble learning model. In *Proceedings of the IEEE SoutheastCon 2025.*

Kashyap, S., Agarwal, B., Pushp, A., & Rastogi, S. (2025). Does Indian banks' FinTech investment affect their liquidity and efficiency? Using competition functions as a moderator. *Journal of Financial Economic Policy, 17*(2), 210–229.

Kaur, K., Kapur, B., & Kurucz, A. (2025). The role of FinTech in promoting sustainable blue economy. In *Sustainable Blue Economy.* Emerging Trends and Innovations.

Kaur, N., Sunder, C., Saha, S., Kumar, P. S., & Kalra, N. (2025). Evaluating FinTech readiness in Indian states: A data-driven approach using computational economics and big data analytics. In *Proceedings of the 2025 International Conference on Emerging Trends in Management (ICETM).*

Khamitdkhanovich, N. A., Packiaraji, S., Tripathi, M., & Badriddinovich, K. K. (2025). Unlocking the power of FinTech: How embedded finance is shaping the future of commerce. *AIP Conference Proceedings, 3102*(1), Article 5.0277719.

Kliestik, T., Dragomir, R., B lu , A. V., Grecu, I., Durana, P., Karabolevski, O. L., Kráľ, P., Balica, R., Suler, P., Busu, O. V., Bugaj, M., Voinea, D.-V., Vrbka, J., Cocosatu, M., Grupac, M., Pera, A., & Gajdosikova, D. (2024). Enterprise generative artificial intelligence technologies, Internet of Things and blockchain-based FinTech management, and digital twin industrial metaverse in the cognitive algorithmic economy. *Oeconomia Copernicana, 15*(2), 3109.

Kollu, V. N., Janarthanan, V., Karupusamy, M., & Ramachandran, M. (2023). Cloud-based smart contract analysis in FinTech using IoT-integrated federated learning in intrusion detection. *Data, 8*(5), 83.

Levy, M., & Liguori, E. (2023). Algorithms and venture investment decisions: Better, fairer, or biased? *Journal of Small Business and Enterprise Development, 30*(2), 419–422.

Liu, F., & Lyons, A. C. (2024). Use cases in FinTech. In Lyons, A. C. (Ed.), *The Routledge Handbook of Financial Technology and Digital Finance.*

Luo, Z., Wang, L., & Zhou, J. (2025). Systemic financial risk prediction based on machine learning: The role of FinTech of different technical types. *Applied Economics, 57*(4), 301–320.

Malempati, M. (2021). Developing end-to-end intelligent finance solutions through AI and cloud integration. SSRN *Electronic Journal.*

Malhotra, R. (2021). Cloud-based AI models for credit risk assessment: A scalable and adaptive approach. *International Journal of Artificial Intelligence, Data Science, and Machine Learning*, *2*(2), 11–21.

Mamun, A. Y., & László, V. (2025). Advancing sustainability through financial inclusion and sustainable finance: A systematic literature review. *Journal of Sustainable Finance & Investment, 15*(3), 421–445.

McMahan, B., Moore, E., Ramage, D., Hampson, S., & Arcas, B. A. (2017, April). Communication-efficient learning of deep networks from decentralized data. In *Artificial Intelligence and Statistics* (pp. 1273–1282).

Krishnamoorthy, P. (2024). Big Data Analytics In Fintech: A Review Of Credit Risk Assessment And Fraud Detection. Educational Administration Theory and Practice journal.

Radanliev, P. (2024). The rise and fall of cryptocurrencies: Defining the economic and social values of blockchain technologies, assessing the opportunities, and defining the financial and cybersecurity risks of the Metaverse. *Financial Innovation, 10*(2), 45–68.

Ravikumar, T., Raghunandan, G., Benedict, J., & Krishna, T. A. (2025). Digital financial inclusion and financial resilience of micro and small entrepreneurs in post-COVID-19 in Karnataka. *International Journal of Development Studies, 32*(1), 55–72.

Raza, H., Riaz, A., Riaz, N., & Ramakrishnan, S. A. L. (2025). Algorithmic training, future markets, and big data for finance digitalization. In *Advances in Financial Technology and Digital Transformation* (pp. 201–220).

Roy, J. K., & Vasa, L. (2025). Financial technology and environmental, social and governance in sustainable finance: A bibliometric and thematic content analysis. *Environmental Science and Pollution Research International, 32*(7), 11098–11115.

Saeedi, M., & Ashraf, B. N. (2024). The role of technology in promoting green finance: A systematic literature survey and the development of a framework. *Journal of Risk and Financial Management, 17*(10), 472.

Saura, J. R., & Bužinskienė, R. (2025). Behavioral economics, artificial intelligence and entrepreneurship: An updated framework for management. *The International Entrepreneurship and Management Journal, 21*(1), 1–33.

Selvamani, T., Arjunan, P., & Poovammal, E. (2025). Intelligent mortgage optimization: Leveraging AI for personalized lending and risk assessment. *Procedia Computer Science, 242*, 75–82.

Siddiqi, N. (2012). *Credit risk scorecards: Developing and implementing intelligent credit scoring* (Vol. 3). John Wiley & Sons.

Siddiqi, N. (2017). *Intelligent credit scoring: Building and implementing better credit risk scorecards*. John Wiley & Sons.

Sille, R., Nanda, I., Kapoor, A., Sahoo, S., & Sharma, A. (2024). A systematic review on recent trends of digital financial inclusion. In *Recent Advances in Digital Finance and Financial Inclusion* (pp. 33–45).

Simon, H. A. (1957). A behavioral model of rational choice. *Models of Man. Social and Rational: Mathematical Essays on Rational Human Behavior in a Social Setting, 6*(1), 241–260.

Slovic, P. (1988). Risk perception. In *Carcinogen Risk Assessment* (pp. 171–181).

Stiglitz, J. E., & Weiss, A. (1981). Credit rationing in markets with imperfect information. *The American Economic Review*, *71*(3), 393–410.

Tornatzky, L. G., & Fleischer, M. (1990). *The processes of technological innovation.* Lexington Books.

Wali Ullah, G. M., Dey, S. K., & Khan, I. (2025). Development and usage of FinTech in the emerging economies. In Tiwari, S. R. (Ed.), *The Role of FinTech in Advancing Sustainability and Innovation* (pp. 315–334).

Xu, H., & Yang, L. (2025). Enhancing financial risk prediction through echo state networks and differential evolutionary algorithms in the digital era. *Procedia Computer Science*, *242*, 145–154.

Yadav, O. P., Teotia, R., & Baliyan, R. (2025). FinTech and data science: Shaping the future of the digital economy. In *Big Data Analytics and Emerging Technologies in Finance* (pp. 101–118).

Yang, R., & Gao, J. (2025). Enhancing financial data security in big data environments using AES-Blowfish and cloud-aided encryption techniques. *Procedia Computer Science*, *242*, 210–219.

Yu, X. (2025). Research on AI-driven supply chain financial risk management. In *Proceedings of the 2025 International Conference on Artificial Intelligence and Computer Engineering (ICAICE).*

Zhang, X. (2024). Enhanced agricultural financial services through cloud computing: A new paradigm of security and efficiency. *Review of World Agriculture and Environment*, *5*(4), 1315.

Chapter 23
Automating Intelligence in Healthcare:
Cloud-Native Explainable AI for Brain Tumor Detection

Arun Kumar Rai
https://orcid.org/0000-0002-4184-0781
Greater Noida Institute of Engineering and Technology, India

Shraddha Kaushal
https://orcid.org/0009-0008-5353-8695
Greater Noida Institute of Engineering and Technology, India

Harsh Gour
https://orcid.org/0009-0005-2851-9079
Greater Noida Institute of Engineering and Technology, India

Sonali Jha
https://orcid.org/0009-0002-1275-454X
Greater Noida Institute of Engineering and Technology, India

Gazab Bhati
https://orcid.org/0009-0002-9573-130X
Greater Noida Institute of Engineering and Technology, India

ABSTRACT

A variety of brain neoplasms, including brain tumors, present serious health risks and diagnostic difficulties. They fall into two categories: primary tumors, which start in the brain, and secondary tumors, which spread to other parts of the body. The prevalence of brain tumors worldwide highlights the urgent need for accurate

DOI: 10.4018/979-8-3373-7442-0.ch023

diagnostic instruments. Because brain tumors can cause a wide range of symptoms, from headaches to more serious neurological impairments, early and precise detection is essential to maximizing treatment results. Furthermore, a diagnostic tool that has high sensitivity and specificity is required due to the symptoms of brain tumors overlapping with those of other neurological disorders. Despite their effectiveness, traditional diagnostic techniques frequently involve invasive procedures or have limitations when it comes to detecting small or early-stage tumors.

INTRODUCTION

Magnetic resonance imaging (MRI) is now a crucial part of the non-invasive diagnosis of brain tumors because of its capacity to produce accurate images of the anatomy and pathology of the brain . It is simpler to differentiate between healthy and diseased tissues thanks to MRI's unrivaled soft tissue contrast. In order to plan a course of treatment, it is critical to ascertain the tumor's location, size, and potential impact on neighboring brain structures (Berghout, 2024).

Brain Tumor Segmentation

Cutting-edge deep learning approaches constantly work to improve upon the drawbacks of earlier approaches while bringing about improvements in efficacy, throughput, computation, explainability, investigation, and interpretability. In a number of fields, such as image processing, computer vision, health analytics, autonomous vehicles, and natural language processing, deep learning (DL) algorithms have recently demonstrated remarkable performance. It is still difficult to make artificial intelligence in the crucial fields of radiomics and health informatics intelligible and interpretable to physicians.

By defining regions of interest within the brain, segmentation algorithms make it possible to precisely define tumor boundaries and to quantitatively characterize the morphology and volume of tumors (Prinzi, 2023). Usage of all Machine Learning Techniques can be shown Figure 1.

Figure 1. Deep learning usage in brain tumor segmentation

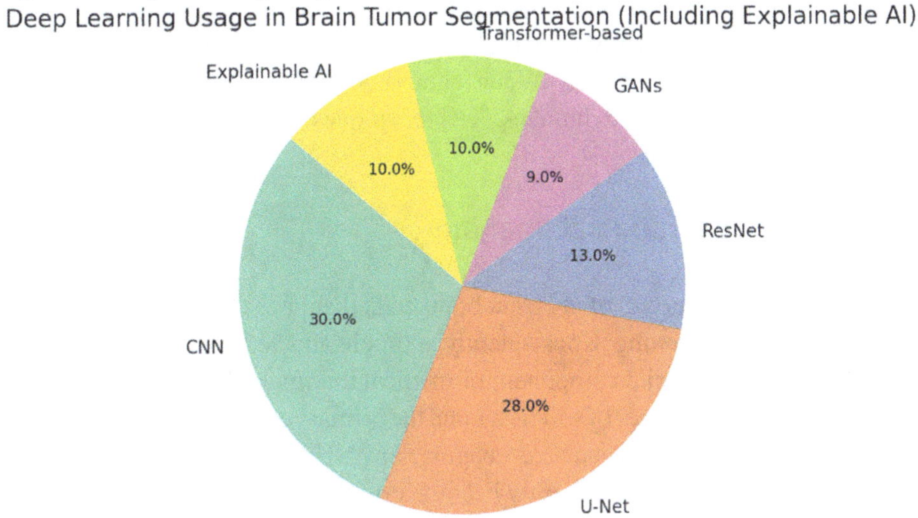

Intersection of AI, Deep learning and Machine Learning

Deep neural networks and machine learning (ML) algorithms are used in many artificial intelligence (AI) systems to solve important issues in precision medicine, bioinformatics, and biomedical informatics. However, it is challenging to comprehend the logic behind the decisions made by sophisticated machine learning models, which are frequently regarded as opaque and black-box techniques. Both AI developers and end users and decision-makers may find this lack of transparency problematic. Explainability and accountability are not only desirable qualities but also legally required for AI systems that can significantly affect human lives in delicate fields like healthcare. Another growing concern is fairness, since algorithmic decisions shouldn't discriminate or show bias against particular groups or individuals based on sensitive attributes (Zafar & Rana, 2025).

Currently, artificial intelligence (AI) is being used in many complex applications, but because many AI models are black-box, it can be difficult to understand and trust their results. Understanding the logic underlying an AI model's decisions is typically crucial. Therefore, there is a need for eXplainable AI (XAI) techniques to increase confidence in AI models. In recent years, XAI has gained popularity as a research topic in the field of artificial intelligence. Existing work have tackled the concepts of XAI, its general terms, and post-hoc explainability methods but there have not been any reviews that have looked at the assessment methods, available tools, XAI datasets, and other related aspects (Gaur & Gaur, 2024).

Artificial Intelligence (XAI) techniques aim to produce ML models with a good interpretability-accuracy tradeoff. When it comes to DNN models and how to make their output understandable to humans, XAI techniques are essential (Gaur & Gaur, 2024).

Explainable AI in medical imaging

A brain tumor is a group of aberrant brain cells that have the potential to spread and become life-threatening. Consequently, a timely and accurate classification of the brain tumor is a crucial component of medical treatment. The primary tool for creating high-quality images of soft tissue and the primary technology for identifying brain tumors is magnetic resonance imaging, or MRI. Computer vision methods like deep learning (DL) have recently become crucial in the classification of brain tumors. The majority of these methods rely on conventional centralized classification models, which are severely hampered by the lack of representative and varied datasets and make it more difficult to create a transparent model (Gaur & Gaur, 2024).

The primary part of the body that is in charge of managing cognitive and physiological processes is the brain. One of the prevalent conditions that significantly lowers quality of life is brain tumors (29, 30). Patients' outcomes and chances of survival can be enhanced by early brain tumor diagnosis. By giving a clear image of the cerebral lesions in the patient's brain, magnetic resonance imaging (MRI) holds promise for effectively diagnosing brain tumors in the field of medical imaging. Using MRI images to accurately classify brain tumors can be extremely helpful in guiding treatment choices and enhancing patient survival rates.

The potential to save lives by promptly and precisely detecting brain tumors served as the driving force behind the investigation and creation of an MRI tumor detection model. The main concept was to realize that if XAI could be utilized to increase prediction accuracy, patients might experience significant advantages. When predicting the survival rate of patients with brain tumors based on MRI scans, XAI seeks to solve the challenge of deciphering the decision-making of an ML algorithm. Brain tumors are recognized to be among the most urgent and severe problems with the body that call for prompt, accurate, and appropriate treatment so that the patient can get back to feeling good. Most of the time, brain tumors are diagnosed by hand using information from medical imaging, such as MRI scans, which renders it greatly time-inefficient and highly susceptible to human errors. But due to the emergence of artificial intelligence (AI) and deep learning (Ayadiet al., 2021; Cheng et al., 2015; Ayadi et al., 2020) methods, there arises a great potential to enhance the precision and productivity of brain tumor diagnosis.

Magnetic resonance imaging (MRI) is the most widely used method in clinical settings for detecting brain tumors because it can produce high-resolution images of soft tissues. However, misdiagnosis cannot be avoided because manual brain MRI analysis requires a high level of skill and experience. Due to notable developments in computer vision and natural language processing, the importance of intelligent healthcare has grown significantly in recent years. Computer-aided diagnosis (CAD) systems, which offer fast, precise medical image analysis with consistent, repeatable results, are the best example of this development. In order to maximize the effectiveness of treatment, these systems significantly help with the early and accurate diagnosis of brain tumors (Geroski & Filipović, 2024).

The application of deep learning in radiology improves radiologists' skills by fusing it with human expertise. Effective brain tumor treatment depends on accurate detection, and deep learning models provide a major benefit in accelerating this process . Biopsies are still the most accurate way to identify particular tumor types, despite their invasiveness and related health risks. Techniques based on non-invasive MRI are becoming more and more dependable substitutes. MRI data is used by machine learning algorithms to distinguish between different kinds of brain tumors. Brain tumors have been successfully classified using MRI data by Convolutional Neural Networks (CNNs) and their variants, including Inception, Xception, MobileNet, and EfficientNet (Alam et al., 2024).

Effective brain tumor treatment depends on accurate detection, and deep learning models provide a major benefit in accelerating this process . Biopsies are still the most accurate way to identify particular tumor types, despite their invasiveness and related health risks. Techniques based on non-invasive MRI are becoming more and more dependable substitutes. MRI data is used by machine learning algorithms to distinguish between different kinds of brain tumors. Brain tumors have been successfully classified using MRI data by Convolutional Neural Networks (CNNs) and their variants, including Inception, Xception, MobileNet, and EfficientNet (Islam et al., 2025).

This chapter examines the use of CNNs coupled with cloud-based systems for classifying brain tumors based on technical feasibility, advantages, and the limitations of this kind of application in the clinical setting. Cloud computing poses immense potential for large-scale, real-time, and the transformation of patient care through collaborative brain tumor diagnosis and outcomes and ultimately transforming the treatment process.

The use of cloud computing has now begun. It is a solid platform that facilitates scaling and deployment of AI models for providing large computing power to process a huge quantity of medical data. The practice of using CNNs for brain tumor classification on the cloud utilizes multiple benefits such as flexibility, scalability, and synergism of complex machine-learning models without requiring on-premise setup. The approach will make healthcare robust by enabling remote communi-cation to gain access to the data center and purchase the most recent diagnostic tools to alleviate the load from healthcare amenities and speed up diagnosis.

HOW EXPLAINABLE AI WORKS

With its ability to provide precise images of the anatomy and pathology of the brain, magnetic resonance imaging (MRI) has become a key component in the non-invasive diagnosis of brain tumors Unmatched soft tissue contrast offered by MRI makes it easier to distinguish between healthy and diseased tissues. It is essential for determining the tumor's location, size, and possible effects on nearby brain structures, all of which are crucial for treatment planning.

Manual Processes to Intelligent Operations

Deep learning, a subset of machine learning, has revolutionized medical image analysis and greatly enhanced the detection and classification of many diseases . Complex MRI data can be analyzed by deep learning algorithms, which can then identify patterns in the data that are not visible to the human eye and use those patterns to predict the presence of brain tumors. These algorithms, especially convolutional neural networks, or CNNs, have demonstrated their ability to increase the speed and accuracy of brain tumor diagnosis by reducing the need for human interpretation and potentially minimizing diagnostic errors.

The term "explainable AI" refers to an AI model's anticipated effects and possible biases. In AI-powered decision making, it aids in describing model accuracy, fairness, transparency, and results. When implementing AI models in production, an organization needs explainable AI to foster confidence and trust. An organization can also embrace a responsible approach to AI development with the aid of AI explainability. The data is used to create these black box models. Furthermore, not even the data scientists or engineers who developed the algorithm are able to comprehend or describe what is going on inside them or how the AI algorithm came to a particular conclusion.

Knowing how an AI-enabled system produced a particular result has several benefits. Explainability may be required to comply with regulatory requirements, assist developers in making sure the system is operating as intended, or be crucial in enabling individuals impacted by a decision to contest or alter the decision's outcome (Rahman et al., 2024).

The term "explainable AI" (XAI) describes strategies and tactics that help humans comprehend the results of AI systems. Its objective is to increase trust, accountability, and transparency in AI systems. The main elements of explainable AI shown in Figure 2.

Figure 2. "Explainable AI" (XAI)

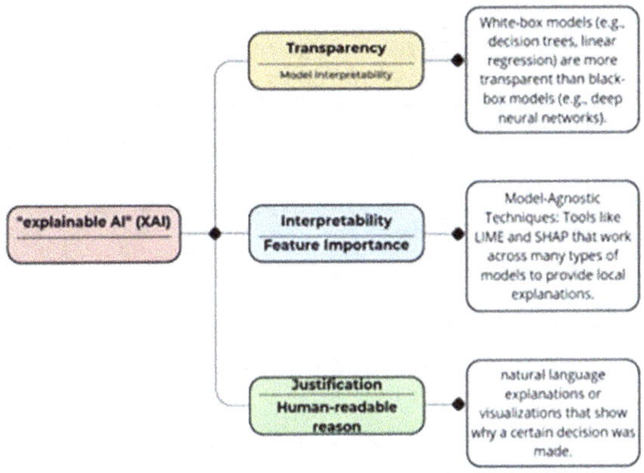

Transparency, interpretability and explainability

In the context of artificial intelligence, the terms explainability, interpretability, and transparency are sometimes used interchangeably and lack a formal definition.

The following is an interpretation of the various concepts in this document:

- Transparency is the capacity to comprehend a particular model. A model is considered transparent in the strictest sense if it is possible to consider the entire model at once. Transparency can be viewed at three different levels: the model as a whole, the level of specific training algorithms, and the level of individual components (such as parameters).

In AI, explainability focuses on offering concise, comprehensible justifications for particular model decisions or predictions. It seeks to provide human-understandable explanations or reasons for a particular result in order to respond to queries such as "Why did the AI system make this particular prediction?" Interpretability is a fundamental component of explainability, but it also considers other domains and areas like ethics, law, and human-computer interaction.

The increasing sophistication of AI makes it harder for humans to understand and trace the algorithm's path to a conclusion. The entire computation process is transformed into what is known as a "black box" that is incomprehensible.

BRAIN TUMOR DETECTION WITH XAI: DATA SET

Data set used in research work shown in Table 1:

Table 1. Data set

Patient ID	Age	Gender	Tumor Type	Detected (Yes/No)	AI Confidence (%)	XAI Method	Key Feature (XAI Insight)
PT001	34	Male	Glioma	Yes	92.4	SHAP	Irregular mass, edema area
PT002	58	Female	Meningioma	Yes	88.7	Grad-CAM	Contrast near dura
PT003	45	Male	Pituitary Tumor	No	46.1	LIME	Non-enhancing lesion
PT004	60	Female	Glioblastoma	Yes	96.3	SHAP	Necrotic core, ring enhancement
PT005	27	Male	Meningioma	Yes	84.5	Grad-CAM	Mass near meninges
PT006	39	Female	Glioma	Yes	89.2	SHAP	T2 hyperintensity
PT007	52	Female	Pituitary Tumor	No	41.3	LIME	Lesion not enhancing
PT008	46	Male	Glioblastoma	Yes	94.8	Grad-CAM	Peritumoral edema
PT009	33	Female	Glioma	Yes	91.5	SHAP	Cortical involvement
PT010	61	Male	Meningioma	Yes	87.1	Grad-CAM	Dural tail sign

Gender wise distribution Brain Tumor detection shown in Figure 3.

Figure 3. Gender wise distribution brain tumor detection shown

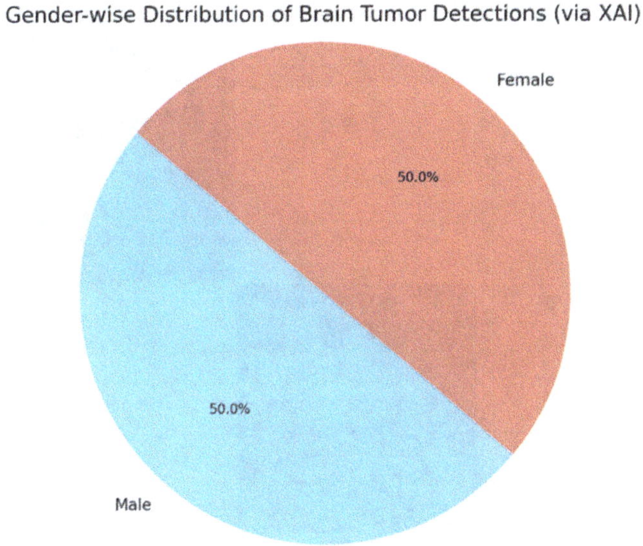

AI confidence Detected shown in Fig 4.

Figure 4. AI confidence detected

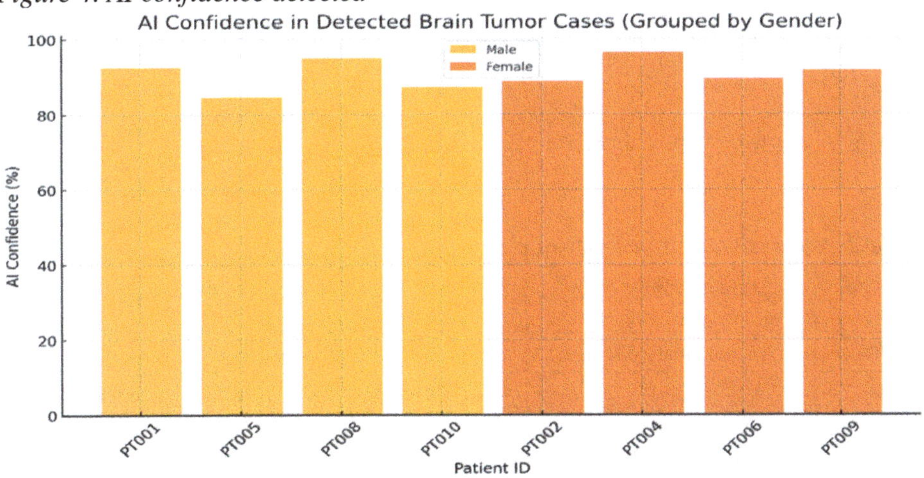

Figure 5. XAI method vs tumor type

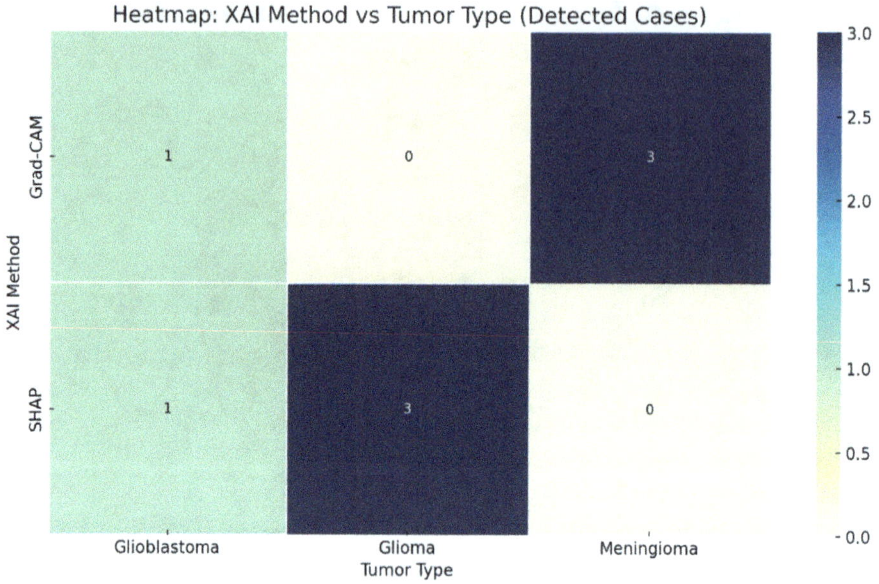

- A dark orange/red cell at “51–60” and “Glioblastoma” would mean many Glioblastoma cases were detected in patients aged 51–60.
- A lighter cell at “20–30” and “Meningioma” might indicate fewer cases in younger individuals.

Age wise tumor types shown in Table 2.

Table 2. Age wise tumor severity and types

Age Group	Common Tumor Types	Likely Insights
20–30	Glioma, Pituitary Tumor	Tumors that occur earlier in life
31–40	Glioma	Some growth, still relatively low numbers
41–50	Mixed	Transition period; detection of more types
51–60	Meningioma, Glioblastoma	Higher detection; age-related tumor risk increases
61–70	Glioblastoma	Peak for aggressive tumors like Glioblastoma

Using Explainable AI (XAI) to analyze synthetic data, the “Age Group vs. Tumor Type” heatmap shows the distribution of brain tumor detections across different age groups and tumor types. The heatmap's cells each indicate the quantity of cases of a particular tumor type found within a given age range. (Abbas et al., 2024) This graphic aids in recognizing patterns in the incidence of various brain tumor types, including gliomas, meningiomas, and glioblastomas, across age groups such as 20–30, 31–40, and up to 61–70 years (Khalighi et al., 2024).

POPULARITY AND ACCURACY OF EXPLAINABLE AI (XAI)

As Explainable AI (XAI) continues to mature, especially in critical domains like healthcare, both its *popularity* and *accuracy* trends show distinct patterns in Table 3.

Table 3.

Trend	*Details*
Increasing Adoption in Healthcare	XAI is gaining significant attention in medical imaging (e.g., brain tumor detection), diagnostics, and genomics. Hospitals and research labs increasingly require transparency from AI systems.
Integration in AI Frameworks	Libraries like SHAP, LIME, and Grad-CAM are now integrated into many ML platforms (TensorFlow, PyTorch, Scikit-learn), making XAI more accessible.
Regulatory Pressure	New regulations, like the EU AI Act, emphasize transparency and fairness, pushing industries to adopt XAI to comply.
Trust and Ethical AI	Organizations want to ensure AI decisions are interpretable and non-biased, leading to a steady rise in XAI integration (Balasubramanyam et al., 2024).

Utility of Popular model

- SHAP – Widely adopted for tabular and deep learning models
- Grad-CAM – Dominant in medical imaging (MRI, CT scans)
- LIME – Useful for local model explanations in NLP and vision

Accuracy

Table 4.

Trend	Details
Complementing Accuracy	XAI doesn't increase raw accuracy but improves **trustworthiness** and **diagnostic confidence**—especially in clinical use cases.
Model Debugging	Developers use XAI to understand *why* a model misclassifies, helping refine training data and architecture.
Better Feature Engineering	Insights from XAI highlight which features influence outcomes, often leading to better feature selection and accuracy.

Benefits of Using XAI in Brain Tumor Detection

- Accurate Diagnosis: XAI helps identify and localize tumor regions, even in complex cases such as low-grade gliomas and small metastatic lesions.
- Enhanced Diagnostic Accuracy: By highlighting crucial regions, XAI can increase the accuracy of AI-driven diagnoses.
- Faster and More Accurate Diagnosis: XAI can reduce the time and effort required for manual examination by automating the identification and classification of tumors.
- Early Detection: XAI may increase the efficacy of treatment by aiding in the early detection of tumors.
- Customized Treatment Plans: XAI can help create more customized treatment plans by providing information about the distinct characteristics of each tumor.

XAI in Brain Tumor Research Examples:

ResNet50 with Grad-CAM: This method creates visual explanations for brain tumor detection by fusing Grad-CAM with the predictive ability of a deep learning model.

UNet with XAISS-BMLBT: This method integrates XAI to improve interpretability by using ResNet50 for feature extraction and UNet for segmentation (Kumar et al., 2024).

XAI is used in the DNA Methylation-Based Brain Tumor Classifier to interpret the output of a machine learning model that categorizes tumors according to DNA methylation patterns.

FastGlioma: This AI tool rapidly and precisely detects cancer cells by combining AI with stimulated Raman histology (Chattopadhyay et al., 2025).

Cloud Based Models

The overall scalability of cloud-based brain tumor classification is one of its greatest benefits. Huge CNNs, particularly those that have numerous layers, are computationally intensive and require loads of power and huge memory. The local systems are mostly unable to supply the capacity for conducting the train and inference operations, especially for the large image data related to medicine. Cloud services like AWS, Google Cloud, or Azure have GPUs or TPUs, and researchers respond and act whenever they wish to enhance their models and corresponding data processing accordingly.

This follows the concept of elasticity machine, where the resources are allocated dynamically according to the requirements of the client. In addition, the use of clouds centralizes the process of distributed data analytics, where the focus is on models is followed by their training on portions of the datasets across various machines, thereby making it possible to reduce the time for training extensively.

Advantage of using Cloud Models

Cloud platforms securely and efficiently share data, an essential component of solid brain tumor classification model establishment. Often, medical image collections are distributed among various individual healthcare professionals; that is why available data to train is sometimes not adequate. Data repositories these datasets, whose killer applications can be accessed, can be collected in the cloud. to healthcare clinicians and researchers from various locations and thereby make it possible for such a collection to work together and share observations. Consequently, the project will entail federated learning since learning will be carried out on dispersed datasets without actually sharing the data. s. This will also protect the privacy of patients. The benefits of the theory model of data sharing include robust generalizability of the model, as well as increased exposure to various tumor morphologies and imaging protocols.

Cloud Model Deployment

Access to classification models of brain tumors is made simpler by cloud-based deployment. A trained model can be deployed as a web API or web service, so clinicians may use it from anywhere they have access to the Internet. It improves advanced diagnostic tools accessible to the public, and generally to lowresource settings. Software as a ser-vice (SaaS) paradigm fits this scenario, where the model is offered for use as a service on the Internet, meaning there is no instal-lation and man-tenance to be dealt with. access on a large scale is not enough. Without justifications to account for model predictions, these systems may be underused or sceptical of clinicians. Therefore, in order for model decisions to be both traceable and explicable, cloud access must be complemented by activity. Models may also be supported by cloud platforms by monitoring and 16 tools for versioning so that clinicians never use outdated or broken version of the model. One advantage of using the cloud services is that it can also serve as a secure refuge for storing data because the disclosure of private medical data could lead to in security and privacy issues. Good-quality data encryption, adequate access policy and compliance with HIPAA and GDPR regulations are wonderful examples. However, a persistent flaw in the existing literature is the absence of practical frameworks that are efficient, privacy-preserving, and easy to understand—especially for systems that are already in operation across institutions or between countries. This emphasizes the necessity even more. to investigate federated learning together with explainability. Cloud major providers also offer security measures including data encryption both at rest and in transit, access controls, and audit logs so that patient data security is guaranteed. Apart from that, compliance and data governance must also be taken into account so that privacy of the consent and agency agreements ensure the safety of patients.

Cloud-supported platforms relate to optimizing artificial intelligence models to detect brain tumors. Measures like compressing, tuning, and pruning can make models smaller and more computationally simple, making them easier to use. execute on such hardware-constrained devices and are faster to infer.

To begin, the theoretical domain of effective deep dive is replicated in this paper. discovering the primary driver in order to produce accurate models that require fewer computational resources. Cloud-based platforms provide tools and high-level frameworks that may be employed to optimize models for research. Consequently, researchers can perform various types of experiments and assign the accuracy-versus-efficiency that will best provide the output.

TECHNOLOGICAL ADVANCES IN CLOUD-BASED BRAIN TUMOR CLASSIFICATION: PROSPECTS FOR THE FUTURE

Integration of Multi-modal Data: MRI images can be combined with other patient data (like genetic or clinical information) to provide a more complete picture of the tumor's characteristics, which can help with a better diagnosis (Onciul et al., 2025). To improve classification accuracy and identify more complex relationships in the data, research into multi-layer transformers and other advanced architectures is recommended.

Classification by Patient: By including patient traits and personal data, brain tumor classification may become more accurate and customized.By adopting XAI, researchers and healthcare providers can develop AI-powered tools for brain tumor detection and treatment that are more dependable, intelligible, and effective. Brain tumor diagnosis remains one of the most difficult tasks in neurology. The development of imaging technologies like CT and MRI established the foundation for contemporary healthcare, and they are still essential tools today. Traditionally, diagnosis relied heavily on expert interpretation. However, recent advancements in cloud computing, artificial intelligence, and data analytics have significantly improved accuracy. and speed of brain tumor detection. Machine learning and artificial intelligence are changing the way we analyze vast amounts of medical imaging data. Cloud-based platforms enable even greater effectiveness, leveraging technologies like deep learning and transfer learning. AI models can now detect subtle patterns in MRI data that might escape human observation. Additionally, techniques for cloud optimization like model compression, quantization, and pruning have significantly enhanced model accuracy and speed. Privacy is essential when dealing with medical data due to its delicate nature. and ensuring security are top priorities. Today's cloud providers implement robust safety measures, including data encryption at rest and in audit logs, strict access controls, and transit, all designed to adhere to regulations like HIPAA and GDPR. Federated learning has emerged as a leading privacy-preserving method, allowing models to be trained on decentralized datasets without the need to share raw data. Additionally, techniques like homomorphic encryption enable computation on encrypted data, providing yet another layer of security.

REFERENCES

Abbas, S., Asif, M., Rehman, A., Alharbi, M., Khan, M. A., & Elmitwally, N. (2024). Emerging research trends in artificial intelligence for cancer diagnostic systems: A comprehensive review.

Ariful Islam, M., Mridha, M. F., Safran, M., Alfarhood, S., & Mohsin Kabir, M. (2025). Revolutionizing Brain Tumor Detection Using Explainable AI in MRI Images. *NMR in Biomedicine*, *38*(3), e70001. PMID: 39948696

Balasubramanyam, A., Ramesh, R., Sudheer, R., & Honnavalli, P. B. (2024). Revolutionizing healthcare: A review unveiling the transformative power of digital twins. *IEEE Access : Practical Innovations, Open Solutions*, *12*, 69652–69676.

Berghout, T. (2024). The Neural Frontier of Future Medical Imaging: A Review of Deep Learning for Brain Tumor Detection. *Journal of Imaging*, *11*(1), 2. PMID: 39852315

Gaur, L., & Gaur, D. (2024). Explainable Artificial Intelligence (XAI) on Neurogenerative Diseases. In *AI and Neuro-Degenerative Diseases: Insights and Solutions* (pp. 63–72). Springer Nature Switzerland.

Geroski, T., & Filipović, N. (2024). Artificial Intelligence Empowering Medical Image Processing. In *In Silico Clinical Trials for Cardiovascular Disease: A Finite Element and Machine Learning Approach* (pp. 179–208). Springer Nature Switzerland.

Khalighi, S., Reddy, K., Midya, A., Pandav, K. B., Madabhushi, A., & Abedalthagafi, M. (2024). Artificial intelligence in neuro-oncology: Advances and challenges in brain tumor diagnosis, prognosis, and precision treatment. *NPJ Precision Oncology*, *8*(1), 80. PMID: 38553633

Kumar, R., Waisberg, E., Ong, J., Paladugu, P., Amiri, D., Saintyl, J., & Tavakkoli, A. (2024). Artificial Intelligence-Based Methodologies for Early Diagnostic Precision and Personalized Therapeutic Strategies in Neuro-Ophthalmic and Neurodegenerative Pathologies. *Brain Sciences*, *14*(12), 1266. PMID: 39766465

Onciul, R., Tataru, C. I., Dumitru, A. V., Crivoi, C., Serban, M., Covache-Busuioc, R. A., & Toader, C. (2025). Artificial intelligence and neuroscience: Transformative synergies in brain research and clinical applications. *Journal of Clinical Medicine*, *14*(2), 550. PMID: 39860555

Prinzi, F. (2023). Innovations in Medical Image Analysis and Explainable AI for Transparent Clinical Decision Support Systems. Chattopadhyay, S., Barman, S., & Lakshmi, D. (2025). The Role of Explainable AI for Healthcare 5.0: Best Practices, Challenges, and Opportunities. *Edge AI for Industry 5.0 and Healthcare 5.0 Applications*, 45-80.

Rahman, A., Sohail, S. S., Alam, M. S., Sharma, A., & Mansoor, W. (2024, November). Detecting Brain Cancer Using Explainable AI. In *2024 7th International Conference on Signal Processing and Information Security (ICSPIS)* (pp. 1-6). IEEE.

Zafar, S., & Rana, N. (2025). The convergence of nanotechnology and artificial intelligence: Unlocking future innovations. *Recent Innovations in Chemical Engineering*.

Chapter 24
Emotion AI and Employee Experience:
Cloud-Native Tools for Human-Centric Management

Jeayaram Subramanian
JGBS, O.P. Jindal Global University, India

Damith Gangodawilage
https://orcid.org/0000-0001-7968-6284
Qasim Ibrahim School of Business, Villa College, Maldives

Thi Minh Ngoc Luu
https://orcid.org/0000-0002-5972-7752
International School, Vietnam National University, Hanoi, Vietnam

Anugamini Priya Srivastava
https://orcid.org/0000-0003-0617-2711
Dnyaan Prasad Global University, Pune, India

ABSTRACT

The integration of Emotion AI (Affective Computing) into employee experience (EX) management represents a paradigm shift towards human-centric, data-driven leadership. This chapter explores how cloud-native tools enable organizations to move beyond lagging indicators like annual surveys by analyzing emotional data from digital communication platforms in real-time. It details use cases across the employee lifecycle, from enhancing onboarding to predicting burnout and fostering inclusive teams. However, this transformative potential is accompanied by significant implications. The chapter provides a critical examination of the ethical, social, and managerial challenges, including privacy, algorithmic bias, and the risk of emotional

DOI: 10.4018/979-8-3373-7442-0.ch024

surveillance. It concludes that the responsible adoption of Emotion AI, governed by strong ethical frameworks, is crucial for creating empathetic, supportive, and productive workplaces.

INTRODUCTION THE NEW FRONTIER OF EMPLOYEE ENGAGEMENT.

The modern office environment can be described as having a deep contradiction. The organizations of the present day already have an unparalleled ability to gather and process data on all aspects of their endeavors, including sales pipelines and supply chain logistics as well as digital employee activity (Kiron & Schrage, 2019). Nevertheless, even with such abundance of data, a lot of companies are puzzled with the continuing levels of employee disengagement, burnout, and attrition. The Great Resignation and the ensuing change to the hybrid work models has given a very clear indication that the old, once-a-year engagement surveys would not be more than a lagging indicator, but would create a post-mortem diagnosis and not real-time and proactive insight into the pulse of the workforce (Gartner, 2022). Such disconnect between information and the real human understanding is a gaping failure in contemporary management practice. It is against this backdrop that a new, potent, and provocative paradigm is formed: the Artificial Intelligence (AI) is used to learn and improve the emotional side of the working process. This chapter presents the idea that the next evolutionary stage of Employee Experience (EX) is a conscious and ethical inversion of the Emotion AI facilitation through the agility and scalability of cloud-native applications to drive the transition of the management paradigm to the proactive, empathetic, and human-centered one.

Emotion AI, or Affective Computing, is the subdivision of AI that allows the ability of machines to detect, understand, imitate, and act on human feelings (Picard, 1997). In addition to the basic sentiment analysis, which categorizes the text as positive, negative, or neutral, the Emotion AI will distinguish between complicated emotional conditions like frustration, joy, anxiety, and ambivalence with the help of the text, voice, and visual information. Combined with the digital tools that constitute the modern workplace, this technology provides the opportunity to measure the subjective, emotional back-currents that have always flavored the morale, teamwork and output of the employees, but are not visible to the data-driven controlling machinery. The promise is a more reactive organization, an organization that will be able to spot symptoms of team conflict or individual burnout in near real-time and encourage managers to pursue supportive interventions before problems turn into resignations or disengagement.

But this promise is tied up inextricably with great dangers. Introduction of technology that claims to identify human emotions poses a pressing ethical, social and legal concern over such aspects as privacy, consent, discrimination and even the definition of the employer-employee relationship. Thus, this chapter is going to not only discuss the potential of Emotion AI technology, but also critique it thoroughly. We are going to start with the conceptual background, which entails defining Emotion AI and the contemporary interpretation of Employee Experience. Then we will explore the engine behind this revolution one which is the cloud-native tools and architectures that enable large-scale, integrated Emotion AI to become viable. Its transformative potential will be demonstrated by the in-depth discussion of practical applications at each of the employee lifecycle stages. Lastly, and most importantly, the chapter will also provide significant space to the examination of the complex set of implications, which organizations have to tread, to essentially exploit this power in a responsible way, where the journey towards a data-driven EX will not compromise on human dignity and trust.

SECTION I: THE BACKGROUND: THE INTRODUCTION OF EMOTION AI TO EMPLOYEE EXPERIENCE.

1.1. What is Emotion AI (Affective Computing)?

Emotion AI at its simplest is an interdisciplinary science that cuts across computer science, psychology, and cognitive science to fill the emotional divide between humans and computers. The basic assumption, determined by the works of Rosalind Picard (1997), in her classic work, is that to be really intelligent and able to communicate with humans in a natural manner, computers have to know how to identify and react to human emotion. Early implementations used to be crude and used simple textual spotting of key words. Nevertheless, with the development of machine learning (ML), specifically, deep learning and natural language processing (NLP), the level of these systems has soared significantly. The state of the art of Emotion AI no longer just classifies sentiment, but tries to predict discrete emotional states based on a set of diverse data modalities all of which have their own weaknesses and challenges.

The three major modalities of emotion recognition are the textual, vocal and visual modalities. The most common and readily incorporated modality of knowledge-work setting is textual analysis. More sophisticated NLP systems, including transformers (e.g., BERT, GPT), can turn written communication, whether in emails, Slack messages, project management systems, or open-ended survey responses, to identify the subtexts of emotion, e.g., sarcasm, confusion, enthusiasm, or stress, which more basic systems do not (Devlin et al., 2018). As an example, an employee can write in a chat Great, another last-minute request; a sentiment analyzer will distinguish it as a positive image because of such a word as great but a complex system of Emotion AI will understand the sarcasm and define it as the inability to meet expectations or anger. Analysis of the vocal tone or paralinguistic analysis is an analysis of non-lexical aspects of speech, which includes pitch, pace, volume, and timbre to conclude about the emotional state of the speaker. It is especially applicable in customer service phone call facilities to check the welfare of the agents and in virtual meetings to determine the interest of the participants (El Ayadi et al., 2011). The manager may also get a dashboard warning that the stress indicators in the vocal tone of his or her weekly sync in a team have been increasing notably over the past month, and there was the need to proactively review workloads.

The most controversial modality is facial expression analysis, which involves computer vision to analyze the movements of muscles on the face (Action Units) into the emotion according to models such as the Facial Action Coding System by Paul Ekman (Ekman and Friesen, 1978). Its popularity in the workplace is highly controversial in terms of ethical issues of surveillance and cultural peculiarities of facial expression, despite its technical sophistication. Lastly, indicators of stress and arousal levels can be measured with the help of physiological data collected by wearable devices such as smartwatches or biometric sensors (e.g. heart rate variability (HRV), galvanic skin response (GSR), temperature, etc.) (Healey and Picard, 2005). Although the approach provides objective physiological data, it is the most privacy-raising and informationally-sensitive method that demands the most explicit and informed consent. The science of these technologies consists of the training of machine learning models on large, labeled datasets of human expressions. An important caveat here is however that these systems do not feel or understand emotion; the systems are statistical pattern recognition engines which attempt to predict an emotional label based on correlated inputs, a point with far reaching consequences to their accuracy and use.

1.2. The Contemporary Employee Experience (EX) Journey.

Employee Experience (EX) has become a prevalent model in human resource management and organisational development that implies an overall perspective of the employer-employee relationship. The concept of EX is determined by the aggregate of all the perceptions that an employee perceives regarding their interaction with an organization in three main environments: physical space (e.g., office design), cultural environment (e.g., leadership, values, belonging), and technological environment (e.g., tools and systems they can use to do their job) (Morgan, 2017). It is a multidimensional construct, embracing all the touchpoints, starting with a candidate seeing a career page, to the interview, to the exit-interview, and so on. This is the sharp contrast with older models that were extremely narrow in their measurement i.e. employee satisfaction or employee engagement which was usually measured separately. Positive EX is closely related to key business results, such as increased productivity, increased innovation, greater customer satisfaction, and reduced voluntary turnover by a significant margin (Deloitte, 2020).

The end-to-end employee lifecycle must be mapped to comprehend which points and areas Emotion AI can interfere with. This road starts with Recruitment and Onboarding when a candidate develops his/her initial impressions about the company in terms of its efficiency, communication, and culture. Bad application procedures which are non-transparent, unclear, or cold emotionally may send away the best talent. The process then moves to the deep stages of Performance and Development that entail constant work, goal setting, feedbacks, manager relationships, team work, learning opportunities and career development. This is the most lasting and significant phase of the EX, during which the emotional experience of an employee (being recognized, supported, and psychologically safe (or the opposite) on a daily basis) gradually provides an impact on his or her commitment. Lastly, the lifecycle ends with Retention and Exit where the organization can get to know why the individual has left by conducting exit interviews with them and ensuring that there is a positive relationship with the alumni.

The basic data-lag in the classical EX management consists in the fact that it uses infrequent, self-reported and frequently biased data-collections. A typical example of a lagging indicator is the annual engagement survey; by the time that the survey is put together and analyzed, the underlying problems might have contributed months, and the emotional background of the responses will be lost. Also, survey data are affected by recall bias and social desirability bias where employees are not safe enough to give honest and critical feedbacks (Podsakoff et al., 2003). This develops a responsive management cycle, and leaders are continually problem solving the issues of yesterday. The new, hybrid workplace, which is premised on digital communicative tools, has merely made this divide even wider as much of the informal, water-cooler chats that managers had a hunch about regarding team morale have vanished. Emotion AI suggests addressing this gap by offering an active but passive and contextual flow of information about the emotional climate of the organization and transforming the subjective art of management into a science-based one.

1.3. The Confluence: Why Emotion AI is a revolutionary EX Tool.

The intersection between Emotion AI and the current EX model constitutes the paradigm shift in the way organizations comprehend and utilize their human capital.

Emotion AI is the most important among them because it helps in converting lagging indicators into leading ones. Rather than having to wait until a yearly survey revealed the existence of burnout six months ago, an Emotion AI system would allow identifying a persistent rise in the use of language that indicated the existence of stress and fatigue within the communication channels of a particular team in real-time. This enables pre-emptive action like a redistribution of workload, the provision of more resources or a supportive conversation before the situation causes a drop in performance, sick-leave, or resignation. This shifts the organization off a reactive stance and to a proactive and preventative one.

Secondly, Emotion AI provides an effective tool of measuring the subjective and in many cases intangible elements of workplace culture. Other such concepts as psychological safety, which can be defined as a collective belief in the safety of a team in terms of interpersonal risk-taking (Edmondson, 1999), are both vital to team performance and in fact, very hard to quantify. Emotion AI can be used to provide proxy measures of psychological safety by examining communication related patterns, including, but not limited to: the equality of conversational turns-taking, the ratio of questions to statements, and emotional tone of feedback. Equally, it could be useful in gauging the emotional response of an organizational change, say the merger or the implementation of a new policy, by tracing the collective mood swings throughout the digital footprint of the organization. This empirical methodology will enable leaders to transcend their feelings and anecdotes in making important decisions on culture and strategy.

Finally, the potential of Emotion AI to transform the world of management is its ability to scale the human-centric one. In big companies, senior leaders cannot have a personal caring relationship with all workers. Emotion AI systems can also become a force multiplier, which will supplement the perception of a manager by pointing to previously invisible indicators of distress or disengagement in his or her teams. They can also give managers suggestions that can be proactively acted upon, like, Your team had an increase of 40 percent in anxiety related language this week and Employee X has been exceptionally silent. You can consider visiting them. This does not take the place of human empathy of the manager but it provides him/her with the contextual knowledge of how to employ it in a more effective and fair manner among his/her entire team. Emotion AI can transform workplaces into more productive and resilient, as well as more humane and caring towards the health and well-being of its employees by making the emotional state of the workforce apparent and addressable.

PART II: THE TECHNOLOGICAL PROPULSION: EMOTION AI CLOUD-NATIVE TOOLS.

2.1. The Capabilities of a Cloud-Native Architecture.

The application of Emotion AI in practice at the organizational level would not be technologically possible without the infrastructure of cloud computing. Applications written to take advantage of cloud environments, i.e. a cloud-native architecture, is not only a convenience of the Emotion AI, but a necessity. This is owed to some natural data and processing requirements. Scalability is the first thing. The amount of information that is produced by the digital communications of a multinational corporation is overwhelming. The number of employees sending emails, messages, and documents, and engaging in video calls, is growing at a rate of thousands daily to generate terabytes of unstructured data. Cloud-native technology is able to automatically scale its computational and storage capabilities up or down to accommodate this load, which a legacy on-premise server infrastructure would not be well-equipped to handle without providing a bottleneck in performance or over-provisioning it (Villamizar et al., 2015).

Another important strength is agility and integration. Cloud native applications are developed in the form of microservices- small, independent and loosely coupled services that perform a specific functionality. This implies that the Emotion AI text analysis service on Slack can be a distinct service, which is not the one that analyses vocal patterns during a Zoom meeting. These microservices can also interact through well-established APIs (Application Programming Interfaces), and it is much easier to add Emotion AI to the already existing mosaic of HR technology, which is referred to as the HR Tech stack. As an example, an Emotion AI microservice can retrieve anonymized data through Microsoft Teams API, analyze it, and send insights and alerts straight onto a manager's dashboard of Salesforce or Microsoft Viva in a smooth, automated process. This microservices approach that is API-driven, will not make the Emotion AI system another data silo, and its insights will be directly integrated into the tools where managers are already operating.

Also, the necessity to process data in real-time is a major motivation to the use of clouds of the native. The more timely an insight on increasing team stress is provided, the greater its value. Cloud computing transforms streaming data into the high-performance computing capabilities required to execute complicated AI models to achieve real-time analysis and feedback. This makes interventions possible in real-time i.e. in a virtual meeting a facilitator can receive a prompt when it appears that the engagement level is dropping, which may prompt them to ask questions. Lastly, security and compliance, which are commonly viewed as a cloud hazard, are in numerous aspects improved by cloud vendors. The major cloud providers such as Amazon Web Services (AWS), Google Cloud Platform (GCP) and Microsoft Azure spend billions of dollars on security infrastructure, providing high-quality encryption of data stored and transferred, advanced identity and access management (IAM) controls, and compliance certifications with a broad set of regulations such as GDPR and HIPAA. In the case of an organization managing the highly sensitive information of employee feelings, using this enterprise level security base can be more secure than developing and sustaining a similar on-premise system.

2.2. Elements of an emotional AI Stack in the clouds.

An entirely implemented, cloud-native Emotion AI system of EX is not an app and instead a stack of networked layers with a specific purpose. This stack is important to learn about both technically speaking and managerially speaking. It is based on the Data Ingestion Layer. This layer will be in charge of connection to different data sources in the digital ecosystem of the organization. It retrieves raw, unstructured data based on its email provider (e.g. outlook, Gmail), collaboration tools (e.g. Slack, Microsoft Teams), HR Information Systems (e.g. Workday, SAP SuccessFactors), survey tools and even calendar systems using a library of pre-written connectors and APIs. An important role of this layer is to anonymise or pseudonymise the data upon ingestion to defend the privacy of the individual prior to any processing taking place. As an illustration, it can substitute the name of employees with random and unique numbers so that they could be analyzed within a group but not reveal identities of the individual employees.

AI/ML Processing Layer is the core of the system. This is the place where the raw data is converted to the emotional insights. This layer is based on cloud-based machine learning systems such as AWS SageMaker, Google Vertex AI, or Azure Machine Learning. It contains a set of AI models: a natural language understanding model of text, an audio processing model of a vocal tone, and possibly a computer vision model of video data (when utilized within a firm set of ethical principles). They are typically trained on large and general datasets, but can be tailored with the specific anonymized data of a given organization to be more accurate and tailored to the specifics of a company in terms of its jargon and culture. The data it produces, which is structured (timestamps, emotional values: e.g., frustration: 0.8, enthusiasm: 0.2), and confidence degrees, is placed into a cloud data warehouse or data lake.

The Data Analytics and Visualization Layer is sitting above the processing layer. This element combines the tabular emotional information and summarizes it to defend the privacy of individuals and uncover significant patterns. It has the ability to produce reports and interactive dashboards according to various stakeholders. The organizational-level dashboard could display to an HR business partner emotional trends per department and the individual department dashboard may display to a team manager aggregate team mood and anonymous notifications regarding any drastic emotional changes. The usefulness of this data is mostly associated with its visualization; it has to be user-friendly, practical, and should not allow it to be misunderstood or misused.

Lastly, the Action and Intervention Layer completes the loop, and changes knowledge into action. Business logic is written here in order to invoke automated and human-centric workflows. This layer can take actions based on a predefined set of rules or even an additional AI model. As an example, when the system identifies a consistent pattern of language linked to burnout within the communication of a team, it might automatically: (1) privately send an anonymous notification to the manager indicating that he/she needs to check in on the team to verify their well-being, (2) encourage the HR system to send a list of mental health resources to the entire team, or (3) remind the manager to cancel a non-essential meeting in order to reduce cognitive load. It is this layer that renders the system proactive and guarantees that once the emotional insights have been acquired, the results reflect in a form of tangible support and not were they merely added to a dashboard.

SECTION III: USE CASES IN ACTION: THEORY TO PRACTICE.

3.1. Improving Recruitment and Onboarding.

Recruitment is the start of the employee lifecycle and first impressions are paramount. Through emotion AI, the language and tone that is employed in communication with the candidates during the hiring process can be analyzed. Evaluating the emotional feeling of email correspondence and scheduling emails, organizations may see the issues of friction or misunderstanding that make candidates drop out or develop a negative vision about the company atmosphere (Black et al., 2021). As an illustration, an overall negative or anxious tone in communications with candidates may be used to show an ineffective or disrespectful process to make some correction by the recruiters. Moreover, in video interviews (only with express and knowledgeable permission), it is possible to utilize Emotion AI to give the hiring managers feedback on their own behavior, whether they are displaying a warm, engaging, and welcoming presence, and this will strengthen the employer brand of the company.

The onboarding period is a critical period when an organization can promote engagement and retention when the candidate accepts an offer. Emotion AI will be able to track the online profile of new employees, how they socialize in their onboarding systems, how they communicate with their coworkers and supervisors, and how they engage in initiation conferences. The system can help in recognizing the struggling employee through the mapping of emotional trajectory of new workforce and those who are experiencing isolation, confusion, or are having imposter syndrome. The manager or HR partner can be alerted that could be in the form of a message indicating that communication indicates that new Hire A feels less confident and more confused over the past week, and to offer the supportive check-in to offer reassurance and resources. Such a proactive support could go a long way in making a successful integration more likely and avoiding the early-stage attrition, which is as expensive as it gets and disruptive.

3.2. In Favor of Manager-Employee Relationships.

One's relationship with his or her immediate manager is always reported to be the biggest driver of employee engagement and retention (Harter, 2022). Emotion AI can also be useful to complement the emotional intelligence of a manager and increase their ability to be empathetic. One of the main uses is in the form of AI-driven nudges. The system can identify macro-level changes in emotions by predicting and analyzing aggregate, anonymized data on digital communications within a team. A manager can be provided with a weekly digest which said: This week, the general morale of your team did not change, however, the signs of frustration on the subject of project deadlines grew by 25 percent. Think about consideration of the project timeline with the team. This will give the manager a contextual and data-based understanding on what is happening to their team, so that they are able to resolve any problems prior to the formalization of the same.

A more high-technology, and ethically responsible, one example is the one-on-one meeting conversation analysis. The audio of manager-employee check-ins may be analysed (usually in real-time) with the consent of both parties in order to provide the manager with feedback. It is not capturing the conversation and transcribing it to be revisited at a later date, but analysing the paralinguistic characteristics (tone, pace, interruption patterns) to create insights. The manager could get a confidential report after the call informing him that: You were talking 70 percent of the meeting. The voice range exhibited some hesitation by the employee when talking about career objectives. The second time, you should use more open-ended questions. This is an effective management development tool as leaders can become more self aware and effective coaches which allows them to build deeper trust and more productive relationships with the reports.

3.3. Developing Psychologically Safe and Inclusive Teams.

Diversity and inclusion are not only moral requirements but also business requirements, which are associated with superior decision-making and innovation (Rock and Grant, 2016). Emotion AI is capable of offering objective information that can be used to achieve these objectives. During virtual or hybrid meetings, the pattern of participation can be observed with the help of AI tools that trace the speaking time and turn-taking. It will be able to see whether some demographics (e.g., women or junior employees) are talking less or interrupted faster than the others. A real-time, non-public nudge can be automatically sent to the meeting facilitator indicating that he/she should encourage input by the less vocal speakers in the discussion in order to impose a more participatory discussion pattern.

Moreover, as stated above, Emotion AI may be used as an indicator of psychological safety. Through the analysis of the language employed in the team channels and meetings, the system may identify the extent to which employees are comfortable to voice their divergent views, to confess mistakes or to present new ideas. There is an excess use of the wording that relates risk-taking, including, I might be wrong but... or What if we tried..., together with affirmative, supportive replies by colleagues, would signify high psychological safety. On the contrary, the absence of such language or high rate of keywords of blame and criticism may be an indication of a problem. This enables the leaders to quantify and positively control the psychological environment of their teams and provide a climate in which innovation can be facilitated.

3.4. Preventing and predicting Burnout.

Burnout (exhaustion, cynicism and decline of professional efficacy) is a widespread problem that has dire human and organizational implications (Maslach et al., 2001). Emotion AI is a powerful instrument of its detection and prevention. Through continuous observance of the long-term signs of language-based burnout, including increased manifestations of fatigue, cynicism (what's the point), and feelings of ineffectiveness, throughout the communications of an individual employee or a group (team) over time, the system will be able to provide an early warning to at-risk employees before it is too late. A continual rise in these metrics, particularly in combination with such data as long working hours (calculated based on email and calendar data) can generate a high-priority notification.

However, the intervention should be planned well to ensure that the issue is not worsened. An employee should not receive the message in an email, which says, Our AI has found out that you are burned out. Rather the alert is to be forwarded to an identified HR or the well-trained manager. The given step could be a sympathetic, non-disclosure talk, which aims to support: I have noticed that you have been working more overtime on Project X of late, and I wanted to ask how you are doing. Is there anything we can do to rearrange your work load or offer more support? Such a human-centered approach where AI is informed enables the organization to refer struggling workers to resources such as counseling services, allow them to get time off, and fix the root causes of the problem such as unrealistic deadlines, which are leading to burnout among teams, and thus, promote a culture of well-being and sustainable performance.

PART IV: FINDING YOUR WAY THROUGH THE IMPLICATIONS: A CRITICAL ANALYSIS.

4.1. Ethical Implications (The Paramount Concern)

Ethical issues that accompany the use of Emotion AI in the work environment should be taken as seriously as possible. Privacy and consent is the immediate issue of concern. Workplace as such is a power imbalanced and the concept of voluntary consent is questionable when a boss, who also determines ones livelihood, requests to access emotional information (Martin & Freeman, 2003). Employees can be forced to agree on it because they are afraid of being punished in the event they decline. Thus, the transparency is a must. Organizations need to articulate the information about what information is being gathered, how the data will be analyzed, who can access the information and how the data will be used. Minimalism of data needs to be utilized--gathering only the data that is absolutely necessary. Individual-level analysis should be an exception, and anonymization and aggregation are to be the norm, with the sole exception of critical and support-focused applications, such as burnout prevention, and only done with explicit and opt-in permission.

Another far-reaching ethical issue is prejudice and reasonableness. The models of AI are trained using data, and in case the data is biased or reflects the biases in the past or society, the models will reproduce it and even enhance it (Mehrabi et al., 2021). As an example, an Emotion AI model, which is mostly trained on data on one particular demographic sample, can fail or fail to make accurate conclusions about people who are in other cultural backgrounds and who express emotions in a different way. What one model may read as someone being passive, quiet, and disengaged, may actually be an illustration of a society where a person is supposed to be calm and quiet, without any emotion expression. To overcome this risk, rigorous bias testing and auditing and various training datasets are necessary. Moreover, the process of measuring emotion in itself can be reductive and not capturing the complexity and cultural particularity of the human emotional experience.

There is also a great danger of dehumanization and manipulation. In case managers get too dependent on algorithmic evaluations of how their team feels, they can outsource their own empathy and internal feeling, causing a transactional and fake management approach. It may also pose the threat that this technology might not be applied to support, but to manipulate, e.g. to get the emotional condition of customer service agents as high as possible to sell more, without consideration of their true well-being. This explains why a robust ethical code should regulate the application of Emotion AI, so that it can be applied to complement human intelligence and promote well-being, rather than dictate and manipulate employees.

4.2. Social and Organizational Implication.

The societal consequences of the massive use of Emotion AI are far-ranging. The fear of the Big Brother is real and tangible. To the extent that a workforce thinks it is being emotionally spied upon most of the time, it is unlikely to be innovative, genuine or even trusting. This may result in a so-called emotional labor where workers feel compelled to act or repress some emotions in order to obtain a good result on the algorithm (Hochschild, 1983). They may also be taught to employ happy words in their messages even when they are not feeling happy and this results in a homogenized and fake culture of work place. The nature of work relationships may also shift because the relations will become performative actions aimed at an algorithmic audience instead of real human relations.

The technology is also one that poses a threat to the creation of a new digital divide. Major and well-equipped companies will be the first to implement and enjoy the advanced Emotion AI systems, which may end up providing an enormous talent management and productivity advantage. Small and medium-sized businesses can also be left out, as they cannot afford these tools, and this will further enhance the existing disparities in the business environment. Domestically, new type of emotion workers or analysts might become a reality and overload to regulate their emotional information might result in a new stressor and anxiety factor among the common working population. It would have a fundamental change in the social contract of work, which should be provided with new norms and protection of autonomy and psychological space of employees.

4.3. Applications to Practice and Management.

Practically, Emotion AI implementation is a complicated organizational change process and not a technology installation. There is a serious discrepancy in the level of skills; the HR specialists and the line managers are not usually taught on how to read emotional data dashboards or respond to AI-based nudges in a positive, non-invasive way. The process of gaining knowledge about the insights and turn them into beneficial discussions instead of punitive measures and micromanagement needs to be thoroughly trained. The managers also have to be instructed on the constraints of the technology that it offers probabilistic indications rather than definite truths and that human judgment on their part is supreme.

Change management is thus very essential. Implementing an Emotion AI system without proper communication and engagement among the employees is the formula of a backlash, mistrust, and failure. Trust and legitimacy can be established through a co-creation strategy, which entails the participation of employees and representatives to design and governance of the system. To prove the value of the technology and eliminate fears, it could be tested in pilot locations (such as well-being) and piloted by groups of volunteers. Moreover, it is not an easy task to define a clear Return on Investment (ROI) yet it has to be done. The measure of success should not be based on the conventional productivity indicators but enhanced human outcomes: a decrease in voluntary turnover, fewer absenteeism rates, increased psychological safety and inclusion survey scores, and the ratings of manager effectiveness. Lastly, it is important that a strong governance structure is put in place. This must have a cross-functional ethics board that includes the HR, legal, IT and such like the employees to go through the use cases, to monitor the data handling and to settle any disagreements so that the technology can be used in a responsible manner and in the interest of all the stakeholders.

4.4. Law and Non-legal Implications.

The Emotion AI legal environment is new and quickly changing. Most jurisdictions will probably have emotional data categorized as biometric or sensitive personal information and create strong regulatory demands. The General Data Protection Regulation (GDPR) of the European Union and the California Consumer Privacy Act (CCPA) as amended by the CPRA set the strict regulations on the processing of sensitive information as it requires explicit opt-in consent, rights to access and ability to delete, and the implementation of an effective security system (GDPR, 2016; CCPA, 2018). The application of Emotion AI on an unlawful basis may lead to huge fines and a negative reputation.

Another significant area of law grey is liability. Who is liable to what extent, in the event an Emotion AI system makes an incorrect recommendation that results in a manager making a harmful decision (ex: denying somebody a promotion because of a mistaken evaluation of their low engagement)? The software vendor, the manager or the employer? There has to be clear terms of service and internal policies that set out that the AI is a decision-support rather than a decision-maker and that the human manager will be responsible over ultimate decisions that involve people-management. Since laws cannot always keep up with new technologies, companies need to take a precautionary principle, actively practicing the utmost levels of transparency, agreement, and information security to reduce legal risk and establish a base of trust.

CONCLUSION: A RESPONSIBLE AND HUMAN-CENTRIC FUTURE.

The introduction of the Emotion AI into the context of Employee Experience is a landmark in the history of the management. The possible advantages are too great to overlook: the possibility to make the workplaces more responsive, supportive, and engaging, making the emotional conditions of the working force visible and impractical. The scalable integrated engine that will turn this vision to a viable reality is offered by cloud-native tools, by improving onboarding to preventing burnout and building inclusive teams. The analytical and informational perspectives that this technology provides can empower the managers to become more efficient coaches and leaders, extending human-centric practice to even the largest international organizations.

But this is a dangerous road. The very technology that can help people be more empathetic can be used to facilitate surveillance, the technology that can facilitate inclusion can be used to strengthen prejudice, the technology that can help well-being can weaken trust and authenticity. The ultimate conclusion made in this chapter is that the success of Emotion AI will not be determined by the sophistication of its algorithms but rather the wisdom, ethics, and empathy with which it is controlled and implemented. Technology should be serving human values and not the other way round. It is to be employed in supplementing and enriching human managers, not to substitute their judgment or their ability to have authentic contact.

Responsible innovation is the way forward. It implies unequivocal openness, effective staff approval, stringent bias alleviation, and effective ethical governance. It demands managers to continue being at the center of the system with the insights derived with AI serving as the kickstarter of the compassionate conversation, not its replacement. With caution and integrity, though, Emotion AI could assist us in creating organizations not just more efficient and resilient but also more humane, gratifying and deserving of the individuals behind which they run. It would be incorrect to think that the future of work is either at the risk of taking the choice between data and humanity or using data to serve our common humanity.

REFERENCES

Black, J. S., van Esch, P., & Johnson, R. D. (2021). AI-enabled recruiting: What is it and how should it be used? *Business Horizons*, *64*(6), 763–774.

California Consumer Privacy Act (CCPA), Cal. Civ. Code § 1798.100 et seq. (2018).

Deloitte. (2020). *Global Human Capital Trends*. Deloitte Insights.

Devlin, J., Chang, M. W., Lee, K., & Toutanova, K. (2018). BERT: Pre-training of deep bidirectional transformers for language understanding. *arXiv preprint arXiv:1810.04805*.

Edmondson, A. C. (1999). Psychological safety and learning behavior in work teams. *Administrative Science Quarterly*, *44*(2), 350–383.

Ekman, P., & Friesen, W. V. (1978). *Facial Action Coding System: A technique for the measurement of facial movement*. Consulting Psychologists Press.

El Ayadi, M., Kamel, M. S., & Karray, F. (2011). Survey on speech emotion recognition: Features, classification schemes, and databases. *Pattern Recognition*, *44*(3), 572–587.

Gartner. (2022). *Gartner HR Research Finds 2021 Was the Year of the Manager*. Gartner.

General Data Protection Regulation (GDPR). (2016). Regulation (EU) 2016/679 of the European Parliament and of the Council.

Harter, J. (2022). *The Manager's Role in Employee Engagement*.

Healey, J. A., & Picard, R. W. (2005). Detecting stress during real-world driving tasks using physiological sensors. *IEEE Transactions on Intelligent Transportation Systems*, *6*(2), 156–166.

Hochschild, A. R. (1983). *The managed heart: Commercialization of human feeling*. University of California Press.

Kiron, D., & Schrage, M. (2019). Strategy for and with AI. *MIT Sloan Management Review*.

Martin, K. E., & Freeman, R. E. (2003). Some problems with employee monitoring. *Journal of Business Ethics*, *43*(4), 353–361.

Maslach, C., Schaufeli, W. B., & Leiter, M. P. (2001). Job burnout. *Annual Review of Psychology*, *52*, 397–422.

Mehrabi, N., Morstatter, F., Saxena, N., Lerman, K., & Galstyan, A. (2021). A survey on bias and fairness in machine learning. *ACM Computing Surveys*, *54*(6), 1–35.

Morgan, J. (2017). *The Employee Experience Advantage: How to Win the War for Talent by Giving Employees the Workspace They Want, the Tools They Need, and a Culture They Can Celebrate*. Wiley.

Picard, R. W. (1997). *Affective computing*. MIT Press.

Podsakoff, P. M., MacKenzie, S. B., Lee, J. Y., & Podsakoff, N. P. (2003). Common method biases in behavioral research: A critical review of the literature and recommended remedies. *The Journal of Applied Psychology*, *88*(5), 879.

Rock, D., & Grant, H. (2016). Why diverse teams are smarter. *Harvard Business Review*, *94*(10), 74–81.

Villamizar, M., Garces, O., Castro, H., Verano, M., Salamanca, L., Casallas, R., & Gil, S. (2015, October). Evaluating the monolithic and the microservice architecture pattern to deploy web applications in the cloud. In *2015 10th Computing Colombian Conference (10CCC)* (pp. 583-590). IEEE.

Chapter 25
The Smart HR Cloud: Automating Workforce Planning and Skill Mapping

Mohit Yadav
https://orcid.org/0000-0002-9341-2527
JGBS, O.P. Jindal Global University, India

Minakshi
College of Computer Science, King Khalid University, Saudi Arabia

Tanupriya Choudhury
https://orcid.org/0000-0002-9826-2759
School of Computer Science, University of Petroleum and Energy Studies, India

Narayanage Jayantha Dewasiri
https://orcid.org/0000-0002-5908-8890
Faculty of Management Studies, Sabaragamuwa University of Sri Lanka, Sri Lanka & Qasim Ibrahim School of Business, Villa College, Maldives

ABSTRACT

The contemporary business landscape, characterized by rapid technological change and VUCA conditions, has rendered traditional, static HR processes obsolete. This chapter delineates the strategic imperative for organizations to adopt a "Smart HR Cloud"—an integrated, data-driven system for automating workforce planning and skill mapping. It explores the foundational elements, from building a dynamic skills inventory to leveraging AI for predictive analytics and gap identification. The chapter details how these insights translate into actionable strategies for talent development, internal mobility, and strategic hiring. Critically, it addresses the

DOI: 10.4018/979-8-3373-7442-0.ch025

essential change management, ethical considerations, and practical implications of this transformation, providing a comprehensive blueprint for building an adaptive, future-proof organization grounded in a continuous skills-based strategy.

INTRODUCTION: THE END OF STATIC HR

Modern business environment is characterized by uncertainty, volatility, complexity, and ambiguity (VUCA). Human capital is the key source of competitive advantage in an organization in such an environment. However, even now, most organizations are still managing their workforce in old-fashioned, stagnant ways of spreadsheets, siloed databases and annual performance reviews, which are essentially out of step with the dynamic character of the contemporary markets. This dislocation poses a serious weakness: a widespread and increasing skills gap. The half-life of professional skills is quickly fading as technological acceleration, usually in the form of generative AI, automation, and data analytics, is turning the industry upside down (Deloitte, 2023). The position occupied by an ideally suited person now might be half or even completely out of existence in 18 months. This shapes a burning platform of change in which organizations are forced to stop being reactive in their talent management. The answer is the strategic implementation of the so-called Smart HR Cloud- an integrated, data-driven, and intelligent system which changes Human Resources into the administrative role to a strategic nerve center. This chapter will outline the path to be taken between the old fashioned, manual-intensive workforce planning to a more agile, predictive and automated model. It will examine the fundamental aspects of data architecture, the dynamics of cloud-based automation, the essential human aspects of implementation, and the measurable success metrics, it will in the end offer a roadmap of how to make a genuinely adaptive and future-proof organization.

SECTION 1: THE "WHY" - THE BURNING PLATFORM FOR CHANGE

The reason to embrace a Smart HR Cloud does not just lie in the technological development, but there is a strategic need under the influence of overwhelming macroeconomic and social forces. The most urgent of them is the international skills shortage. According to the estimates of the World Economic Forum (2023), 44% of core skills of workers will be disrupted by the year 2027, and six out of ten workers will need training before that time. Nonetheless, it is observed that only half of the employees are provided with sufficient training opportunities in the present day. Such a disconnect is a direct challenge to productivity, innovation and growth. Firms can no longer implement strategic projects due to lack of internal talent required to drive the projects and end up delaying their projects or missing market opportunities, depending on external contractors who are expensive and in most cases scarce.

The price of doing nothing in this field is immense. Companies that are stuck with manual workforce planning have a major absence of time between the need and the satisfaction. A conventional recruitment procedure can take many months whereby by the time the business requirement has changed or disappeared. On the inside, the managers are blind to the hidden talents that lie within their own organization, so they end up hiring external staff to do the job that the current employees could do. It does not only raise the cost of recruiting but also leads to staff turnover since employees who want to grow move to organizations that can provide them with improved internal mobility (Gartner, 2023). What is needed is a fundamental change in approach, which is a proactive approach instead of a reactive approach, where vacancies are being filled as they appear. The concept of proactive workforce planning entails that future requirements of the skills are projected in accordance with the strategic business goals and talent that is continuously nurtured through the current workforce. It is a change over the perception of HR as an assisting element to the realization of the HR as the driver of strategic implementation. The technological solution that allows making this proactive, strategic approach possible, efficient, and scalable is called the Smart HR Cloud.

SECTION 2: LAYING THE FOUNDATION - DATA AS THE NEW Oil

Intelligence is as smart as data that drives a Smart HR Cloud. The first step in this change is a shift in the unstructured, qualitative descriptions of jobs to a structured, quantitative ontology of skills. This starts with the creation of a sound skill taxonomy. A skill taxonomy is a systematic approach that classifies and systematizes the skills, competencies and knowledge needed in an organization (Tippins et al., 2021). It goes beyond the generic job description that a Marketing Manager can do this or that, such as SEO/SEM Strategy, Marketing Automation Platform Management, Data-Driven Customer Segmentation, and Stakeholder Influence.

This process of developing this taxonomy is an urgent but not an easy one. It is the process of breaking down all the positions in the organization into the essential elements. The best practice is to begin with a hybrid model, using the current public frameworks (such as O*NET) and industry-specific models, and tailoring it to the specific situation and strategic focus of the organization. The idea is to build a living breathing library of skills that can be readily updated with the advent of new technologies and methodologies. After determining the taxonomy, the second step will be to create a dynamic skills inventory. This is the database that is actively updated, which associates individual employees with the skills in the taxonomy. The advantage of the Smart HR Cloud is that it allows automating the process of populating this inventory and updating it. It achieves this by the multi-source data ingestion. These encompass the extraction of structured information on Human Resource Information Systems (HRIS) on tenure and formal roles, analysis of performance reviews and project results on achieved competencies, integration with Learning Management Systems (LMS) on completed certifications and courses, and self-assessments and peer recommendations of internal collaboration tools such as Microsoft Viva or Slack. This gives a rich, multi-dimensional, and ever-changing image of the overall organizational capacity, shifting away, as the organization was hired to do, to one that is dynamic as to what employees are able to do.

SECTION 3: THE ENGINE ROOM - HOW SMART CLOUD AUTOMATION WORKS

Having a clean, structured and dynamic inventory of skills, the analytical capabilities of the Smart HR Cloud can be put to work. This is where data is converted into actionable intelligence using two main automated processes workforce analytics and skill mapping.

3.1 Automated Workforce Analytics:

This feature transforms the HR cloud into a strategic planning application. A supply and demand visualization is one of the key features, typically displayed in informative dashboards and heat maps. The existing skills inventory is the supply side. Strategic business plans, project pipelines, and financial forecasts provide food to the demand side. The system automatically compares the two showing visually the areas of excess (skills that are underutilized), balance, and critical deficit. To give an example, a heat map may depict a vivid red circle in the Data Science category in comparison to the AI Product Launch strategic pillar, which is a direct, unambiguous visual indication of a threat to a critical business initiative. Moreover, Predictive Attrition Modeling is based on machine learning models of historical data (e.g., engagement survey scores, promotion history, compensation changes, even anonymized patterns of collaboration) to predict employees who are at high risk of voluntary turnover (Bauer et al., 2021). This enables the managers and HR business partners to hold proactive retention discussions and make succession plans before a resignation letter is ever presented. Lastly, Scenario Planning tools enable executives to simulate the workforce effect of different scenarios that are referred to as what if. As an example, they can replicate the skill gaps that would arise as a result of joining a new geographic market, introducing a new line of products, or going through a merger and acquisition. This changes workforce planning into an ongoing strategic discussion and not an annual budgetary exercise.

3.2 Intelligent Skill Mapping & Gap Analysis:

It is converting the analytical understanding into particular talent action. The system does Real-time Gap Identification which means the supply-and-demand information is constantly scanned, critical skill gaps automatically identified and flagged. The Head of Engineering may be alerted that according to the Q3 product roadmap, the skill of the Internet of Things (IoT) Cybersecurity will be short of 40% in 90 days. A more advanced system makes use of The Power of Adjacency. This is done by applying AI and graph theory to map skill relations. Assuming that one of the employees already knows how to operate the Python Programming environment, the AI could discover that Data Engineering with Apache Spark is a nearby skill that has a high probability of success in acquisition because the underlying logical structures and paradigms are connected (Tippins et al., 2021). This is past the process of finding gaps to the proactive discovery of the most efficient ways of bridging them. The end result of the process is the development of a Skills DNA of the organization a full dynamic and searchable and map of the cumulative capability, potential and aspirations of the whole workforce, which offers a strategic insight never before seen.

SECTION 4: FROM INSIGHT TO ACTION - CLOSING THE SKILLS GAP

The real worth of the Smart HR Cloud becomes tangible only in case the insights can be translated into specific steps that will seal the gaps in skills. The initial and the strongest lever is the establishment of Personalized Learning Pathways. The system has the ability to suggest particular courses, micro-credentials, mentorship partnerships, and stretch assignments to individual employees automatically based on the gaps found in their team or function and their career ambitions. As an example, an accountant with high potential indicated and a nearby skill match could be suggested a sequence of financial data analytics courses and a mentorship with a senior FP&A manager, which would produce a custom reskilling pathway.

In the case of skills that are inefficiently developed within the organization, Strategic Talent Acquisition is notified through the system. The exact gap analysis data will enable recruiters to create hyper-specific job descriptions that will focus on the candidates who have the exact skills that the organization is lacking. Moreover, the skills taxonomy makes it possible to apply a more advanced screening process, where AI scans profile of the candidates based on the needed competencies, improving the quality of the hire and decreasing the time-to-fill. The ability to facilitate an Internal Talent Marketplace perhaps is the most transformational result. It is a platform, which is driven by the skills inventory, where managers post projects, gigs, and open roles, and employees can browse and apply to them according to their verified skills and interests. It democratizes career development, makes employees more engaged as they are challenged, and makes organizational agility much more dramatic as talent can easily move to the work of the greatest priority, without the drag of the traditional internal process of transferring talent. It is the final manifestation of an organization that is dynamic and skills based.

SECTION 5: THE HUMAN TOUCH - CHANGE MANAGEMENT AND ETHICS

The deployment of Smart HR Cloud is not a technical project per se, rather, it is an extensive organizational change initiative that cannot be successful without the close consideration of the human factor. The most advanced system will not work in case employees and managers lack trust and awareness of its value. The most important thing is to gain Employee Trust. This involves total openness regarding the nature of data being gathered, its usage and accessibility. The system should be placed as an empowerment device aimed at assisting employees to maneuver and build their careers rather than a surveillance system that can be used to detect non-productive employees and fire them. There should be no compromise on strong data governance and communication regarding privacy protection.

At the same time, Upskilling Managers is imperative. Managers are the cogs of this mechanism. They need to be conditioned towards a command and control thinking to that of a coach and career facilitator. They must be educated on how to read the data dashboards, how to have development discussions grounded on skill gap analyses and how to enable their team members to be involved in the internal talent marketplace. Lastly, organizations should be keen to prevent Algorithmic Bias. Historical data is used to train AI models, which may be biased on gender, race, age, or background by humans (Raghavan et al., 2020). Unchecked, an AI that suggests skill adjacencies or finds high-potential employees may also contribute to these biases. It is necessary to have a constant audit of the AI recommendations concerning fairness and inclusiveness. These include the diversification of the data sets, imposing fairness limits on the algorithms and human controls in critical decision making processes.

SECTION 6: MEASURING SUCCESS - KPIS FOR THE SMART HR CLOUD

The effects of the Smart HR Cloud need to be quantified using the corresponding Key Performance Indicators (KPIs) to justify the investment and help sustain the process of continuous improvement. These measures must go beyond the conventional HR metrics to reflect the strategic value of the system. Key KPIs include:

* Time-to-Productivity of New Roles: This will measure how long it takes an employee (internal or external) to become fully knowledgeable in a new job. This should be minimized with an efficient system by better matching and pre-onboarding development.
* Internal Hire Rate / Internal Mobility Rate: The proportion of vacancies that were filled by an internal applicant. Increasing rate means that the system is effectively supporting internal career development and decreasing the use of external hiring, which saves money and increases retention.
* Skill Gap Closure Rate: This will be tracking the rate at which critical skill gaps are reduced over time as a percentage. This is a direct indicator of the effectiveness of the system in creation of future-ready capabilities.
* Employee Engagement Scores: In particular, career development, learning opportunities, and internal mobility scores. These areas should be significantly enhanced through a successful implementation.
* Strategic Initiative Success Rate: Keeping track of the on-time, on-budget completion of the major projects which relied on the reskilling or redeployment of the talent identified by the system.

Practical Implications

The shift towards a Smart HR Cloud has a number of tangible, practical implications on the HR functional work and the organization in general. To begin with, the HR Technology Stack should be integrated and rationalized. It frequently means the replacement of a best-of-breed strategy of having numerous unconnected systems with a more connected cloud-based HCM stack or the creation of a best-of-breed stack with a heavy emphasis on API-led connectivity to guarantee the smooth flow of data. Secondly, there will be a need to change the HR Operating Model. The HR Business Partner (HRBP) position is no longer an administrative specialist and an advocate on behalf of employees but a more data-savvy strategic consultant capable of explaining the workforce analytics to their business executives. The Learning and Talent Acquisition Centers of Excellence should redefine their processes to be powered by the real-time data provided by the cloud, and transition to tailor-made and on-demand solutions instead of standardized programs. Thirdly, Data Governance emerges as an important discipline. An interdisciplinary team should be created to own the skill taxonomy, determine data quality standards, access control, and be in compliance with regulations (e.g., GDPR or CCPA). This involves the introduction of new positions, including HR data analysts and HR data stewards, who authenticate the transition between HR domain expertise and data science.

Managerial Implications

The Smart HR Cloud is a potent instrument and a change of roles to the line managers. Managers acquire Data-Based Decision-Making. They do not have to use their gut feeling to find out what is missing in team skills or the person to bring on board as a successor; they have objective and data-rich dashboards that give them evidence to support their decisions about people, whether to hire, promote or restructure a team. This results in a Coaching-Oriented Leadership style. As the system takes the administrative load of monitoring skills and providing recommendations on where to develop the skills, the managers will be able to concentrate on the human aspects: mentoring their employees, meaningful career discussions, using the information, and assist their team members in participating in gigs and projects within the organization. Agreement on Talent Development Accountability also requires a greater level in this system. Managers can no longer say that they had no idea that there were any skill gaps or flight risks. The information gives early warning signs that are clear and therefore leaders have the responsibility of developing their team proactively and reducing risks associated with talents in the business. This brings people management as a business responsibility to a core part.

CONCLUSION: DEVELOPING AN ADAPTIVE ORGANIZATION

The trip to a Smart HR Cloud deployment is a business change that reinvents the relationship between an organization and its talent. It is a transition between the stagnant, bureaucratic paradigms of the 20 th century to the dynamic, agile and people-centered paradigm of the 21 st century. This chapter has described the critical path: a solid business case cannot be denied, a foundation of clean, structured skills information should be established, cloud automation should be used to make predictive analytics and intelligent skill map, and, last but not least, the whole initiative must be based on solid change management and ethical values. The result is not a more efficient HR department, but a more adaptive organization in general. With such an organization, one can expect a rapid restructuring of its talent to respond to new challenges, predict the market changes, and constantly develop the talent it requires to survive in an uncertain future. Ultimately, the Smart HR Cloud cannot be considered a people management system; it is the technology to create a strong and innovative enterprise that is resilient and future-proof.

REFERENCES

Bauer, T., Erdogan, B., & Taylor, S. (2021). *The Oxford handbook of leader-member exchange*. Oxford University Press.

Deloitte. (2023). *2023 Global Human Capital Trends*. Deloitte Insights. Retrieved from https://www2.deloitte.com/us/en/insights/focus/human-capital-trends.html

Gartner. (2023). *Gartner HR Research Finds 3 Must-Haves to Win the Intensifying Race for Talent*. Gartner Newsroom. Retrieved from https://www.gartner.com/en/newsroom/press-releases/2023-04-04-gartner-hr-research-finds-3-must-haves-to-win-the-intensifying-race-for-talent

Raghavan, M., Barocas, S., Kleinberg, J., & Levy, K. (2020). Mitigating bias in algorithmic hiring: Evaluating claims and practices. In *Proceedings of the 2020 Conference on Fairness, Accountability, and Transparency* (pp. 469–481).

Tippins, N. T., Oswald, F. L., & McPhail, S. M. (2021). *The handbook of personnel assessment and selection*. Wiley-Blackwell.

World Economic Forum. (2023). *Future of Jobs Report 2023*. World Economic Forum. Retrieved from https://www.weforum.org/reports/the-future-of-jobs-report-2023/

Chapter 26

AI-Driven Personalization in Cloud-Powered Fashion:

E-Commerce Customer Experience

Priti Gupta
https://orcid.org/0000-0002-3000-9968
Department of Electrical Engineering, Greater Noida Institute of Technology, Greater Noida, India

Neha Arora
https://orcid.org/0000-0001-5935-2658
Amity School of Fashion Technology, Amity University, Noida, India

Akanksha Kathuria
School of Business, Manav Rachna University, Faridabad, India

Neeraj Chopra
https://orcid.org/0000-0002-0969-3574
School of Business, Manav Rachna University, Faridabad, India

ABSTRACT

Artificial intelligence (AI)-driven personalization has emerged as a transformative force in fashion e-commerce, redefining how consumers engage with online retail ecosystems and elevating customer experience (CX) through tailored interactions. Coupled with the scalability and flexibility of cloud-based infrastructure, modern

DOI: 10.4018/979-8-3373-7442-0.ch026

fashion platforms can analyze extensive customer data in real time, optimize content delivery, and provide individualized shopping experiences that significantly enhance satisfaction, loyalty, and conversion metrics.

INTRODUCTION

Increased revenues, longer client loyalty, and improved purchase outcomes are just a few of the benefits that physical, automated personalization tactics for e-commerce platforms offer. Businesses may provide their customers with a natural and simple purchasing experience because to AI's ability to anticipate client wants. Through its Chabot systems, AI personalization improves inventory forecasts and supply chain efficiency while also assisting customers with virtual purchasing. These technical developments improve business productivity and emphasize customer-focused e-commerce delivery. Artificial Intelligence (AI) has a revolutionary effect on the e-commerce sector since it changes customer-business communications and their personalized buying experiences. Businesses utilize AI-driven personalization to evaluate consumer behavior, product preferences, and previous purchases through machine learning and prediction algorithms with big data analytics. Businesses may create individualized product recommendations, flexible pricing models, marketing directing tactics, and customized user interfaces that increase consumer engagement and loyalty thanks to organization-driven data processing. The development of AI technology over this period guarantees a major advancement in e-commerce personalization standards. Emerging trends such as hyper-personalization, AI-powered voice commerce, augmented reality (AR) retail experiences, and generative AI in content production are transforming the sector. Businesses must strike a balance between advancing technology development and upholding ethical standards if they hope to maintain customer trust.

This chapter looks at the fundamental AI and cloud technologies that enable personalization, their e-commerce applications, and the advantages they provide to both businesses and consumers. The chapter examines the barriers to AI technology adoption while providing real-world instances of its commercial success and outlining potential market trends that could influence AI-enhanced e-commerce customization.

PERSONALIZATION MAKES A DIFFERENCE IN FASHION E-COMMERCE

The fashion industry is fiercely competitive, and consumers are becoming more picky. Fashion firms must come up with ways to differentiate themselves, boost productivity, and please consumers.Providing a flawless shopping experience is now just as important as having the best products. One way to overcome this difficulty is through personalization. Before, during, and after the purchase are all phases of the consumer journey where personalization is present.

Personalization is still crucial during activation and retention, although brands tend to concentrate on everything during user acquisition.

It all comes down to providing a high-quality user experience that will captivate the client and motivate them to act.

When a customer has an emotionally satisfying experience tailored to his evolving demands, he is more likely to purchase from and recommend the company.

It is more challenging to replicate the pleasant in-store buying experience, which includes fitting rooms, the ability for customers to touch the clothing, and salespeople to offer advice.

Indeed, most e-shops have a rather static customer journey: a simple navigation from the category page, to the purchase page, through the product page, which impacts the perception of the brand for its e-commerce customers. This standardization of the e-commerce customer journey is driving consumers to new brands that offer more innovative experiences (Huang & Rust, 2021).

Customers expect more and more personalization of the experience Customers are naturally attracted to brands that value them, that make them feel like they are being listened to, understood and cared for. The Covid-19 pandemic has accentuated this feeling even more, as consumers have only been shopping online for more than a year, and their level of demand has risen sharply. Brand choice now depends heavily on personalization. According to a study conducted by e-commerce platform Shopify 75% of customers value tailored brand content.

- When offered content is irrelevant to them, 74% of customers report feeling annoyed as shown in Figure 1.

*Figure 1. This reflects broader data showing a strong desire for personalized, relevant content — and parallel frustration when brands *fail* to deliver it*

Fashion firms must consequently adopt a hyper-personalized strategy, providing each consumer with a distinctive experience.

CUSTOMIZATION OFFERS INFORMATION TO SATISFY E-COMMERCE CLIENTS' NEEDS

The results that customisation produces for brands are just one of its many advantages. Brands get a lot of information about their consumers in addition to engagement and conversion (habits, preferences, searches, order history, purchase values, etc.).

With the use of this information, they are able to get to know each consumer well, address their specific demands, and even anticipate what they might want from a future purchase.

Veesual's Mix&Match allows marketers to display customers the most relevant photos based on their preferences (based on their interactions with the experience), highlight looks that complement the clothing they've consulted, and modify the models to maximize consumer engagement.

AI CUSTOMIZATION OPERATIONAL FRAMEWORK

Real-time processing, machine learning algorithms, and data collection are all necessary for AI-driven customisation. AI collects and analyzes user data from various sources, including as website visits, past purchases, social media activity, and consumer feedback. Behavioral analysis: By identifying patterns in consumer behavior, machine learning models forecast preferences and future purchases. Personalized Recommendations (Wedel & Kannan, 2016).AI creates personalized product recommendations based on contextual information, past interactions, and user preferences. Continuous Learning and Optimization: Over time, AI systems adjust and improve personalization tactics, increasing precision and efficacy. Operational Framework shown in Figure 2.

Figure 2. Operational framework

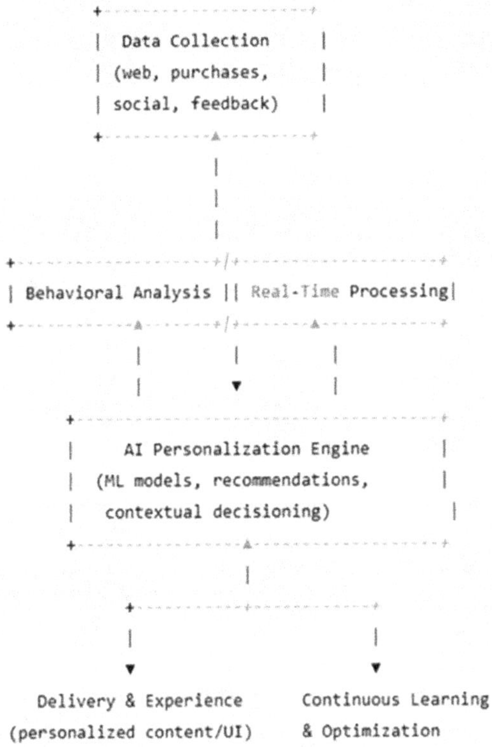

AI is revolutionizing customer experience (CX) through intelligent automation, predictive engagement, and hyper-personalized interactions. To stay ahead in 2025, companies will need to overcome new obstacles like labor upskilling, AI governance, and data privacy. This essay examines the major changes in the market, geographical differences, and practical methods that businesses can use to leverage AI for more intelligent, human-centered CX. Artificial Intelligence (AI) is actively changing how companies engage with their clients; it is no longer a futuristic idea. By improving automation, personalization, and engagement, generative artificial intelligence (GenAI) is leading the way in the transformation of the consumer experience (CX) in 2025.

Businesses have a rare chance to improve customer interactions thanks to AI's expanding influence. CX is becoming quicker, smarter, and more effective thanks to automation, AI-driven chatbots, and predictive analytics. But there are drawbacks to this change as well. Business executives navigating the AI-driven CX landscape are particularly concerned about issues like data protection, legal compliance, and skill gaps in the workforce (Davenport et al., 2020).

The main market changes, difficulties, and tactics that companies need to adopt in order to remain competitive in the AI-powered customer experience will all be covered in this chapter.Time line for traditional brick-and-mortar to e-commerce, cloud adoption, and AI-driven personalization shown in in Table 1.

Table 1. Timeline for transformation

Era / Years	Milestone	Description
1970s–1980s	Traditional Retail	Dominance of physical brick-and-mortar stores with in-person shopping and limited digital engagement.
1990s	Early E-Commerce	The first secure online sales begin. Major platforms like Amazon and eBay launch, pioneering e-commerce. (LinkedIn)
2000s	E-Commerce Growth	E-commerce expands with secure payment systems and increased consumer adoption; online marketplaces flourish. (BigCommerce)
2010s	Omni-Channel & Mobile Commerce	Retail evolves into online + offline integration; mobile commerce grows with smartphones and apps. (Contentful)
Mid-2010s	Cloud Adoption	Retail systems shift to cloud infrastructure for scalability, analytics, and real-time operations. (Contentful)
Late 2010s–2020s	AI & Personalization	AI becomes embedded (recommendations, analytics, dynamic pricing, automation), reshaping customer experiences. (Contentful)
2020s+	AI-First Retail Era	AI goes beyond pilots to core operations — real-time personalization, predictive analytics, and automation drive retail strategy. (Contentful)
Future	Intelligent Commerce	Emerging technologies (AR/VR, voice commerce, autonomous fulfillment) elevate personalized, immersive experiences.

Customer engagement has entered a new era where AI plays a central role in enhancing interactions. Businesses are using AI to analyze customer behavior, provide real-time recommendations, and deliver seamless omnichannel experiences. This not only improves customer satisfaction but also drives higher conversion rates. Marketing and customer service teams are particularly benefiting from AI, with a reported 40% increase in productivity across these roles. AI is taking over repetitive tasks, allowing human employees to focus on more complex, high-value interactions.

AI MEETS CLOUD

Two of the most revolutionary technologies of our day are cloud computing and artificial intelligence (AI), which are changing sectors, spurring innovation, and opening up new business prospects. Gartner projects that the global market for public cloud services will reach $723.4 billion in 2025, a growth of 21.5%. According to Grand View Research, the AI market is expected to reach $1,811.75 billion by 2030, growing at a compound annual growth rate (CAGR) of 36.6% between 2024 and 2030. These numbers highlight how both technologies are being more widely used and how interdependent they are becoming.

However, what happens when cloud computing and AI are combined? While AI improves cloud services with intelligence, automation, and predictive capabilities, cloud computing offers the scalable infrastructure and processing power required for AI applications. Healthcare, banking, retail, and entertainment are just a few of the sectors that this potent synergy is transforming. The combination of AI and cloud computing is propelling previously unheard-of breakthroughs, from AI-powered recommendation engines used by Netflix and Spotify to predictive analytics in healthcare that enhance patient outcomes.

It can be difficult for people who are unfamiliar with IT and AI to comprehend how these technologies interact. AI gives machines the ability to think, learn, and make choices. It encompasses technologies that enable systems to identify patterns, comprehend language, and automate intricate procedures, such as computer vision, machine learning (ML), and natural language processing (NLP) (Verhoef et al., 2021)

Cloud computing is the provision of computer services via the internet, including networking, processing power, and storage. Companies can now acquire computing resources on-demand from cloud providers like AWS, Azure, and GCP, eliminating the need to maintain physical servers.The predicted forecast shown in Figure 3.

Figure 3. Public cloud services spending (2024–2025 forecast) and AI market size (2024–2030 forecast)

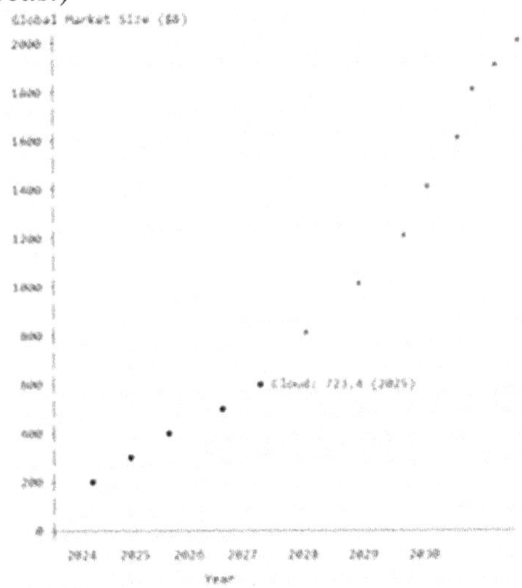

CLOUD COMPUTING ENABLES SCALE & SPEED

In cloud computing, cloud scalability is the capacity to expand or contract IT resources in response to shifting demand. One of the key characteristics of the cloud and the main factor contributing to its rapidly rising popularity among enterprises is scalability.

With the current cloud computing infrastructure, networking, processing power, and data storage capacity can all be expanded. Even better, scaling may be done fast and simply, usually with little to no downtime or disturbance. In the past, scaling with on-premises physical infrastructure could take weeks or months and be very expensive; however, third-party cloud providers have all the infrastructure already in place. Scalable business models enable successful companies to expand rapidly and adapt to shifting market conditions. Their IT is no different. The benefits of cloud scalability enable firms to maintain their agility and competitiveness.

One of the main motivators for moving to the cloud is scalability. Organizations can respond appropriately and economically to boost storage and performance with a scalable cloud solution, regardless of whether traffic or workload demands increase abruptly or gradually over time.

Companies can use public, private, or hybrid clouds to set up a customized, scalable cloud solution. Vertical scaling, sometimes referred to as "scaling up" or "scaling down," is the process of increasing or decreasing the memory (RAM), storage, or processing capacity (CPU) of an already-existing cloud server. This typically indicates that scaling has an upper limit determined by the server's or machine's capability; scaling beyond that frequently necessitates downtime.

By adding additional resources, like as servers, to your system, you can expand horizontally (also known as scaling in or out). This boosts performance and storage capacity by distributing the burden among multiple machines. For companies with high availability services that need little downtime, horizontal scalability is particularly crucial (Zhang et al., 2019).

CLOUD-BASED AI WORKFLOW AUTOMATION: INTELLIGENT BUSINESS OPERATIONS

From simple job execution to intelligent systems that adapt, optimize, and function independently, AI workflow automation has come a long way. Organizations may automate complicated business operations without being constrained by legacy hardware or siloed systems thanks to this change, which is driven by cloud infrastructure.

Cloud-based AI automation improves decision-making, guarantees operational consistency, and yields quantifiable productivity increases. Industries ranging from customer operations to finance are depending more and more on these technologies to efficiently and precisely handle high-volume procedures.At a compound annual growth rate (CAGR) of 17.85%, the worldwide cloud workflow market is expected to increase from USD 3.41 billion in 2023 to USD 14.81 billion by 2032. Global Cloud Workflow Market (2023–2032) shown in Figure 4 and Table 2.

Figure 4. Global cloud workflow market (2023–2032)

Table 2. Global cloud workflow market

Metric	Value
2023 Market Size (Base Year)	USD 3.41 billion
2032 Projected Market Size	USD 14.81 billion
Forecast Period	2024–2032
Compound Annual Growth Rate (CAGR)	17.85 % (2024–2032)
Key Growth Drivers	Cloud-based automation adoption, AI integration, digital transformation
Major Regions Covered	North America, Europe, Asia-Pacific, MEA, Latin America

The use of artificial intelligence (AI) integrated with cloud-hosted workflow automation platforms to manage, optimize, and carry out business processes without human interaction is known as “cloud-based AI workflow automation.” It improves operational efficiency, decision-making, and task automation by fusing the intelligence of AI with the scale and flexibility of cloud computing.

PERSONALITY VS. PERSONALIZATION" IN AI SYSTEMS

AI conversational technologies are extremely individualized. Businesses are concentrating on providing individualized experiences that are customized to users' tastes, behaviors, and virtual and physical environments across all industries. These include general-purpose LLMs and the quickly expanding market for corporate assistants, educational aids, and AI companions driven by LLMs. There are distinct trends in this general focus, including the expansion of short- and long-term "memory," increased access to systems, systems with increasingly distinct "personalities," and systems that are more personalized to individual users through the collection and inference of personal information. These developments are pushing the boundaries of tort, product responsibility, consumer protection, and data protection regulations, all of which have new implications for US law.

There is a difference between two trends in this first piece of a multi-part blog post series: "personality" and "personalization." Both have practical applications; future blog entries will delve deeper into these and examine specific hazards as well as possible risk mitigation for each category(Ricci et al., 2022).

Features of AI systems that adjust to a specific user's preferences, behavior, history, or surroundings are referred to as personalization. Conversational AI systems raise important privacy, transparency, and consent issues as their capacity to infer and retain information grows through a number of mechanisms (such as larger context windows, memory, system prompts, and retrieval augmented generation) and as they are granted more access to data and content.

The human-like characteristics and actions (such as being amiable, succinct, funny, or skeptical) that are becoming more prevalent in conversational systems are referred to as personality. The more human-like characteristics of interacting AI systems might raise new concerns, such as manipulation, over-reliance, and emotional dependency, which in extreme situations has resulted in delusional conduct or self-harm, even in the absence of memory or data-driven customisation. While a smaller context window might make it more difficult for the model to manage lengthier prompts or retain coherence during protracted talks, a wider context window enables the model to comprehend and react to more complicated and lengthy cues (Grewal, Roggeveen, & Nordfält, 2017).Meta AI helps to retain certain information about you, such as your love of travel and language acquisition, and it can also identify significant details based on context. For instance, suppose you ask Meta AI for some breakfast ideas because you're hungry. You reply in the chat window to inform Meta AI that you are vegan when it recommends an omelette or a gourmet frittata. Meta AI is able to retain such data and utilize it to guide recipe suggestions in the future as shown in Figure 5.

Figure 5. Personality vs. personalization

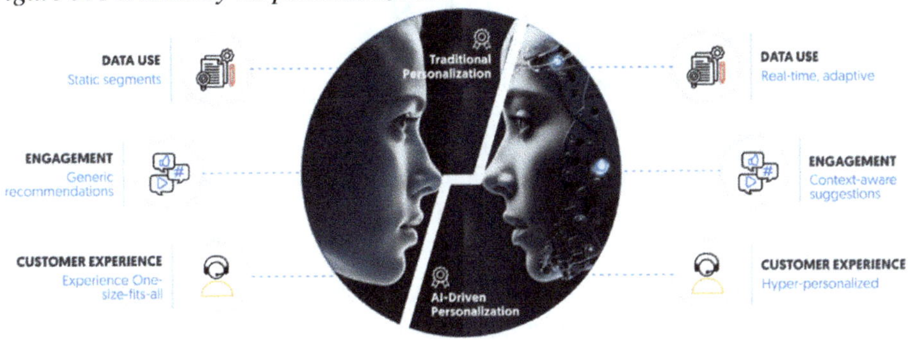

In order to provide experiences, services, and interactions that are genuinely tailored to each user, AI personalization, also known as personalization by artificial intelligence, is based on a methodology based on the thorough analysis of customer data.

Creating memorable moments in accordance with client expectations and particular behaviours is the aim. This technology makes use of artificial intelligence's remarkable capabilities, particularly machine learning, to examine and interpret customer preferences and behaviours across a wide range of channels and touchpoints (Armbrust et al., 2010).

AI personalization uses a variety of data types, including transactional, behavioural, psychographic, and demographic, rather than just simple segmentation. Boosted Personalization shown in Figure 6.

Figure 6. AI boosted personalization

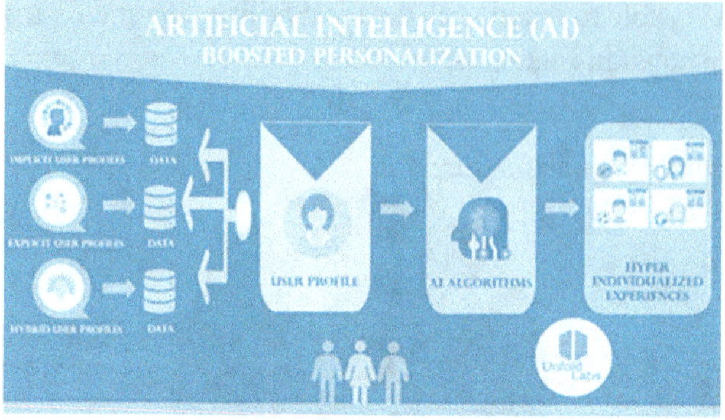

In order to provide experiences, services, and interactions that are genuinely tailored to each user, AI personalization, also known as personalization by artificial intelligence, is based on a methodology based on the thorough analysis of customer data.

Creating memorable moments in accordance with client expectations and particular behaviors is the aim. This technology makes use of artificial intelligence's remarkable capabilities, particularly machine learning, to examine and interpret customer preferences and behaviours across a wide range of channels and touchpoints (Marston et al., 2011).

AI personalization uses a variety of data types, including transactional, behavioural, psychographic, and demographic, rather than just simple segmentation.

AI PERSONALISATION WORKFLOW

A well-oiled system based on a number of crucial phases underlies the seeming smoothness of AI-personalized experiences. Data collecting is the first step in the process. Websites, mobile apps, social networks, e-commerce platforms, CRMs, and many other sources can provide this valuable yet unprocessed data. They create a raw material that AI will progressively mold.

After being centralized, machine learning techniques are used to process this data. This is where the magic happens: AI recognizes implicit preferences, finds recurrent patterns, and decodes weak signals that occasionally evade traditional human analysis. It is a dynamic correlation of behaviors, situations, and intentions rather than a straightforward sorting procedure (Kim & Forsythe, 2008).

This comprehension mostly depends on machine learning methods like unsupervised learning, when the objective is to identify groups or typical profiles without prior indication, or supervised learning, which uses labelled samples to create predictions.

Another option is reinforcement learning, in which an algorithm improves its suggestions while experimenting with various actions and monitoring the outcomes, akin to a system that learns by trial and error.AI personalization is radically changing how businesses communicate with their customers—not through set routes or pre-made profiles, but through encounters that are nearly always customized (Blázquez, 2014).

The AI assistant goes beyond straightforward responses by honing its responses over exchanges. It predicts, modifies speech, identifies implicit preferences, and conforms to stated—or suggested—needs. In a world where prospects are constantly being contacted and where it is necessary to both attract and hold their attention, such a resource shows to be quite pertinent.

Offering something genuinely different from what rivals provide can be challenging, depending on the market. AI and thus enhanced personalization could be the differentiator. In order to be competitive, this new dynamic should not be disregarded (McKinsey & Company, 2023).

HYPER-PERSONALISATION THE KEY TO THE FUTURE

Hyper-personalisation today goes beyond a trend and is gradually becoming an operational necessity. Far exceeding traditional customer segmentation, it allows marketing and customer relations professionals to enter into an almost individualised interaction logic, where every message, every offer, every product recommendation seems designed for a person—not for a customer category. Concretely, this means the ability to adjust messaging in real time, based on weak signals emitted by the user—whether it's a purchase history, browsing behaviour, or even simple silence. This level of subtlety, impossible to achieve without the support of well-trained AI, redefines the standards of a successful customer experience.

The influence of AI personalization on sales is among the first obvious consequences. Businesses promote recurring, frequently more impulsive, and occasionally emotional purchases by tailoring offers to customers' expectations and preferences. Repetition fosters connection, a relationship with the brand that transcends simple cost-effectiveness.

In order to build a long-lasting relationship, it is now crucial to provide each person with an experience that is similar to them, or at least feels tailored to them. Customers are more likely to return when they feel understood, not only because of the product but also because they felt that the encounter was attentive. This holds true for both B2C and many B2B situations, where difference now boils down to the specifics. Personalization becomes a sign of maturity as it is included into procedures. Early adopters gain a lead that is difficult to overcome. While others continue to use generic methods, they produce unforgettable encounters. According to McKinsey, high-growth businesses credit sophisticated personalization for up to 40% of their extra revenue.

Sharper, more useful consumer knowledge

Data quality is a deeper issue that lies behind these apparent advantages. When correctly powered, AI turns data collection into intelligence that can be put to use. It enables businesses to watch, contrast, and comprehend. Both marketing initiatives and more general strategy choices are improved by these insights. You can anticipate changes in the market more accurately if you know your customers better.

TYPE OF DATA

Customer data, in its most basic form, is whatever information you are aware of about your clients, be they persons, companies, or organizations. Contact details, demographic information, consumer preferences, internet interactions, and more are all included in this.

Gathering and evaluating consumer data gives you a deeper understanding of your target market. It makes their needs and desires clear, as well as the beliefs that motivate them. This gives you the ability to make more data-driven choices for your sales, marketing, and promotional initiatives in order to draw in and keep more clients.

Characteristics like age, gender, income, and educational attainment that help you identify your clientele are referred to as demographic data.

Knowing who your customers are and what they could require is much easier with this kind of information. Additionally, it can be used to segment your clientele for full-fledged campaigns and marketing plans.

For instance, you could want to concentrate your marketing efforts on the age groups who are most likely to buy your anti-wrinkle cream. Your advertising will be more focused and produce better results if you collect demographic information about your clientele.

Your selling strategy and messaging can be influenced by psychographic data, which includes a customer's values, personality traits, opinions, attitudes, beliefs, and way of life (Amazon Web Services, 2022)

Using demographic information to target a segment of the market is no longer sufficient. Consider psychographic information as the next level. It's straightforward: psychographic information helps you appeal to the emotional aspect of a customer's purchasing decision, whereas demographic information may help you appeal to their logical or analytical side.

Purchase history, online interactions, and subscribing to your newsletter in exchange for a discount are all examples of behavioral data. Put otherwise, it refers to the behaviors and acts that your consumers engage in.

This kind of information enables you to comprehend how your clients have previously engaged with your company, which may offer insights into their future demands and desires.

Continuing with the wrinkle cream scenario above, for instance, individuals of a particular age who visit your homepage and browse through some of the featured product pages have both buyer intent and some degree of brand familiarity. BBuyer intent insights strengthen when you track user behavior data and see that they came back a day later to sign up for your brand newsletter. (Sun, Teng, & Chen, 2017).

Information regarding the transactions your clients have made with your company is known as transactional data. This can contain information such as the quantity and frequency of purchases, the time it takes the client to make those purchases, and the number of returned items.

This kind of information aids marketers in comprehending consumer purchasing trends and patterns. It's particularly crucial when new factors join the purchasing equation, new technology is introduced, and market pressures change (Pantano, Pizzi, Scarpi, & Dennis, 2020).

Constant support via virtual assistants and chatbots

While occasionally covert, intelligent chatbots are frequently incredibly successful. They are among the customer ecosystem's most obvious AI access points. Their promise: immediate answers, 24/7, to an ever-increasing number of requests. However, these assistants do more than just automate responses; they also learn, adapt, and contextualize. They consider recent behaviors, preferences, and past experiences.This kind of information might be useful in figuring out what drives your target audience and how to best meet their demands.

For instance, let's say you have a website that sells holiday cards but doesn't target any specific faith. Knowing a customer's religion allows you to customize your messaging, product recommendations, and promotional communications to reflect their values. Then, when a Christian holiday like Easter approaches, for instance, you can send a promotional eblast to clients you know are Christians, who might be motivated to buy something, without unintentionally offending or turning off clients who follow other religions (Accenture, 2021).

AR VIRTUAL TRY-ON IMPROVES RETAIL EXPERIENCES

From Snapchat effects that make you look like your favorite animal, augmented reality (AR) has advanced significantly. Although AR was present in stores before to the pandemic, 67% of consumers say they now shop differently than they did prior to COVID-19.

Retailers who are ready to cater to these customers will expand their online sales offerings and adopt the technologies required to provide customers with an enjoyable shopping experience.

Today's consumers look for virtual solutions for convenience and to escape big crowds. A large reduction in overhead might also be advantageous to retailers. However, sophisticated technologies that let customers see exactly what they're purchasing are needed for a virtual shopping experience that can compete with a physical store in the retail industry. Virtual try-on capabilities come into play here.

AR is compatible with many different devices. Consumers can try on anything, including cosmetics, special occasion attire, and basics like eyeglasses (Pillai, Sivathanu, & Dwivedi, 2020).

AR-powered virtual try-on

Smartphones, tablets, wearables, laptops, TVs, virtual mirrors, and even AR fitting rooms are just a few of the many devices that can use augmented reality. AR uses the camera and other technology on a portable device, such as a smartphone or tablet, to collect data about the surrounding actual world or, in the instance of a virtual try-on, a selfie or video. Then, by superimposing more contextual data that depicts the product the buyer tries on, the software enhances the real surroundings Chattaraman et al., 2019).

Customers who want to benefit from the ease of online buying will benefit from the virtual try-on experience.A captivating experience is offered by cutting-edge technology that lets customers virtually try on everything from shoes and clothes to eyewear. Additionally, it enables businesses to use a range of marketing strategies to connect with new clients.

Retailers can target devoted fans with gift cards or promotions that showcase new products, time-limited offers, or top sellers through interactive advertisements on social media and search pages, as well as brand websites devoted to personalizing unique products. Take a look at these three ways virtual try-on technology can increase your online retail sales and enhance the customer experience.The majority of consumers have had the regrettable experience of buying something that appears fantastic on a model only to discover that it isn't even close to the ideal size, color, or form for them. Because of this, online shopping may seem like a risky endeavor.

Maybe this explains why 63% of consumers claimed that augmented reality will enhance their shopping experience.

Customers may view how specific products appear on them from the comfort of their own homes thanks to virtual try-on technology (Gao, Li, Liu, & Fang, 2021).

CONCLUSION

Personalized customer experiences, dynamic pricing strategies, improved security, and faster operations are all made possible by the integration of AI and Cloud, which represents a revolutionary moment in the digital retail sector. In order to promote responsible and inclusive AI research, ethical concerns about bias, data privacy, and job displacement must be addressed head-on, even while the potential benefits are indisputable. E-commerce is reliable, effective, and tailored for everyone. To create strong ethical frameworks for AI in e-commerce, further study is required, with an emphasis on reducing bias, protecting data privacy, and enabling human-in-the-loop monitoring. In order to shape the ethical development and application of AI and cloud and guarantee that this revolutionary technology helps both enterprises and society, collaborative initiatives including researchers, companies, legislators, and consumers are essential.

REFERENCES

Accenture. (2021). *Personalization at scale: AI-driven customer experience.* https://www.accenture.com

Amazon Web Services. (2022). *Machine learning for personalization in retail.* https://aws.amazon.com

Armbrust, M., Fox, A., Griffith, R., Joseph, A. D., Katz, R., Konwinski, A., & Zaharia, M. (2010). A view of cloud computing. *Communications of the ACM, 53*(4), 50–58. DOI: 10.1145/1721654.1721672

Blázquez, M. (2014). Fashion shopping in multichannel retail: The role of technology in enhancing the customer experience. *International Journal of Electronic Commerce, 18*(4), 97–116. DOI: 10.2753/JEC1086-4415180404

Chattaraman, V., Kwon, W.-S., Gilbert, J. E., & Ross, K. (2019). Should AI-based, conversational digital assistants employ social- or task-oriented interaction style? *Computers in Human Behavior, 90*, 315–330. DOI: 10.1016/j.chb.2018.08.048

Davenport, T. H., Guha, A., Grewal, D., & Bressgott, T. (2020). How artificial intelligence will change the future of marketing. *Journal of the Academy of Marketing Science, 48*(1), 24–42. DOI: 10.1007/s11747-019-00696-0

Gao, B., Li, X., Liu, S., & Fang, D. (2021). How power distance affects online hotel ratings: The moderating role of AI-based personalization. *Tourism Management, 82*, 104204. DOI: 10.1016/j.tourman.2020.104204

Grewal, D., Roggeveen, A. L., & Nordfält, J. (2017). The future of retailing. *Journal of Retailing, 93*(1), 1–6. DOI: 10.1016/j.jretai.2016.12.008

Huang, M.-H., & Rust, R. T. (2021). Artificial intelligence in service. *Journal of Service Research, 24*(1), 3–18. DOI: 10.1177/1094670520902266

Kim, J., & Forsythe, S. (2008). Adoption of virtual try-on technology for online apparel shopping. *Journal of Interactive Marketing, 22*(2), 45–59. DOI: 10.1002/dir.20113

Marston, S., Li, Z., Bandyopadhyay, S., Zhang, J., & Ghalsasi, A. (2011). Cloud computing—The business perspective. *Decision Support Systems, 51*(1), 176–189. DOI: 10.1016/j.dss.2010.12.006

McKinsey & Company. (2023). *The state of AI in retail and fashion.* https://www.mckinsey.com

Pantano, E., Pizzi, G., Scarpi, D., & Dennis, C. (2020). Competing during a pandemic? Retailers' ups and downs during the COVID-19 outbreak. *Journal of Business Research, 116*, 209–213. DOI: 10.1016/j.jbusres.2020.05.036

Pillai, R., Sivathanu, B., & Dwivedi, Y. K. (2020). Shopping intention at AI-powered automated retail stores (AIPARS). *Journal of Retailing and Consumer Services, 57*, 102207. DOI: 10.1016/j.jretconser.2020.102207

Ricci, F., Rokach, L., & Shapira, B. (2022). *Recommender systems handbook* (3rd ed.). Springer., DOI: 10.1007/978-1-0716-2197-4

Sun, J., Teng, J. T. C., & Chen, Y. (2017). Product recommendation system for e-commerce websites: An example of clothing retailing. *Decision Support Systems, 92*, 57–65. DOI: 10.1016/j.dss.2016.09.008

Verhoef, P. C., Broekhuizen, T., Bart, Y., Bhattacharya, A., Dong, J. Q., Fabian, N., & Haenlein, M. (2021). Digital transformation: A multidisciplinary reflection and research agenda. *Journal of Business Research, 122*, 889–901. DOI: 10.1016/j.jbusres.2019.09.022

Wedel, M., & Kannan, P. K. (2016). Marketing analytics for data-rich environments. *Journal of Marketing, 80*(6), 97–121. DOI: 10.1509/jm.15.0413

Zhang, S., Yao, L., Sun, A., & Tay, Y. (2019). Deep learning based recommender system: A survey and new perspectives. *ACM Computing Surveys, 52*(1), 1–38. DOI: 10.1145/3285029

Compilation of References

Abbas, S., Asif, M., Rehman, A., Alharbi, M., Khan, M. A., & Elmitwally, N. (2024). Emerging research trends in artificial intelligence for cancer diagnostic systems: A comprehensive review.

Abeysekera, L., & Dawson, P. (2015). Motivation and cognitive load in the flipped classroom: Definition, rationale and a call for research. *Higher Education Research & Development*, *34*(1), 1–14. DOI: 10.1080/07294360.2014.934336

Abu Hashima, M., & Al-Azab, M. (2024). Entrepreneurship in the age of artificial intelligence: A unique set of essential skills and competencies. *International Journal of Artificial Intelligence in Education and Training*, *4*(2).

Accenture. (2021). *Personalization at scale: AI-driven customer experience*. https://www.accenture.com

Adadi, A., & Berrada, M. (2018). Peeking inside the black-box: A survey on explainable artificial intelligence (XAI). *IEEE Access : Practical Innovations, Open Solutions*, *6*, 52138–52160.

Adler, N. J., & Gundersen, A. (2008). *International dimensions of organizational behavior* (5th ed.). Thomson South-Western.

Afzal, S. (2019). *Recommender systems for technology enhanced learning*. Springer International Publishing.

Agarwal, S., & Hauswald, R. (2010). Distance and private information in lending. *Review of Financial Studies*, *23*(7), 2757–2788.

Agrawal, A., Gans, J., & Goldfarb, A. (2018). *Prediction machines: The simple economics of artificial intelligence*. Harvard Business Review Press.

Agüera, Y., & Arcas, B. (2022). Do Large Language Models Understand Us? *Daedalus*, *151*(2), 183–197. DOI: 10.1162/DAED_a_01909

Ahmed, H. O. K. (2016). Flipped learning as a new educational paradigm: An analytical critical study. *European Scientific Journal, 12*(10).

Ahuja, I. P. S., & Khamba, J. S. (2008). Total productive maintenance: Literature review and directions. *International Journal of Quality & Reliability Management, 25*(7), 709–756. DOI: 10.1108/02656710810892786

Ahuja, J., Yadav, M., & Sergio, R. (2022). Green leadership and pro-environmental behaviour: A moderated mediation model with rewards, self-efficacy and training. *International Journal of Ethics and Systems, 39*. Advance online publication. DOI: 10.1108/IJOES-02-2022-0041

Akerlof, G. A., & Spence, M. (1970). L'asymétrie d'information. *The American Economic Review, 60*(3), 393–410.

Amazon Web Services. (2022). *Machine learning for personalization in retail.* https://aws.amazon.com

Amodei, D., Olah, C., Steinhardt, J., Christiano, P., Schulman, J., & Mané, D. (2016). Concrete problems in AI safety. *arXiv preprint arXiv:1606.06565.*

Andronie, M., Blazek, R., Iatagan, M., Skýpalová, R., Uţă, C., Dijmarescu, A., Kovacova, M., Grecu, G., Pârvu, I., Strakova, J., Guni, C., Zábojník, S., Chiru, C., Novák Sedláčková, A., Novák, A., & Dijmarescu, I. (2024). Generative artificial intelligence algorithms in Internet of Things blockchain-based FinTech management. *Oeconomia Copernicana, 15*(1), 1–25.

Aneesh, A. (2015). *Neutral accent: How language, labor, and life become global.* Duke University Press.

Angrave, D., Charlwood, A., Kirkpatrick, I., Lawrence, M., & Stuart, M. (2016). HR and analytics: Why HR is set to fail the big data challenge. *Human Resource Management Journal, 26*(1), 1–11.

Ansoff, H. I. (1965). *Corporate strategy: An analytic approach to business policy for growth and expansion.* McGraw-Hill.

Ariful Islam, M., Mridha, M. F., Safran, M., Alfarhood, S., & Mohsin Kabir, M. (2025). Revolutionizing Brain Tumor Detection Using Explainable AI in MRI Images. *NMR in Biomedicine, 38*(3), e70001. PMID: 39948696

Armbrust, M., Fox, A., Griffith, R., Joseph, A. D., Katz, R., Konwinski, A., & Zaharia, M. (2010). A view of cloud computing. *Communications of the ACM, 53*(4), 50–58. DOI: 10.1145/1721654.1721672

Arora, P. (2024). *From pessimism to promise: Lessons from the Global South on designing inclusive tech.* MIT Press.

Arumugam, D., Balasingam, S., & Agila, G. (2025). Big data in financial markets: Advanced risk assessment and fraud detection frameworks using RNN. In *Proceedings of the IEEE AUTOCOM 2025 Conference*.

Balasubramanyam, A., Ramesh, R., Sudheer, R., & Honnavalli, P. B. (2024). Revolutionizing healthcare: A review unveiling the transformative power of digital twins. *IEEE Access : Practical Innovations, Open Solutions, 12*, 69652–69676.

Ball, K. (2021). Electronic monitoring and surveillance in the workplace: Literature review and policy recommendations. *Journal of Business Ethics, 170*(4), 609–625.

Bannya, A., Bainbridge, H., & Chan-Serafin, S. (2022). Hr practices and work relationships: A 20 year review of relational hrm research. *Human Resource Management, 62*(4), 391–412. DOI: 10.1002/hrm.22151

Barney, J. (1991). Firm resources and sustained competitive advantage. *Journal of Management, 17*(1), 99–120.

Barocas, S., & Selbst, A. D. (2016). Big data's disparate impact. *California Law Review, 104*, 671–732.

Barrett, L. F., Adolphs, R., Marsella, S., Martinez, A. M., & Pollak, S. D. (2019). Emotional expressions reconsidered: Challenges to inferring emotion from human facial movements. *Psychological Science in the Public Interest, 20*(1), 1–68. DOI: 10.1177/1529100619832930 PMID: 31313636

Bauer, T., Erdogan, B., & Taylor, S. (2021). *The Oxford handbook of leader-member exchange*. Oxford University Press.

Benchsci. (2025). *benchsci*. www.benchsci.com

Benjamin, R. (2023). Race after technology. In *Social Theory Re-Wired* (pp. 405–415). Routledge.

Benlian, A., Klumpe, J., & Hinz, O. (2020). Mitigating the intrusive effects of smart home assistants by using anthropomorphic design features: A multimethod investigation. *Information Systems Journal, 30*(6), 1010–1042.

Berg, F., Kölbel, J. F., & Pavlova, A. (2022). The ESG-returns relationship: A meta-analysis. *Journal of Financial Economics, 144*(1), 1–32.

Berghout, T. (2024). The Neural Frontier of Future Medical Imaging: A Review of Deep Learning for Brain Tumor Detection. *Journal of Imaging*, *11*(1), 2. PMID: 39852315

Berg, T., Burg, V., Gombović, A., & Puri, M. (2019). On the rise of FinTechs: Credit scoring using digital footprints. *Review of Financial Studies*, *32*(5), 1785–1820.

Bersin, J. (2017). *The disruptive, integrated, and personalized learning systems are here*. Josh Bersin Academy. https://joshbersin.com/2017/12/the-disruptive-integrated-and-personalized-learning-systems-are-here/

Bersin, J. (2019). *Learning in the flow of life*. Josh Bersin Academy. https://joshbersin.com/2019/03/learning-in-the-flow-of-life/

Bersin, J. (2020). *The rise of the empathetic leader: How the pandemic is changing HR and leadership forever*. Josh Bersin Academy.

Bessen, J. E. (2019). AI and jobs: The role of demand. In *The Economics of Artificial Intelligence: An Agenda*. University of Chicago Press.

Besson, M. (2019). *The Upskilling Crisis: How to Build a Future-Ready Workforce*. Gartner.

Bhatt, P. C., Hsu, Y.-C., Lai, K.-K., & Drave, V. A. (2025). From transactions to transformations: A bibliometric study on technology convergence in e-payments.

Bijker, W. E. (1994). Sociohistorical technology studies. *Handbook of science and technology studies*, 229-256.

Birhane, A. (2020). Algorithmic colonization of Africa. *Script-ed*, *17*, 389.

Bishop, J. L., & Verleger, M. A. (2013). *The flipped classroom: A survey of the research*. In *Proceedings of the 2013 ASEE Annual Conference & Exposition*. American Society for Engineering Education.

Black, J. S., van Esch, P., & Johnson, R. D. (2021). AI-enabled recruiting: What is it and how should it be used? *Business Horizons*, *64*(6), 763–774.

Blanc-Brude, V., & Defélix, C. (2024). Of chips and men: When working in Industry 4.0 is more human than expected. *Annales Des Mines - Gérer & Comprendre, N° 157*(3), 22–31. https://doi.org/DOI: 10.3917/geco1.157.0022

Blázquez, M. (2014). Fashion shopping in multichannel retail: The role of technology in enhancing the customer experience. *International Journal of Electronic Commerce*, *18*(4), 97–116. DOI: 10.2753/JEC1086-4415180404

Boddington, P. (2017). *Towards a code of ethics for artificial intelligence*. Springer.

Bogue, R. (2016). Growth in warehouse robotics: The success stories continue. Industrial Robot. *International Journal (Toronto, Ont.)*, *43*(6), 583–587. DOI: 10.1108/IR-07-2016-0184

Bolaños, F., Salatino, A., Osborne, F., & Motta, E. (2024). Artificial intelligence for literature reviews: opportunities and challenges. In *Artificial Intelligence Review* (Vol. 57, Issue 9). Springer Netherlands. https://doi.org/DOI: 10.1007/s10462-024-10902-3

Borenstein, J., & Howard, A. (2021). Emerging challenges in AI and the need for AI ethics education. *AI and Ethics*, *1*(1), 61–65. DOI: 10.1007/s43681-020-00002-7 PMID: 38624388

Boudreau, J. W., & Rice, S. (2015). *The new organization: Different by design.* McGraw Hill Professional.

Bousdekis, A., Magoutas, B., Apostolou, D., & Mentzas, G. (2017). Predictive maintenance using data-driven approaches: A review. *Procedia CIRP*, *67*, 268–273. DOI: 10.1016/j.procir.2017.03.015

Boushey, H., & Glynn, S. J. (2012). *The economic benefits of family and medical leave insurance.* Center for American Progress.

Bowker, G., Star, S. L., Gasser, L., & Turner, W. (2014). *Social science, technical systems, and cooperative work: Beyond the great divide.* Psychology Press.

Brougham, D., & Haar, J. (2018). Smart technology, artificial intelligence, robotics, and algorithms (STARA): Employees' perceptions of our future workplace. *Journal of Management & Organization*, *24*(2), 239–257.

Brown, T. B., Mann, B., Rydcr, N., Subbiah, M., Kaplan, J., Dhariwal, P., Neelakantan, A., Shyam, P., Sastry, G., Askell, A., Agarwal, S., Herbert-Voss, A., Krueger, G., Henighan, T., Child, R., Ramesh, A., Ziegler, D. M., Wu, J., Winter, C., . . . Amodei, D. (2020). Language Models are Few-Shot Learners Tom. *Advances in Neural Information Processing Systems, NeurIPS*. https://commoncrawl.org/the-data/

Brynjolfsson, E., & McAfee, A. (2014). *The second machine age: Work, progress, and prosperity in a time of brilliant technologies.* WW Norton & Company.

Brynjolfsson, E., & McAfee, A. (2017). *Machine, platform, crowd: Harnessing our digital future.* W. W. Norton & Company.

Bubeck, S., Chandrasekaran, V., Eldan, R., Gehrke, J., Horvitz, E., Kamar, E., . . . Zhang, Y. (2023). Sparks of artificial general intelligence: Early experiments with GPT-4. *arXiv preprint arXiv:2303.12712.*

Buhmann, A., & Fieseler, C. (2022). Deep learning meets deep democracy: Deliberative governance and responsible innovation in artificial intelligence. *Business Ethics Quarterly*, *33*(1), 146–179. DOI: 10.1017/beq.2021.42

Buolamwini, J., & Gebru, T. (2018). Gender shades: Intersectional accuracy disparities in commercial gender classification. *Proceedings of the 1st Conference on Fairness, Accountability and Transparency, 81*, 77-91. http://proceedings.mlr.press/v81/buolamwini18a.html

Buolamwini, J., & Gebru, T. (2018, January). Gender shades: Intersectional accuracy disparities in commercial gender classification. In *Conference on fairness, accountability and transparency* (pp. 77-91). PMLR.

Burt, A., & Dworak, W. (2021). *The AI-powered enterprise: How to build an AI-driven organization.* Life Tree Press.

California Consumer Privacy Act (CCPA), Cal. Civ. Code § 1798.100 et seq. (2018).

Calzada, I. (2025). Decentralizing AI Economics for Poverty Alleviation: Web3 Social Innovation Systems in the Global South. *AI*, *6*(12), 309.

Cantner, U., et al. (2021). Entrepreneurial lifecycle and innovation dynamics.

Cappellani, F., Card, K. R., Shields, C. L., Pulido, J. S., & Haller, J. A. (2024). Reliability and accuracy of artificial intelligence ChatGPT in providing information on ophthalmic diseases and management to patients. *Eye (Basingstoke)*, *38*(7), 1368–1373. DOI: 10.1038/s41433-023-02906-0 PMID: 38245622

Cappelli, P., & Keller, J. R. (2017). The historical context of talent management. In Collings, D. G., Mellahi, K., & Cascio, W. F. (Eds.), *The Oxford handbook of talent management*. Oxford University Press.

Castelvecchi, D. (2016). Can we open the black box of AI? *Nature*, *538*(7623), 20–23. PMID: 27708329

Ceci, S. J. (1996). On Intelligence. *On Intelligence*, 1–2. https://doi.org/DOI: 10.4159/9780674029316

Chalmers, D.. (2021). Generative AI and startup transformations. *Entrepreneurship Theory and Practice*, *45*(3), 567–589.

Chattaraman, V., Kwon, W.-S., Gilbert, J. E., & Ross, K. (2019). Should AI-based, conversational digital assistants employ social- or task-oriented interaction style? *Computers in Human Behavior*, *90*, 315–330. DOI: 10.1016/j.chb.2018.08.048

Chen, Y., Wang, Y., & Chen, N.-S. (2014). Is FLIP enough? Or should we use the FLIPPED model instead? *Computers & Education, 79*, 16–27. DOI: 10.1016/j.compedu.2014.07.004

Chernyavskiy, A., Ilvovsky, D., & Nakov, P. (2021). Transformers: "The End of History" for NLP? *Joint European Conference on Machine Learning and Knowledge Discovery in Databases*. https://arxiv.org/abs/2105.00813

Chesney, R., & Citron, D. (2019). Deep fakes: A looming challenge for privacy, democracy, and national security. *Lawfare Research Paper Series, 1*(1), 1-25.

Choudhary, S., Yadav, M., & Srivastava, A. P. (2021). Cognitive Biases Among Millennial Indian Investors: Do Personality and Demographic Factors Matter? *FIIB Business Review, 13*(1), 106–117. DOI: 10.1177/23197145211057343

Choudhary, V., Marchetti, A., Shrestha, Y., & Puranam, P. (2023). Human-ai ensembles: When can they work? *Journal of Management, 51*(2), 536–569. DOI: 10.1177/01492063231194968

Choudhury, P., Foroughi, C., & Larson, B. (2020). Work-from-anywhere: The productivity effects of geographic flexibility. *Strategic Management Journal, 42*(4), 655–683. DOI: 10.1002/smj.3251

Chowdhury, G. G. (2003). Natural language processing. *Annual Review of Information Science & Technology, 37*(1), 51–89.

Chowdhury, S., Budhwar, P., & Wood, G. (2024). Generative artificial intelligence in business: Towards a strategic human resource management framework. *British Journal of Management, 35*(4), 1680–1691. DOI: 10.1111/1467-8551.12824

Chowdhury, S., Dey, P., Joel-Edgar, S., Bhattacharya, S., Rodríguez-Espíndola, O., Abadie, A., & Truong, L. (2023). Unlocking the value of artificial intelligence in human resource management through ai capability framework. *Human Resource Management Review, 33*(1), 100899. DOI: 10.1016/j.hrmr.2022.100899

Cito, J., Leitner, P., Fritz, T., & Gall, H. C. (2017). The making of cloud applications: An empirical study on software development for the cloud. In *Proceedings of the 2017 11th Joint Meeting on Foundations of Software Engineering* (pp. 728-739).

Couldry, N., & Mejias, U. A. (2019). Data colonialism: Rethinking big data's relation to the contemporary subject. *Television & New Media, 20*(4), 336–349.

Cowgill, B. (2018). *Bias and productivity in humans and algorithms: Theory and evidence from resume screening*. Columbia Business School Research Paper No. 18-19. https://ssrn.com/abstract=3181120

Cummings, M. L. (2004). Automation bias in intelligent time critical decision support systems. *American Institute of Aeronautics and Astronautics*, 1-6.

Dassault Systems. (2025). *K'Watch*. K'WATCH. https://www.3ds.com/3dexperiencelab/portfolio/pkvitality

Dastin, J. (2018, October 10). Amazon scraps secret AI recruiting tool that showed bias against women. Reuters.

Davenport, T. H., Guha, A., Grewal, D., & Bressgott, T. (2020). How artificial intelligence will change the future of marketing. *Journal of the Academy of Marketing Science*, *48*(1), 24–42. DOI: 10.1007/s11747-019-00696-0

Davenport, T. H., Harris, J., & Shapiro, J. (2019). Competing on talent analytics. *Harvard Business Review*, *88*(10), 52–58.

Davenport, T. H., & Kirby, J. (2016). *Only humans need apply: Winners and losers in the age of smart machines*. HarperCollins.

Davenport, T. H., & Ronanki, R. (2018). Artificial intelligence for the real world. *Harvard Business Review*, *96*(1), 108–116.

Dellermann, D., Calma, A., Lipusch, N., Weber, T., Weigel, S., & Ebel, P. (2019). The future of human-AI collaboration: A taxonomy of design knowledge for hybrid intelligence systems. In *Proceedings of the 52nd Hawaii International Conference on System Sciences*.

Dellermann, D., Ebel, P., Söllner, M., & Leimeister, J. M. (2019). Hybrid intelligence. *Business & Information Systems Engineering*, *61*(5), 637–643. DOI: 10.1007/s12599-019-00595-2

Deloitte. (2020). *Global Human Capital Trends*. Deloitte Insights.

Deloitte. (2020). *Global Human Capital Trends*. Deloitte Insights. https://www2.deloitte.com/us/en/insights/focus/human-capital-trends.html

Deloitte. (2023). *2023 Global Human Capital Trends*. Deloitte Insights. Retrieved from https://www2.deloitte.com/us/en/insights/focus/human-capital-trends.html

Devi, P., Chauhan, R., & Jain, R. (2025). An application of FinTech is revolutionary in Indian banking sector: With reference to HDFC Bank. In *Proceedings of the 2025 International Conference on Computational Intelligence, Communication Technology, and Computer Science*.

Devlin, J., Chang, M. W., Lee, K., & Toutanova, K. (2018). BERT: Pre-training of deep bidirectional transformers for language understanding. *arXiv preprint arXiv:1810.04805.*

Dineen, B., Yu, K., & Stevenson-Street, J. (2022). Recruitment in personnel psychology and beyond: Where we've been working, and where we might work next. *Personnel Psychology*, *76*(2), 617–650. DOI: 10.1111/peps.12548

Doung-In, S. (2017). Flip your classroom: Reach every student in every class every day. *Walailak Journal of Learning Innovations*, *3*(2), 71–78.

Duan, Y., Edwards, J. S., & Dwivedi, Y. K. (2019). Artificial intelligence for decision making in the era of Big Data – evolution, challenges and research agenda. *International Journal of Information Management*, *48*, 63–71.

Du-Harpur, X., Watt, F. M., Luscombe, N. M., & Lynch, M. D. (2020). What is AI? Applications of artificial intelligence to dermatology. *British Journal of Dermatology*, *183*(3), 423–430. DOI: 10.1111/bjd.18880 PMID: 31960407

Dutordoir, M., & Struyfs, K. (2024). The impact of a compressed workweek on shareholder value: An event study analysis of belgium's 4-day workweek legislation. *Human Resource Management*, *63*(6), 903–917. DOI: 10.1002/hrm.22238

Ebert, C., Gallardo, G., Hernantes, J., & Serrano, N. (2016). DevOps. *IEEE Software*, *33*(3), 94–100.

Eccles, R. G., Ioannou, I., & Serafeim, G. (2014). The impact of corporate sustainability on organizational processes and performance. *Management Science*, *60*(11), 2835–2857.

Edmondson, A. C. (1999). Psychological safety and learning behavior in work teams. *Administrative Science Quarterly*, *44*(2), 350–383.

Edwards, P. N. (2003). Infrastructure and modernity: Force, time, and social organization in the history of sociotechnical systems. *Modernity and technology, 1*, 185-226.

Einola, K., & Khoreva, V. (2022). Best friend or broken tool? exploring the co-existence of humans and artificial intelligence in the workplace ecosystem. *Human Resource Management*, *62*(1), 117–135. DOI: 10.1002/hrm.22147

Ekman, P., & Friesen, W. V. (1978). *Facial Action Coding System: A technique for the measurement of facial movement.* Consulting Psychologists Press.

El Ayadi, M., Kamel, M. S., & Karray, F. (2011). Survey on speech emotion recognition: Features, classification schemes, and databases. *Pattern Recognition*, *44*(3), 572–587.

Elahi, M., Afolaranmi, S. O., Martinez Lastra, J. L., & Perez Garcia, J. A. (2023). A comprehensive literature review of the applications of AI techniques through the lifecycle of industrial equipment. In *Discover Artificial Intelligence* (Vol. 3, Issue 1). Springer International Publishing. https://doi.org/DOI: 10.1007/s44163-023-00089-x

Enfield, J. (2013). Looking at the impact of the flipped classroom model of instruction on undergraduate multimedia students at CSUN. *TechTrends*, *57*(6), 14–27. DOI: 10.1007/s11528-013-0698-1

Eubanks, V. (2018). *Automating inequality: How high-tech tools profile, police, and punish the poor*. Macmillan+ ORM.

Eubanks, V. (2018). *Automating inequality: How high-tech tools profile, police, and punish the poor*. St. Martin's Press.

European Commission High-Level Expert Group on Artificial Intelligence. (2019). *Ethics guidelines for trustworthy AI*. European Commission.

European Commission. (2021). *Proposal for a regulation of the European Parliament and of the Council laying down harmonised rules on artificial intelligence (Artificial Intelligence Act)*. COM(2021) 206 final.

Fehling, C., Leymann, F., Retter, R., Schupeck, W., & Arbitter, P. (2014). *Cloud computing patterns: Fundamentals to design, build, and manage cloud applications*. Springer.

Felzmann, H., Villaronga, E. F., Lutz, C., & Tamo-Larrieux, A. (2020). Transparency you can trust: Transparency requirements for artificial intelligence between legal norms and contextual concerns. *Big Data & Society*, *6*(1), 2053951719860542.

Feng, Q. (2024). Data statistical analysis of the impact of financial technology on bank credit risk. In *Proceedings of the 2024 9th International Conference on Big Data and Computing*.

Ferrati, F., & Muffatto, M. (2021). Reviewing equity investors' funding criteria: A comprehensive classification and research agenda. *Venture Capital*, *23*(2), 157–178.

Ferreira, J.. (2025). The transformative role of AI in modern entrepreneurship. *Business Horizons*, *68*(2), 134–146.

Field, J., Achten, N., Hilligoss, H., Nagy, A., & Srikumar, M. (2020). Principled artificial intelligence: A map of ethical and rights-based approaches. *Berkman Klein Center for Internet & Society Research Publication*, 2020-1.

Field, J., Achten, N., Hilligoss, H., Nagy, A., & Srikumar, M. (2020). *Principled artificial intelligence: Mapping consensus in ethical and rights-based approaches to principles for AI*. Society Research Publication. Berkman Klein Center for Internet.

Fildes, R., Ma, S., & Kolassa, S. (2019). Retail forecasting: Research and practice. *International Journal of Forecasting, 35*(1), 1–22. DOI: 10.1016/j.ijforecast.2018.08.004 PMID: 36217499

Følstad, A., Araujo, T., Law, E. L. C., Brandtzaeg, P. B., Papadopoulos, S., Reis, L., & Luger, E. (2021). Future directions for chatbot research: An interdisciplinary research agenda. *Computing, 103*(12), 2915–2942.

Forsgren, N., Humble, J., & Kim, G. (2018). *Accelerate: The science of lean software and DevOps: Building and scaling high performing technology organizations*. IT Revolution Press.

Foster, A., & Shah, M. (2020). Principles for advancing game-based learning in teacher education. *Journal of Digital Learning in Teacher Education, 36*(2), 84–95.

Fourcade, M., & Healy, K. (2017). Seeing like a market. *Socio-economic Review, 15*(1), 9–29.

Frey, C. B., & Osborne, M. A. (2017). The future of employment: How susceptible are jobs to computerisation? *Technological Forecasting and Social Change, 114*, 254–280. DOI: 10.1016/j.techfore.2016.08.019

Frost, J., Gambacorta, L., Huang, Y., Shin, H. S., & Zbinden, P. (2019). BigTech and the changing structure of financial intermediation. *BIS Working Papers No. 779*. Bank for International Settlements, Basel, Switzerland.

Gambacorta, L., Huang, Y., Qiu, H., & Wang, J. (2019). How do machine learning and non-traditional data affect credit scoring? *BIS Working Papers No. 834*. Bank for International Settlements.

Gander, P., Hartley, L., Powell, D., Cabon, P., Hitchcock, E., Mills, A., & Popkin, S. (2011). Fatigue risk management: Organizational factors at the regulatory and industry/company level. *Accident; Analysis and Prevention, 43*(2), 573–590. DOI: 10.1016/j.aap.2009.11.007 PMID: 21130218

Gao, B., Li, X., Liu, S., & Fang, D. (2021). How power distance affects online hotel ratings: The moderating role of AI-based personalization. *Tourism Management, 82*, 104204. DOI: 10.1016/j.tourman.2020.104204

Gardner, H. (2011). *The Theory of Multiple Intelligences*. Harvard Resources. https://pz.harvard.edu/resources/the-theory-of-multiple-intelligences

Gartner. (2022). *Gartner HR Research Finds 2021 Was the Year of the Manager*. Gartner.

Gartner. (2023). *Gartner HR Research Finds 3 Must-Haves to Win the Intensifying Race for Talent*. Gartner Newsroom. Retrieved from https://www.gartner.com/en/newsroom/press-releases/2023-04-04-gartner-hr-research-finds-3-must-haves-to-win-the-intensifying-race-for-talent

Gaur, L., & Gaur, D. (2024). Explainable Artificial Intelligence (XAI) on Neurogenerative Diseases. In *AI and Neuro-Degenerative Diseases: Insights and Solutions* (pp. 63–72). Springer Nature Switzerland.

Gautam, A., Chirputkar, A., & Pathak, P. (2022). Opportunities and challenges in the application of Artificial Intelligence-based technologies in the healthcare Industry. *International Interdisciplinary Humanitarian Conference for Sustainability, IIHC 2022 - Proceedings*, 1521–1524. https://doi.org/DOI: 10.1109/IIHC55949.2022.10059767

General Data Protection Regulation (GDPR). (2016). Regulation (EU) 2016/679 of the European Parliament and of the Council.

Geroski, T., & Filipović, N. (2024). Artificial Intelligence Empowering Medical Image Processing. In *In Silico Clinical Trials for Cardiovascular Disease: A Finite Element and Machine Learning Approach* (pp. 179–208). Springer Nature Switzerland.

Gilboy, M. B., Heinerichs, S., & Pazzaglia, G. (2015). Enhancing student engagement using the flipped classroom. *Journal of Nutrition Education and Behavior*, *47*(1), 109–114. DOI: 10.1016/j.jneb.2014.08.008

Gindert, M., & Müller, M. L. (2024). The impact of generative artificial intelligence on ideation and the performance of innovation teams. *arXiv*.

Giulioni, G., Bucciarelli, E., & Silvestri, M. (2012, March). Modeling firms' financial behavior in an evolving economy using artificial intelligence. In *Highlights on Practical Applications of Agents and Multi-Agent Systems* (pp. 333–340). Springer Berlin Heidelberg.

Glikson, E., & Woolley, A. W. (2020). Human trust in artificial intelligence: Review of empirical research. *The Academy of Management Annals*, *14*(2), 627–660. DOI: 10.5465/annals.2018.0057

Goodman, B., & Flaxman, S. (2017). European Union regulations on algorithmic decision-making and a "right to explanation". *AI Magazine*, *38*(3), 50–57.

Gray, M. L., & Suri, S. (2019). *Ghost work: How to stop Silicon Valley from building a new global underclass*. Harper Business.

Greipl, S., Moeller, K., & Ninaus, M. (2020). Potential and limits of game-based learning. *International Journal of Technology Enhanced Learning*, *12*(4), 363–389.

Grewal, D., Roggeveen, A. L., & Nordfält, J. (2017). The future of retailing. *Journal of Retailing*, *93*(1), 1–6. DOI: 10.1016/j.jretai.2016.12.008

Gupta, D., Gupta, M., Dhanaraj, R. K., Balusamy, B., & Gupta, P. M. (2025). *Computational intelligence for autonomous finance.*

Gupta, R., & Gupta, R. (2023). Lost in the perilous boulevards of gig economy: Making of human drones. *SA Journal of Human Resource Management*, *10*(1), 85–106.

Gwagwa, A., Kazim, E., & Hilliard, A. (2022). The role of the African value of Ubuntu in global AI inclusion discourse: A normative ethics perspective. *Patterns (New York, N.Y.)*, *3*(4). PMID: 35465235

Hadhli, F., & Sheikh, H. (2022). Applications of artificial intelligence in design and its relationship to creativity and innovation: The Japanese experience as a model. *Journal of Legal and Economic Studies*, *5*(1).

Hall, E. T. (1976). *Beyond culture*. Anchor Books.

Harris, M., Qi, A., Jeagal, L., Torabi, N., Menzies, D., Korobitsyn, A., Pai, M., Nathavitharana, R. R., & Khan, F. A. (2019). A systematic review of the diagnostic accuracy of artificial intelligence-based computer programs to analyze chest x-rays for pulmonary tuberculosis. *PLoS One*, *14*(9), 1–19. DOI: 10.1371/journal.pone.0221339 PMID: 31479448

Harter, J. (2022). *The Manager's Role in Employee Engagement.*

Hasan Tantawi, K., Sokolov, A., & Tantawi, O. (2024). Advances in Industrial Robotics: From Industry 3.0 Automation to Industry 4.0 Collaboration. *IEEE Proceedings*.

Hashem, I. A. T., Yaqoob, I., Anuar, N. B., Mokhtar, S., Gani, A., & Khan, S. U. (2015). The rise of "big data" on cloud computing: Review and open research issues. *Information Systems*, *47*, 98–115.

Hazan, E., Kelly, G., Khan, H., Spillecke, D., & Yee, L. (2021). *The top trends in tech*. McKinsey & Company.

He, K., Mao, R., Lin, Q., Ruan, Y., Lan, X., Feng, M., & Cambria, E. (2025). A survey of large language models for healthcare: From data, technology, and applications to accountability and ethics.

Healey, J. A., & Picard, R. W. (2005). Detecting stress during real-world driving tasks using physiological sensors. *IEEE Transactions on Intelligent Transportation Systems*, *6*(2), 156–166.

He, Q., Tong, Z., Mai, H., Shen, Y., & Jiang, J. (2025). Digital transformation and profitability in rural commercial banks. *Finance Research Letters*, *60*, 106985.

Herreid, C. F., & Schiller, N. A. (2013). Case study: Case studies and the flipped classroom. *Journal of College Science Teaching*, *42*(5), 62–66.

Hiba, A. M., & Mohamed, I. (2021). The economic implications of the fourth industrial revolution: Artificial intelligence. *Series of Economic Studies*.

Hidayata, S. E., Musari, K., Masrohatinb, S., & Smolo, E. (2024). The future of Islamic finance. In *The Routledge Handbook of Islamic Finance and Sustainable Development*. Routledge.

Hightower, K., Burns, B., & Beda, J. (2017). *Kubernetes: Up and running*. O'Reilly Media.

Hintze, A. (2016). *Understanding the four types of AI, from reactive robots to self-aware beings*. The Conversation. https://theconversation.com/understanding-the-four-types-of-ai-from-reactive-robots-to-self-aware-beings-67616

Hirschberg, J., & Manning, C. D. (2015). Advances in natural language processing. *Science*, *349*(6245), 261–266. PMID: 26185244

Hochschild, A. R. (1983). *The managed heart: Commercialization of human feeling*. University of California Press.

Hofmann, P., Samp, C., & Urbach, N. (2020). Robotic process automation. *Electronic Markets*, *30*(1), 99–106.

Hofstede, G. (2001). *Culture's consequences: Comparing values, behaviors, institutions, and organizations across nations* (2nd ed.). Sage Publications.

Holmes, J. B., & Gee, E. R. (2016). A framework for understanding game-based teaching and learning. *On the Horizon*, *24*(1), 1–16.

Hoppe, S., & Toussaint, M. (2019). Language Models are Unsupervised Multitask Learners Alec. *OpenAI Blog*. https://arxiv.org/abs/2007.07582

Hoskens, M., & De Boeck, P. (1991). An Implicit Theory of Intelligence-Related Mental Activities. *Journal of Personality*, *59*(4), 793–814. DOI: 10.1111/j.1467-6494.1991.tb00932.x

House, R. J., Hanges, P. J., Javidan, M., Dorfman, P. W., & Gupta, V. (Eds.). (2004). *Culture, leadership, and organizations: The GLOBE study of 62 societies*. Sage Publications.

Housman, M. (2017). Can algorithms reduce bias in the hiring process? *Harvard Business Review*.

Huang, M. H., & Rust, R. T. (2021). A strategic framework for artificial intelligence in marketing. *Journal of the Academy of Marketing Science*, *49*(1), 30–50.

Huang, M.-H., & Rust, R. T. (2021). Artificial intelligence in service. *Journal of Service Research*, *24*(1), 3–41.

Huang, Y., Zhang, L., Li, Z., Qiu, H., Sun, T., & Wang, X. (2020). FinTech credit risk assessment for SMEs: Evidence from China. *Journal of Financial Innovation*, *6*(3), 145–160.

Humble, J., & Farley, D. (2020). *Continuous delivery: Reliable software releases through build, test, and deployment automation*. Addison-Wesley Professional.

Iansiti, M., & Lakhani, K. R. (2020). *Competing in the age of AI: Strategy and leadership when algorithms and networks run the world*. Harvard Business Press.

Ivanov, D., & Dolgui, A. (2020). Viability of intertwined supply networks: Extending the supply chain resilience angles towards survivability. A position paper motivated by COVID-19 outbreak. *International Journal of Production Research*, *58*(10), 2904–2915. DOI: 10.1080/00207543.2020.1750727

Ivanov, D., Dolgui, A., & Sokolov, B. (2019). The impact of digital technology and Industry 4.0 on the ripple effect and supply chain risk analytics. *International Journal of Production Research*, *57*(3), 829–846. DOI: 10.1080/00207543.2018.1488086

Jain, J. (2025). Enhancing FinTech solutions with AI: Predicting income and financial transactions using ensemble learning model. In *Proceedings of the IEEE SoutheastCon 2025*.

Jardine, A. K. S., Lin, D., & Banjevic, D. (2006). A review on machinery diagnostics and prognostics implementing condition-based maintenance. *Mechanical Systems and Signal Processing*, *20*(7), 1483–1510. DOI: 10.1016/j.ymssp.2005.09.012

Jin, D. Y. (2024). The rise of digital platforms as a soft power apparatus in the New Korean Wave era. *Communication and the Public*, *9*(2), 161–177.

Jobin, A., Ienca, M., & Vayena, E. (2019). The global landscape of AI ethics guidelines. *Nature Machine Intelligence*, *1*(9), 389–399.

Kahneman, D. (2011). *Thinking, fast and slow*. Farrar, Straus and Giroux.

Kaplan, A., & Haenlein, M. (2020). Rulers of the world, unite! The challenges and opportunities of artificial intelligence. *Business Horizons, 63*(1), 37–50. DOI: 10.1016/j.bushor.2019.09.003

Kar, S., Yadav, M., & Panda, T. (2023). Inclusive organizational behaviour – the dynamic rules of building new workplaces. *VINE Journal of Information and Knowledge Management Systems, 55*. Advance online publication. DOI: 10.1108/VJIKMS-05-2022-0155

Kashyap, S., Agarwal, B., Pushp, A., & Rastogi, S. (2025). Does Indian banks' FinTech investment affect their liquidity and efficiency? Using competition functions as a moderator. *Journal of Financial Economic Policy, 17*(2), 210–229.

Kaur, K., Kapur, B., & Kurucz, A. (2025). The role of FinTech in promoting sustainable blue economy. In *Sustainable Blue Economy*. Emerging Trends and Innovations.

Kaur, N., Sunder, C., Saha, S., Kumar, P. S., & Kalra, N. (2025). Evaluating FinTech readiness in Indian states: A data-driven approach using computational economics and big data analytics. In *Proceedings of the 2025 International Conference on Emerging Trends in Management (ICETM)*.

Kay, A., Vlachos, P., Tasoulis, K., & Farndale, E. (2024). Fraught expectations: A fairness heuristic process model of the pros and cons of csr for talent acquisition. *Human Resource Management, 64*(2), 465–483. DOI: 10.1002/hrm.22269

Keiper, A., & Schulman, A. N. (2015). The Problem with 'Friendly' Artificial Intelligence. *Science, Virtue, and the Future of Humanity*, 69–77.

Kellogg, K. C., Valentine, M. A., & Christin, A. (2020). Algorithms and the future of work: A research agenda. In *The Future of Work* (pp. 1–40). Emerald Publishing Limited.

Kellogg, K. C., Valentine, M. A., & Christin, A. (2020). Algorithms at work: The new contested terrain of control. *The Academy of Management Annals, 14*(1), 366–410.

Khalighi, S., Reddy, K., Midya, A., Pandav, K. B., Madabhushi, A., & Abedalthagafi, M. (2024). Artificial intelligence in neuro-oncology: Advances and challenges in brain tumor diagnosis, prognosis, and precision treatment. *NPJ Precision Oncology, 8*(1), 80. PMID: 38553633

Khamitdkhanovich, N. A., Packiaraji, S., Tripathi, M., & Badriddinovich, K. K. (2025). Unlocking the power of FinTech: How embedded finance is shaping the future of commerce. *AIP Conference Proceedings, 3102*(1), Article 5.0277719.

Kim, J., & Forsythe, S. (2008). Adoption of virtual try-on technology for online apparel shopping. *Journal of Interactive Marketing*, *22*(2), 45–59. DOI: 10.1002/dir.20113

Kim, P. T., & Schessler, M. (2022). The algorithmic bill of rights: A framework for governing automated employment decisions. *Yale Law & Policy Review*, *40*(2), 345–410.

Kim, S., Khoreva, V., & Vaiman, V. (2024). Strategic human resource management in the era of algorithmic technologies: Key insights and future research agenda. *Human Resource Management*, *64*(2), 447–464. DOI: 10.1002/hrm.22268

Kirkpatrick, J. D., & Kirkpatrick, W. K. (2016). *Kirkpatrick's four levels of training evaluation*. Association for Talent Development.

Kiron, D., & Schrage, M. (2019). Strategy for and with AI. *MIT Sloan Management Review*.

Kliestik, T., Dragomir, R., B lu , A. V., Grecu, I., Durana, P., Karabolevski, O. L., Kráľ, P., Balica, R., Suler, P., Busu, O. V., Bugaj, M., Voinea, D.-V., Vrbka, J., Cocosatu, M., Grupac, M., Pera, A., & Gajdosikova, D. (2024). Enterprise generative artificial intelligence technologies, Internet of Things and blockchain-based FinTech management, and digital twin industrial metaverse in the cognitive algorithmic economy. *Oeconomia Copernicana*, *15*(2), 3109.

Knight, R. (2023). How AI is transforming the role of the manager. *MIT Sloan Management Review*, *64*(2), 1–5.

Koch, M. (2021). The digital transformation of HR: A review of the literature. *German Journal of Human Resource Management*, *35*(1), 3–28.

Kollu, V. N., Janarthanan, V., Karupusamy, M., & Ramachandran, M. (2023). Cloud-based smart contract analysis in FinTech using IoT-integrated federated learning in intrusion detection. *Data*, *8*(5), 83.

Kong, X. (2021). AI in education: A systematic review of the literature. *Journal of Educational Technology Development and Exchange*, *14*(1), 1–20.

Korinek, A. (2023). *Generative AI and the future of work*. Brookings Institution.

Krishnamoorthy, P. (2024). Big Data Analytics In Fintech: A Review Of Credit Risk Assessment And Fraud Detection. Educational Administration Theory and Practice journal.

Kumar, M. Yadav, M., & Sahoo, A. (2025). The Importance of Collaboration Skills in Workforce Preparation From Textbooks to Teamwork. In Revitalizing Student Skills for Workforce Preparation (pp. 333–352). https://doi.org/DOI: 10.4018/979-8-3693-3856-8.ch011

Kumar, M., Dagar, M., Yadav, M., Choudhury, T., & Luu, T. M. N. (2025). Navigating the future: Challenges and opportunities in intelligent RPA implementation. In Intelligent Robotic Process Automation: Development, Vulnerability and Applications (pp. 393–420). IGI Global. DOI: 10.4018/979-8-3693-4365-4.ch015

Kumar, M., Yadav, R., Suresh, A. S., Singh, R., Yadav, M., Balodi, A., Vihari, N. S., & Srivastava, A. P. (2023). Mapping AI-Driven Marketing: Review of Field and Directions for Future Research. 2023 2nd International Conference on Futuristic Technologies (INCOFT), 1–5. DOI: 10.1109/INCOFT60753.2023.10425169

Kumar, M., Nain, N., Dagar, M., Luu, T. M. N., & Kaur, K. (2025). Redefining work and outsourcing in the AI era: Challenges and opportunities. In *Global Work Arrangements and Outsourcing in the Age of AI*. IGI Global. DOI: 10.4018/979-8-3373-1270-5.ch021

Kumar, M., & Yadav, R., Sushma, & Yadav, M. (2025). Investigating the influence of digital and social media marketing on university brand equity: An in-depth literature review and bibliometric study. *Global Knowledge, Memory and Communication*, 1–26. DOI: 10.1108/GKMC-07-2024-0426

Kumar, R., Waisberg, E., Ong, J., Paladugu, P., Amiri, D., Saintyl, J., & Tavakkoli, A. (2024). Artificial Intelligence-Based Methodologies for Early Diagnostic Precision and Personalized Therapeutic Strategies in Neuro-Ophthalmic and Neurodegenerative Pathologies. *Brain Sciences*, *14*(12), 1266. PMID: 39766465

Kumar, S. (2021). *The digital frontier: Infrastructures of control on the global web*. Indiana University Press.

Kumar, S., Singh, S. K., & Sharma, R. (2021). A review of natural language processing in ESG. *Journal of Sustainable Finance & Investment*, •••, 1–18.

Kumar, V., Rajan, B., Venkatesan, R., & Lecinski, J. (2019). Understanding the role of artificial intelligence in personalized engagement marketing. *California Management Review*, *61*(4), 135–155.

Kusetogullari, A.. (2025). Clustering AI research in entrepreneurship: A systematic review. *Journal of Innovation & Knowledge*, *10*(4), 78–89.

Lage, M. J., Platt, G. J., & Treglia, M. (2000). Inverting the classroom: A gateway to creating an inclusive learning environment. *The Journal of Economic Education, 31*(1), 30–43. DOI: 10.1080/00220480009596759

Larcara, M. (2015). Benefits of the flipped classroom model. In *Curriculum design and classroom management: Concepts, methodologies, tools, and applications* (pp. 93–105). IGI Global.

Latour, B. (2005). An introduction to actor-network-theory. *Reassembling the social.*

Le Clair, C., & Cullen, A. (2020). The Forrester Wave™ [Forrester Research.]. *Robotic Process Automation, Q4*, 2020.

Lee, J., Bagheri, B., & Jin, C. (2016). Introduction to cyber manufacturing. *Manufacturing Letters, 8*, 11–15. DOI: 10.1016/j.mfglet.2016.03.001

Lee, J., Bagheri, B., & Kao, H.-A. (2015). A Cyber-Physical Systems architecture for Industry 4.0-based manufacturing systems. *Manufacturing Letters, 3*, 18–23. DOI: 10.1016/j.mfglet.2014.12.001

Lee, J., Lapira, E., Bagheri, B., & Kao, H. A. (2013). Recent advances and trends in predictive manufacturing systems in big data environment. *Manufacturing Letters, 1*(1), 38–41. DOI: 10.1016/j.mfglet.2013.09.005

Lee, S., & Mönch, L. (2019). A cyber-physical approach to predictive maintenance in manufacturing. *Computers & Industrial Engineering, 130*, 332–342. DOI: 10.1016/j.cie.2019.03.014

Levy, M., & Liguori, E. (2023). Algorithms and venture investment decisions: Better, fairer, or biased? *Journal of Small Business and Enterprise Development, 30*(2), 419–422.

LinkedIn Learning. (2023). *Workplace Learning Report*. LinkedIn. https://learning.linkedin.com/resources/workplace-learning-report

Liu, F., & Lyons, A. C. (2024). Use cases in FinTech. In Lyons, A. C. (Ed.), *The Routledge Handbook of Financial Technology and Digital Finance.*

Liu, Q., Li, P., Zhao, W., Cai, W., Yu, S., & Leung, V. C. (2018). A survey on security threats and defensive techniques of machine learning: A data driven view. *IEEE Access : Practical Innovations, Open Solutions, 6*, 12103–12117.

Lo, C. K., & Hew, K. F. (2017). A critical review of flipped classroom challenges in K–12 education: Possible solutions and recommendations for future research. *Research and Practice in Technology Enhanced Learning, 12*, 4. Advance online publication. DOI: 10.1186/s41039-016-0044-2

Luo, Z., Wang, L., & Zhou, J. (2025). Systemic financial risk prediction based on machine learning: The role of FinTech of different technical types. *Applied Economics*, *57*(4), 301–320.

Lu, Y. (2017). Industry 4.0: A survey on technologies, applications and open research issues. *Journal of Industrial Information Integration*, *6*, 1–10. DOI: 10.1016/j.jii.2017.04.005

Lyon, T. P., & Montgomery, A. W. (2015). The means and end of greenwash. *Organization & Environment*, *28*(2), 223–249.

Malempati, M. (2021). Developing end-to-end intelligent finance solutions through AI and cloud integration. SSRN *Electronic Journal.*

Malhotra, R. (2021). Cloud-based AI models for credit risk assessment: A scalable and adaptive approach. *International Journal of Artificial Intelligence, Data Science, and Machine Learning*, *2*(2), 11–21.

Malik, A., Budhwar, P., Mohan, H., & Srikanth, N. (2022). Employee experience –the missing link for engaging employees: Insights from an mne's ai-based hr ecosystem. *Human Resource Management*, *62*(1), 97–115. DOI: 10.1002/hrm.22133

Mamun, A. Y., & László, V. (2025). Advancing sustainability through financial inclusion and sustainable finance: A systematic literature review. *Journal of Sustainable Finance & Investment*, *15*(3), 421–445.

Mann, M., & Daly, A. (2020). Geopolitics, jurisdiction and surveillance.

Mantelero, A. (2014). The future of consumer data protection in the EU: Re-thinking the "notice and consent" paradigm in the new era of predictive analytics. *Computer Law & Security Review*, *30*(6), 643–660.

Mariani, M., & Dwivedi, Y. K. (2024). Generative artificial intelligence in innovation management: A preview of future research developments. *Journal of* Mariani, M., Machado, I., Magrelli, V., & Dwivedi, Y. K. (2023). Artificial intelligence in innovation research: A systematic review, conceptual framework and future research directions. *Technovation*, *122*, 102623.

Marler, J. H., & Boudreau, J. W. (2017). An evidence-based review of HR Analytics. *International Journal of Human Resource Management*, *28*(1), 3–26.

Márquez, J. (2020). *Cloud Computing: Theory and Practice*. Morgan Kaufmann.

Marston, S., Li, Z., Bandyopadhyay, S., Zhang, J., & Ghalsasi, A. (2011). Cloud computing—The business perspective. *Decision Support Systems*, *51*(1), 176–189.

Martin, K. (2019). Ethical implications and accountability of algorithms. *Journal of Business Ethics*, *160*(4), 835–850.

Martin, K., & Freeman, R. E. (2003). Some problems with employee monitoring. *Journal of Business Ethics*, *43*(4), 353–361.

Maslach, C., Schaufeli, W. B., & Leiter, M. P. (2001). Job burnout. *Annual Review of Psychology*, *52*, 397–422.

Mayer, R., & Weinreich, R. (2018). *The cloud native landscape: A systematic mapping study*. IEEE/ACM 11th International Conference on Utility and Cloud Computing (UCC).

McAfee, A., & Brynjolfsson, E. (2012). Big data: The management revolution. *Harvard Business Review*, *90*(10), 60–68. PMID: 23074865

McKinsey & Company. (2020). *AI Index Report*. McKinsey Global Institute.

McKinsey & Company. (2023). *The state of AI in retail and fashion.* https://www.mckinsey.com

McLaughlin, J. E.. (2014). The flipped classroom: A course redesign to foster learning and engagement in a health professions school. *Academic Medicine*, *89*(2), 236–243. DOI: 10.1097/ACM.0000000000000086

McMahan, B., Moore, E., Ramage, D., Hampson, S., & Arcas, B. A. (2017, April). Communication-efficient learning of deep networks from decentralized data. In *Artificial Intelligence and Statistics* (pp. 1273–1282).

Mehrabi, N., Morstatter, F., Saxena, N., Lerman, K., & Galstyan, A. (2021). A survey on bias and fairness in machine learning. *ACM Computing Surveys*, *54*(6), 1–35.

Mell, P., & Grance, T. (2011). The NIST definition of cloud computing. *National Institute of Standards and Technology*, *53*(6), 50.

Meta. (2025). *A.R.I.A.* https://www.projectaria.com/

Mikalef, P., & Gupta, M. (2021). Artificial intelligence capability: Conceptualization, measurement calibration, and empirical study on its impact on organizational creativity and firm performance. *Information & Management*, *58*(3), 103434.

Milan, S., & Treré, E. (2019). Big data from the South (s): Beyond data universalism. *Television & New Media*, *20*(4), 319–335.

Mintzberg, H. (1994). The fall and rise of strategic planning. *Harvard Business Review*, *72*(1), 107–114.

Mitchell, M. (2019). *Artificial Intelligence: A Guide for Thinking Humans*. Farrar, Straus and Giroux.

Mitchell, M., Wu, S., Zaldivar, A., Barnes, P., Vasserman, L., Hutchinson, B., & Gebru, T. (2019, January). Model cards for model reporting. In *Proceedings of the conference on fairness, accountability, and transparency* (pp. 220-229).

Mobley, R. K. (2002). *An Introduction to Predictive Maintenance*. Elsevier., DOI: 10.1016/B978-075067532-3/50003-9

Möhlmann, M. (2021). *The impact of cloud computing on AI innovation*. In *Proceedings of the 54th Hawaii International Conference on System Sciences*.

Möhlmann, M., & Zalmanson, L. (2017). Hands on the wheel: Navigating algorithmic management and Uber drivers' autonomy. In *Proceedings of the International Conference on Information Systems (ICIS)*.

Möhlmann, M., Zalmanson, L., Henfridsson, O., & Gregory, R. W. (2021). Algorithmic management of work on online labor platforms: When matching meets control. *Management Information Systems Quarterly*, *45*(4), 1999–2022.

Mollick, E. (2022). *How to use AI to do practical stuff: A guide*. Retrieved from https://oneusefulthing.substack.com/

Moniz, A., & de Jong, M. (2021). The use of NLP for corporate social responsibility monitoring. In *Proceedings of the International Conference on Web Intelligence* (pp. 450-457).

Morgan, J. (2017). *The Employee Experience Advantage: How to Win the War for Talent by Giving Employees the Workspace They Want, the Tools They Need, and a Culture They Can Celebrate*. Wiley.

Mourtzis, D., Doukas, M., & Psarommatis, F. (2019). Industrial big data as a result of IoT adoption in manufacturing. *Procedia Manufacturing*, *39*, 1613–1620. DOI: 10.1016/j.promfg.2020.01.413

Mullender, S. (2019). *Cloud native transformation: Practical patterns for innovation*. O'Reilly Media.

Musiani, F., Cogburn, D. L., DeNardis, L., & Levinson, N. S. (Eds.). (2016). *The turn to infrastructure in Internet governance* (pp. 268-p). New York: Palgrave Macmillan.

Musiani, F. (2022). Infrastructuring digital sovereignty: A research agenda for an infrastructure-based sociology of digital self-determination practices. *Information Communication and Society*, *25*(6), 785–800.

Naveed, H., Khan, A. U., Qiu, S., Saqib, M., Anwar, S., Usman, M., Akhtar, N., Barnes, N., & Mian, A. (2025). A Comprehensive Overview of Large Language Models. *ACM Transactions on Intelligent Systems and Technology*, *16*(5), 1–72. DOI: 10.1145/3744746

Newman, S. (2015). *Building microservices*. O'Reilly Media.

Newman, S. (2015). *Building Microservices*. O'Reilly Media.

Nguyen, T. B., Kumar, M., Sahoo, D. K., Nain, N., Yadav, M., & Dadhich, V. (2025). Mapping the landscape of women entrepreneurs in micro and small enterprises: Trends, themes, and insights through systematic review and text mining. F1000Research, 14. DOI: 10.12688/f1000research.163030.1

Noble, S. U. (2018). Algorithms of oppression: How search engines reinforce racism. In *Algorithms of oppression*. New York university press.

O'Flaherty, J., & Phillips, C. (2015). The use of flipped classrooms in higher education: A scoping review. *The Internet and Higher Education*, *25*, 85–95. DOI: 10.1016/j.iheduc.2015.02.002

O'Neil, C. (2016). *Weapons of math destruction: How big data increases inequality and threatens democracy*. Crown Publishing Group.

Obschonka, M., & Audretsch, D. (2020). The impact of AI on entrepreneurial activities. *Small Business Economics*, *55*(3), 541–565.

Ogbonnaya, C., Messersmith, J., Schnacker, M., & Laker, B. (2025). Setting the stage: Line managers' HR orientation as an antecedent to perceived HR effectiveness. *Human Resource Management*, *64*(6), 1581–1604. DOI: 10.1002/hrm.70002

Onciul, R., Tataru, C. I., Dumitru, A. V., Crivoi, C., Serban, M., Covache-Busuioc, R. A., & Toader, C. (2025). Artificial intelligence and neuroscience: Transformative synergies in brain research and clinical applications. *Journal of Clinical Medicine*, *14*(2), 550. PMID: 39860555

Pankaj, Yadav, R., Naim, I., Kumar, M., Misra, S., Dewasiri, N. J., Rathnasiri, M. S. H., Yadav, M., Balodi, A., Kar, S., & Vihari, N. S. (2023). An Exploratory Study to Identify Effects of Blockchain Technology on Organizational Change and Practices. 2023 IEEE Technology & Engineering Management Conference - Asia Pacific (TEMSCON-ASPAC), 1–8. DOI: 10.1109/TEMSCON-ASPAC59527.2023.10531372

Pantano, E., Pizzi, G., Scarpi, D., & Dennis, C. (2020). Competing during a pandemic? Retailers' ups and downs during the COVID-19 outbreak. *Journal of Business Research*, *116*, 209–213. DOI: 10.1016/j.jbusres.2020.05.036

Pasquale, F. (2015). *The black box society: The secret algorithms that control money and information.* Harvard University Press.

Picard, R. W. (1997). *Affective computing.* MIT Press.

Pillai, R., Sivathanu, B., & Dwivedi, Y. K. (2020). Shopping intention at AI-powered automated retail stores (AIPARS). *Journal of Retailing and Consumer Services, 57,* 102207. DOI: 10.1016/j.jretconser.2020.102207

Podsakoff, P. M., MacKenzie, S. B., Lee, J. Y., & Podsakoff, N. P. (2003). Common method biases in behavioral research: A critical review of the literature and recommended remedies. *The Journal of Applied Psychology, 88*(5), 879.

Porter, M. E., & Heppelmann, J. E. (2014). How smart, connected products are transforming competition. *Harvard Business Review, 92*(11), 64–88.

Prikshat, V., Malik, A., & Budhwar, P. (2023). Ai-augmented hrm: Antecedents, assimilation and multilevel consequences. *Human Resource Management Review, 33*(1), 100860. DOI: 10.1016/j.hrmr.2021.100860

Prinzi, F. (2023). Innovations in Medical Image Analysis and Explainable AI for Transparent Clinical Decision Support Systems. Chattopadhyay, S., Barman, S., & Lakshmi, D. (2025). The Role of Explainable AI for Healthcare 5.0: Best Practices, Challenges, and Opportunities. *Edge AI for Industry 5.0 and Healthcare 5.0 Applications*, 45-80.

Proud, J. K., Lai, D. T. H., Mudie, K. L., Carstairs, G. L., Billing, D. C., Garofolini, A., & Begg, R. K. (2022). Exoskeleton Application to Military Manual Handling Tasks. *Human Factors, 64*(3), 527–554. DOI: 10.1177/0018720820957467 PMID: 33203237

Provost, F., & Fawcett, T. (2013). Data science and its relationship to big data and data-driven decision making. *Big Data, 1*(1), 51–59. PMID: 27447038

Pulakos, E. D., Hanson, R. M., Arad, S., & Moye, N. (2015). Performance management can be fixed: An on-ramp to high performance. *Business Horizons, 58*(6), 639–646.

Qadri, R., Shelby, R., Bennett, C. L., & Denton, R. (2023, June). Ai's regimes of representation: A community-centered study of text-to-image models in south asia. In *Proceedings of the 2023 ACM Conference on Fairness, Accountability, and Transparency* (pp. 506-517).

Qin, J., Liu, Y., & Grosvenor, R. (2016). A categorical framework of manufacturing for industry 4.0 and beyond. *Procedia CIRP, 52,* 173–178. DOI: 10.1016/j.procir.2016.07.040

Qin, X., Zhou, X., Chen, C., Wu, D., Zhou, H., Dong, X., & Lu, J. (2025). Ai aversion or appreciation? a capability–personalization framework and a meta-analytic review. *Psychological Bulletin, 151*(5), 580–599. DOI: 10.1037/bul0000477 PMID: 40489180

Rabbit, R. (2025). *Research Rabbit.* https://www.researchrabbit.ai/

Rabenu, E., & Baruch, Y. (2024). Cyborging hrm theory: From evolution to revolution – the challenges and trajectories of ai for the future role of hrm. *Personnel Review, 54*(1), 174–198. DOI: 10.1108/pr-02-2024-0111

Radanliev, P. (2024). The rise and fall of cryptocurrencies: Defining the economic and social values of blockchain technologies, assessing the opportunities, and defining the financial and cybersecurity risks of the Metaverse. *Financial Innovation, 10*(2), 45–68.

Raghavan, M., Barocas, S., Kleinberg, J., & Levy, K. (2020). Mitigating bias in algorithmic hiring: Evaluating claims and practices. In *Proceedings of the 2020 Conference on Fairness, Accountability, and Transparency* (pp. 469-481).

Rahman, A., Sohail, S. S., Alam, M. S., Sharma, A., & Mansoor, W. (2024, November). Detecting Brain Cancer Using Explainable AI. In *2024 7th International Conference on Signal Processing and Information Security (ICSPIS)* (pp. 1-6). IEEE.

Raisch, S., & Krakowski, S. (2021). Artificial intelligence and management: The automation–augmentation paradox. *Academy of Management Review, 46*(1), 192–210.

Raji, I. D., Smart, A., White, R. N., Mitchell, M., Gebru, T., Hutchinson, B., & Barnes, P. (2020, January). Closing the AI accountability gap: Defining an end-to-end framework for internal algorithmic auditing. In *Proceedings of the 2020 conference on fairness, accountability, and transparency* (pp. 33-44).

Rakova, B., Yang, J., Cramer, H., & Chowdhury, R. (2021). Where responsible AI meets reality: Practitioner perspectives on enablers for shifting organizational practices. *Proceedings of the ACM on Human-Computer Interaction*, 5(CSCW1), 1-23. https://doi.org/DOI: 10.1145/3449081

Ransbotham, S., Kiron, D., Gerbert, P., & Reeves, M. (2017). Reshaping business with artificial intelligence: Closing the gap between ambition and action. *MIT Sloan Management Review, 59*(1), 1–17.

Rao, U., & Nair, V. (2019). Aadhaar: Governing with biometrics. *South Asia, 42*(3), 469–481.

Ravikumar, T., Raghunandan, G., Benedict, J., & Krishna, T. A. (2025). Digital financial inclusion and financial resilience of micro and small entrepreneurs in post-COVID-19 in Karnataka. *International Journal of Development Studies*, *32*(1), 55–72.

Raza, H., Riaz, A., Riaz, N., & Ramakrishnan, S. A. L. (2025). Algorithmic training, future markets, and big data for finance digitalization. In *Advances in Financial Technology and Digital Transformation* (pp. 201–220).

Ribeiro, M. T., Singh, S., & Guestrin, C. (2016, August). " Why should I trust you?" Explaining the predictions of any classifier. In *Proceedings of the 22nd ACM SIGKDD international conference on knowledge discovery and data mining* (pp. 1135-1144).

Ricaurte, P. (2019). Data epistemologies, the coloniality of power, and resistance. *Television & New Media*, *20*(4), 350–365.

Ricci, F., Rokach, L., & Shapira, B. (2022). *Recommender systems handbook* (3rd ed.). Springer., DOI: 10.1007/978-1-0716-2197-4

Robbins, S., & Van Wynsberghe, A. (2022). Our new artificial intelligence infrastructure: Becoming locked into an unsustainable future. *Sustainability*, *14*(8), 4829.

Rock, D., & Grant, H. (2016). Why diverse teams are smarter. *Harvard Business Review*, *94*(10), 74–81.

Roy, J. K., & Vasa, L. (2025). Financial technology and environmental, social and governance in sustainable finance: A bibliometric and thematic content analysis. *Environmental Science and Pollution Research International*, *32*(7), 11098–11115.

Saeedi, M., & Ashraf, B. N. (2024). The role of technology in promoting green finance: A systematic literature survey and the development of a framework. *Journal of Risk and Financial Management*, *17*(10), 472.

Sato, D., Wider, A., & Windheuser, C. (2019). Continuous delivery for machine learning. Martin Fowler. https://martinfowler.com/articles/cd4ml.html

Saura, J. R., & Bužinskienė, R. (2025). Behavioral economics, artificial intelligence and entrepreneurship: An updated framework for management. *The International Entrepreneurship and Management Journal*, *21*(1), 1–33.

Sayah, W. (2025). The role of artificial intelligence in enhancing innovation in entrepreneurship. *ASJP*, *10*(4), 118–128.

Schoox. (2022). *The LMS is dead, long live the LMS!* https://www.schoox.com/blog/the-lms-is-dead-long-live-the-lms/

Schwab, K. (2016). *The fourth industrial revolution*. World Economic Forum.

Schwartz, R., Dodge, J., Smith, N. A., & Etzioni, O. (2022). Green AI. *Communications of the ACM, 65*(1), 54–63.

Sculley, D., Holt, G., Golovin, D., Davydov, E., Phillips, T., Ebner, D., . . . Dennison, D. (2015). Hidden technical debt in machine learning systems. In *Advances in neural information processing systems* (pp. 2503-2511).

Sculley, D., Holt, G., Golovin, D., Davydov, E., Phillips, T., Ebner, D., . . . Dennison, D. (2015). Hidden technical debt in machine learning systems. In *Advances in Neural Information Processing Systems* (pp. 2503-2511).

Segarra, P., & Williams, K. (2025). Calling in: Mental health in management education and academia. *Management Learning, 56*(1), 83–89. DOI: 10.1177/13505076241284433

Selvamani, T., Arjunan, P., & Poovammal, E. (2025). Intelligent mortgage optimization: Leveraging AI for personalized lending and risk assessment. *Procedia Computer Science, 242*, 75–82.

Shankar, V. (2018). How artificial intelligence is transforming marketing. *GfK Marketing Intelligence Review, 10*(2), 34–39.

Sharma, S., Bhardwaj, S., & Gupta, B. (2023). Algorithmic control: a disruption to motivation of gig workers? A critical review. *Multidimensional and strategic outlook in digital business transformation: human resource and management recommendations for performance improvement*, 1-14.

Shi, Z. (2022). Generative Design: A Paradigm Shift in the Design Process. *Proceedings of the 2022 International Conference on Engineering Design.*

Shi, W., Cao, J., Zhang, Q., Li, Y., & Xu, L. (2016). Edge computing: Vision and challenges. *IEEE Internet of Things Journal, 3*(5), 637–646.

Shneiderman, B. (2020). Human-centered artificial intelligence: Reliable, safe & trustworthy. *International Journal of Human-Computer Interaction, 36*(6), 495–504.

Shrestha, Y. R., Ben-Menahem, S. M., & von Krogh, G. (2019). Organizational decision-making structures in the age of artificial intelligence. *California Management Review, 61*(4), 66–83.

Siddiqi, N. (2012). *Credit risk scorecards: Developing and implementing intelligent credit scoring* (Vol. 3). John Wiley & Sons.

Siddiqi, N. (2017). *Intelligent credit scoring: Building and implementing better credit risk scorecards.* John Wiley & Sons.

Sille, R., Nanda, I., Kapoor, A., Sahoo, S., & Sharma, A. (2024). A systematic review on recent trends of digital financial inclusion. In *Recent Advances in Digital Finance and Financial Inclusion* (pp. 33–45).

Simon, H. A. (1957). A behavioral model of rational choice. *Models of Man. Social and Rational: Mathematical Essays on Rational Human Behavior in a Social Setting*, *6*(1), 241–260.

Singh, R., Kumar, U., & Gupta, M. (2018). Internet of Things (IoT) in manufacturing: Review and challenges. *International Journal of Advanced Manufacturing Technology*, *102*, 511–526. DOI: 10.1007/s00170-019-03731-3

Sinha, N. (2024). Aadhaar, AI, and Identity: Negotiating Power and Surveillance in the Global South. *Социологическое обозрение, 23*(4), 80-112.

Slovic, P. (1988). Risk perception. In *Carcinogen Risk Assessment* (pp. 171–181).

Society for Human Resource Management. (2022). *The cost of a bad hire*. SHRM.

Song, L., Yasin, M., & Mokhtar, M. M. (2024). The significance of the virtue of friendliness for moral education in the age of. [InjET]. *International Journal of Education and Training*, *10*(2), 1–12.

Spotify. (2018). *Spotify's failed #SquadGoals model*. [Blog post]. Retrieved from https://www.spotify.com/intl/ie/blog/spotify's-failed-squadgoals-model/

Sreenivasan, A., & Suresh, M. (2024). Can we unlock deep-tech in Indian startups for long-term success? *Technological Sustainability*, *3*(1), 68–75.

Srnicek, N. (2017). *Platform capitalism*. John Wiley & Sons.

Stahl, B. C., Brooks, L., Hatzakis, T., & Santiago, N. (2023). The ethics of artificial intelligence in international business: A review of the literature and a research agenda. *Journal of Business Ethics*, *182*, 1–22. DOI: 10.1007/s10551-022-05239-2

Star, S. L., & Bowker, G. C. (2006). How to infrastructure. *Handbook of new media: Social shaping and social consequences of ICTs*, 230-245

Star, S. L. (1999). The ethnography of infrastructure. *The American Behavioral Scientist*, *43*(3), 377–391.

Stiglitz, J. E., & Weiss, A. (1981). Credit rationing in markets with imperfect information. *The American Economic Review*, *71*(3), 393–410.

Strayer, J. F. (2012). How learning in an inverted classroom influences cooperation, innovation and task orientation. *Learning Environments Research*, *15*, 171–193. DOI: 10.1007/s10984-012-9108-4

Strubell, E., Ganesh, A., & McCallum, A. (2020). Energy and policy considerations for deep learning in NLP. *Proceedings of the 58th Annual Meeting of the Association for Computational Linguistics*, 3645–3653.

Suchman, L. A. (2007). *Human-machine reconfigurations: Plans and situated actions*. Cambridge university press.

Sun, J., Teng, J. T. C., & Chen, Y. (2017). Product recommendation system for e-commerce websites: An example of clothing retailing. *Decision Support Systems, 92*, 57–65. DOI: 10.1016/j.dss.2016.09.008

Sweller, J. (2011). Cognitive load theory. In Mestre, J. P., & Ross, B. H. (Eds.), *Psychology of learning and motivation* (Vol. 55, pp. 37–76). Academic Press.

Talbert, R., & Bergmann, J. (2023). *Flipped learning: A guide for higher education faculty*. Routledge.

Tallon, P. P., Queiroz, M., Coltman, T., & Sharma, R. (2019). Information technology and the search for organizational agility: A systematic review with future research possibilities. *The Journal of Strategic Information Systems, 28*(2), 218–237.

Tambe, P., Cappelli, P., & Yakubovich, V. (2019). Artificial intelligence in human resources management: Challenges and a path forward. *California Management Review, 61*(4), 15–42.

Tao, F., Qi, Q., Liu, A., & Kusiak, A. (2018). Data-driven smart manufacturing. *Journal of Manufacturing Systems, 48*, 157–169.

Tao, F., Zhang, M., & Nee, A. Y. C. (2018). *Digital twin driven smart manufacturing*. Academic Press.

Tarafdar, M., Page, X., & Marabelli, M. (2019). Algorithms as co-workers: Human-algorithm role interactions in professional work. *Journal of the Association for Information Systems, 20*(9), 1–38. DOI: 10.17705/1jais.00565

Tayal, R., Upadhyay, R., Yadav, M., & Singh, R. (2019). The Development and Validation of the Organizational Effectiveness Scale Using Confirmatory Factor Analysis. *Global Business Review, 22*, 097215091881172. DOI: 10.1177/0972150918811720

Tilley, S., & Sussna, M. (2020). *Cloud Native: Using containers, functions, and data to build next-generation applications*. O'Reilly Media.

Tippins, N. T., Oswald, F. L., & McPhail, S. M. (2021). *The handbook of personnel assessment and selection*. Wiley-Blackwell.

Tornatzky, L. G., & Fleischer, M. (1990). *The processes of technological innovation.* Lexington Books.

Treveil, M.. (2020). *Introducing MLOps: How to scale machine learning in the enterprise.* O'Reilly Media.

Trivedi, S., Singh, A., Yadav, M., Kumar, A., & Srivastava, P. (2023). Intention to Use Social Media in Online and Blended Learning: Potential Predictors and Mediation of Collectivism. IEEE Transactions on Engineering Management, PP, 1–14. https://doi.org/DOI: 10.1109/TEM.2023.3257422

Trivedi, S., Vishnu, S., Singh, A., & Yadav, M. (2023). Research Trends in Sustainable E-Payment Systems: A Study Using Topic Modeling Approach. IEEE Transactions on Engineering Management, PP, 1–15. https://doi.org/DOI: 10.1109/TEM.2023.3280216

Truong, H. T. (2023). Closing the velocity gap: Governing AI in the cloud era. *Journal of Enterprise Technology*, *45*(2), 112–125.

Tu n, L. (2024). Fostering public sector employees' citizen-oriented boundary-spanning behaviors through human resource flexibility and employee perceptions of normative public values. *Personnel Review*, *53*(8), 2087–2108. DOI: 10.1108/pr-02-2022-0078

Tubaro, P., Coville, M., Le Ludec, C., & Casilli, A. A. (2022). Hidden inequalities: The gendered labour of women on micro-tasking platforms. *Internet Policy Review*, *11*(1), 1–26.

Tune, J. D., Sturek, M., & Basile, D. P. (2013). Flipped classroom model improves graduate student performance in cardiovascular, respiratory, and renal physiology. *Advances in Physiology Education*, *37*(4), 316–320. DOI: 10.1152/advan.00091.2013

Tunio, M. N., Chaudhry, I. S., Shaikh, S., Jariko, M. A., & Brahmi, M. (2021). Determinants of the Sustainable Entrepreneurial Engagement of Youth in Developing Country—An Empirical Evidence from Pakistan. In Sustainability (Vol. 13, Issue 14). https://doi.org/DOI: 10.3390/su13147764

Tunio, M. N., & Tunio, S. (2025). *Navigating Digital Transformation: A Systematic Review of SMEs' Transitions and Growth in the Digital Age.* Sukkur IBA Journal of Management and Business., DOI: 10.30537/sijmb.v11i2.1509

van der Aalst, W. M. P. (2016). *Process mining: Data science in action.* Springer.

van der Aalst, W. M. P. (2016). *Process Mining: Data Science in Action.* Springer.

van der Aalst, W. M. P., Bichler, M., & Heinzl, A. (2018). Robotic process automation. *Business & Information Systems Engineering, 60*(4), 269–272.

Van Dis, E. A., Bollen, J., Zuidema, W., van Rooij, R., & Bockting, C. L. (2023). ChatGPT: Five priorities for research. *Nature, 614*(7947), 224–226. DOI: 10.1038/d41586-023-00288-7 PMID: 36737653

Verhoef, P. C., Broekhuizen, T., Bart, Y., Bhattacharya, A., Dong, J. Q., Fabian, N., & Haenlein, M. (2021). Digital transformation: A multidisciplinary reflection and research agenda. *Journal of Business Research, 122*, 889–901. DOI: 10.1016/j.jbusres.2019.09.022

Vihari, N., Yadav, M., & Panda, T. K. (2021). Impact of soft TQM practices on employee work role performance: role of innovative work behaviour and initiative climate. The TQM Journal, ahead-of-print. https://doi.org/DOI: 10.1108/TQM-03-2021-0092

Villamizar, M., Garces, O., Castro, H., Verano, M., Casallas, R., Gil, S., & Salamanca, L. (2015). Evaluating the monolithic and the microservice architecture pattern to deploy web applications in the cloud. In 2015 10th Computing Colombian Conference (10CCC) (pp. 583–590). IEEE. https://doi.org/DOI: 10.1109/ColumbianCC.2015.7333476

Villamizar, M., Garces, O., Castro, H., Verano, M., Salamanca, L., Casallas, R., & Gil, S. (2015, October). Evaluating the monolithic and the microservice architecture pattern to deploy web applications in the cloud. In *2015 10th Computing Colombian Conference (10CCC)* (pp. 583-590). IEEE.

Voigt, P., & Von dem Bussche, A. (2017). *The EU General Data Protection Regulation (GDPR): A practical guide*. Springer.

Vygotsky, L. S. (1978). *Mind in society: The development of higher psychological processes*. Harvard University Press.

Wali Ullah, G. M., Dey, S. K., & Khan, I. (2025). Development and usage of FinTech in the emerging economies. In Tiwari, S. R. (Ed.), *The Role of FinTech in Advancing Sustainability and Innovation* (pp. 315–334).

Wang, A., Pruksachatkun, Y., Nangia, N., Singh, A., Michael, J., Hill, F., Levy, O., & Bowman, S. R. (2019). SuperGLUE: A stickier benchmark for general-purpose language understanding systems. *Advances in Neural Information Processing Systems, 32*(July), 1–15.

Wang, K., Ma, Y., Yang, J., Liu, Z., & Kang, R. (2018). Data-driven fault diagnosis and prognosis for manufacturing equipment: A review. *IEEE Transactions on Industrial Informatics, 15*(6), 3312–3325. DOI: 10.1109/TII.2018.2799931

Wang, S., Attal, D., Rafiq, M., & Hubner-Benz, K. (2022). Leveraging GenAI for entrepreneurial decision-making. *Journal of Business Research, 138*, 336–347.

Wedel, M., & Kannan, P. K. (2016). Marketing analytics for data-rich environments. *Journal of Marketing, 80*(6), 97–121. DOI: 10.1509/jm.15.0413

White, M.. (2023). Prompt engineering: Maximizing large language model outputs for entrepreneurial use. *AI & Society, 38*(3), 45–60.

Willcocks, L. P., & Lacity, M. C. (2016). *Service Automation: Robots and the Future of Work.* Steve Brookes Publishing.

Wilson, H. J., & Daugherty, P. R. (2018). Collaborative intelligence: Humans and AI are joining forces. *Harvard Business Review, 96*(4), 114–123.

Wong, J. (2020). The cost of ESG reporting. *Journal of Applied Corporate Finance, 32*(2), 1–8.

World Economic Forum. (2023). *Future of Jobs Report 2023.* https://www.weforum.org/reports/the-future-of-jobs-report-2023/

World Economic Forum. (2023). *Future of Jobs Report 2023.* World Economic Forum.

World Economic Forum. (2023). *Future of Jobs Report 2023.* World Economic Forum. Retrieved from https://www.weforum.org/reports/the-future-of-jobs-report-2023/

Xu, H., & Yang, L. (2025). Enhancing financial risk prediction through echo state networks and differential evolutionary algorithms in the digital era. *Procedia Computer Science, 242*, 145–154.

Xu, L. D., Xu, E. L., & Li, L. (2018). Industry 4.0: State of the art and future trends. *International Journal of Production Research, 56*(8), 2941–2962. DOI: 10.1080/00207543.2018.1444806

Yadav, O. P., Teotia, R., & Baliyan, R. (2025). FinTech and data science: Shaping the future of the digital economy. In *Big Data Analytics and Emerging Technologies in Finance* (pp. 101–118).

Yadav, M., Kumar, M., Alyousef, M. I., Dagar, M., Dewasiri, N. J., & Amoozegar, A. (2025). Mapping Research Trends on Gender Diversity, Work Stress, and Job Satisfaction in Higher Education: A Bibliometric Analysis. *International Journal of Learning. Teaching and Educational Research*, *24*(8), 444–472. DOI: 10.26803/ijlter.24.8.19

Yadav, M., & Rangnekar, S. (2016). Role Clarity and Organizational Citizenship Behaviour: Does Tenure Matter? A Study on Indian Power Sector. *Global Business Review*, *17*(3, suppl), 207S–224S. DOI: 10.1177/0972150916631216

Yadav, M., & Vihari, N. S. (2021). Employee Experience: Construct Clarification, Conceptualization and Validation of a New Scale. *FIIB Business Review*, *12*(3), 328–342. DOI: 10.1177/23197145211012501

Yadav, R., Yadav, M., & Vihari, N. (2023). High-performance work system and learning orientation in offline, online, and hybrid workplaces: The mediating role of affective commitment. *The Learning Organization*, *31*. Advance online publication. DOI: 10.1108/TLO-10-2022-0118

Yalenios, J., & d'Armagnac, S. (2022). Work transformation and the hr ecosystem dynamics: A longitudinal case study of hrm disruption in the era of the 4th industrial revolution. *Human Resource Management*, *62*(1), 55–77. DOI: 10.1002/hrm.22114

Yang, R., & Gao, J. (2025). Enhancing financial data security in big data environments using AES-Blowfish and cloud-aided encryption techniques. *Procedia Computer Science*, *242*, 210–219.

Yeung, K. (2018). Algorithmic regulation: A critical interrogation. *Regulation & Governance*, *12*(4), 505–523.

Yu, X. (2025). Research on AI-driven supply chain financial risk management. In *Proceedings of the 2025 International Conference on Artificial Intelligence and Computer Engineering (ICAICE)*.

Zafar, S., & Rana, N. (2025). The convergence of nanotechnology and artificial intelligence: Unlocking future innovations. *Recent Innovations in Chemical Engineering*.

Zainuddin, Z., & Halili, S. H. (2016). Flipped classroom research and trends from different fields of study. *International Review of Research in Open and Distance Learning*, *17*(3), 313–340. DOI: 10.19173/irrodl.v17i3.2274

Zeng, J., Khan, Z., & De Silva, M. (2022). The role of artificial intelligence in innovation: A systematic literature review. *Technological Forecasting and Social Change*, *179*, 121–158.

Zhai, X., Chu, X., Chai, C. S., Jong, M. S. Y., Istenic, A., Spector, M., Liu, J. B., Yuan, J., & Li, Y. (2021). A Review of Artificial Intelligence (AI) in Education from 2010 to 2020. *Complexity, 2021*. Advance online publication. DOI: 10.1155/2021/8812542

Zhang, S., Yao, L., Sun, A., & Tay, Y. (2019). Deep learning based recommender system: A survey and new perspectives. *ACM Computing Surveys, 52*(1), 1–38. DOI: 10.1145/3285029

Zhang, W., Yang, D., & Wang, H. (2019). Data-driven methods for predictive maintenance of industrial equipment: A survey. *IEEE Systems Journal, 13*(3), 2213–2227. DOI: 10.1109/JSYST.2018.2874340

Zhang, X. (2024). Enhanced agricultural financial services through cloud computing: A new paradigm of security and efficiency. *Review of World Agriculture and Environment, 5*(4), 1315.

Zhou, K., & Yang, S. (2016). Understanding the power of smart energy management. *Renewable & Sustainable Energy Reviews, 57*, 1–10.

Zuboff, S. (2019). *The age of surveillance capitalism: The fight for a human future at the new frontier of power*. PublicAffairs.

Zuboff, S. (2022). Surveillance capitalism or democracy? The death match of institutional orders and the politics of knowledge in our information civilization. *Organization Theory, 3*(3), 26317877221129290.

Zwitter, A. (2021). The rise of the digital divide: AI and the global south. Journal of Cyber Policy, 6(2), 147-162.

About the Contributors

Minakshi, Senior Member, IEEE, received her B.Tech. in Computer Science and Engineering from Maharishi Dayanand University, India (2005), her M.Tech. in Computer Science and Engineering from Maharishi Dayanand University, India (2010), and her Ph.D. in Computer Science and Engineering from Gyan Vihar University, Jaipur, India (2019). With over 18 years of experience in teaching and research, she is currently an Assistant Professor in Computer Science at King Khalid University, Abha, Saudi Arabia. She has previously served at several prestigious institutions, including Uttaranchal University, Dehradun, where she was Professor and Head of Department, CSE (2019-2023), Tulas Institute as Head of Department, CSE (2018-2019), and various academic roles at World College of Technology and Management, Dronacharya College of Engineering, and Apeejay College of Engineering. Her research interests include Artificial Intelligence, Machine Learning, Computer Networks, Multimedia Systems, and Network Security Management. A prolific innovator, she has filed 16 patents covering healthcare technology, blockchain applications, and smart systems, and has published over 75 research papers in SCI-indexed journals and international conferences. Dr. Minakshi holds memberships with IEEE (Senior Member), CSTA, SDIWC, IAENG, CSI, and Indian Science Congress Association. She has served in leadership roles including Chair of Subsection Coordination Outreach and Strategic Planning under IEEE UP Section, Convenor of IEEE SIGHT Group, and currently serves as Principal Investigator on multiple funded research projects at King Khalid University focusing on multimedia forensics, medical image segmentation, and natural language processing applications.

Madhulika Bhatia is working as Software Development Officer with Digital Healthcare Wales, UK. She holds a Diploma in Computer Science and Engineering, B.E in Computer Science and Engineering, MBA in Information Technology, M. Tech in Computer Science and PhD from Amity University, Noida. She has a total of 18 years of teaching experience. She published almost 58 Research Papers in National, International conferences and Journals. She is also the author of two books. She Filed two Provisional Patent. She attended and organized many workshops, Guest Lectures, seminars. She is also a member of many technical societies like IET, ACM, UACEE, IEEE She reviewed for Elsevier-Heliyon, IGI, IEEE ACCESS, and as well as did Editorial for Springer Nature, Switzerland. She has guided 12 M.Tech. Thesis and around 50 B. Tech. Major and Minor Projects and guiding PhD scholars, Research collaborations. She delivered Expert talks and hands on sessions in AICTE Sponsored workshops in collaboration with IIT Roorkie. She was also invited for Expert talk by NIFT, Patna.She has international exposure on various Projects and Research collaborations with University of Cape Town, South Africa and Lancaster University, UK. She was also invited as Visiting researcher by Cardiff Metropolitan University, UK. She is pursuing her Post doctorate Federal University of Ceira, Brazil.She is Tableau and Agile Scrum certified.

Gaurav Dubey is a Professor of Computer Science and Engineering with over 24 years of teaching and research experience. He earned his Ph.D. in Computer Science and Engineering from Amity University, India, and holds M.Tech. and B.Tech. degrees in the same discipline. Dr. Dubey has published more than 51 research papers in reputed SCIE and Scopus-indexed journals, contributed to several Springer-indexed book chapters, and presented widely at national and international conferences. He has more than 30 patents published and 1 design patent granted.His key research areas include machine learning, cloud computing, sentiment analysis, and data mining. In addition to his research contributions, Dr. Dubey has actively participated in curriculum design, peer reviewing, and academic leadership roles, and has served as a session chair and keynote speaker at international forums. His commitment to advancing scholarly work and supporting knowledge dissemination aligns well with IGI Global's mission to promote quality research worldwide.

Tanupriya Choudhury, Senior Member, IEEE, received his B.Tech. in Computer Science and Engineering from West Bengal University of Technology, Kolkata (2004–2008), and his M.Tech. in Computer Science and Engineering from Dr. M.G.R. University, Chennai (2008–2010). He earned his Ph.D. from Jagannath University, Jaipur, in 2016. With over 15 years of experience in teaching and research, he is currently a Professor at the University of Petroleum and Energy Studies (UPES), Dehradun, India, and a Visit ng Professor at Daffodil International University, Bangladesh. He has previously served at several prestigious institutions, including Symbiosis International Deemed University, Graphic Era Hill University (as Research Professor), and Amity University, where he was Assistant Professor (Grade-III) and International Department Head. He also held academic and industry roles at Dronacharya College of Engineering, Lingaya's University, BBDIT, and Syscon Solutions Pvt. Ltd., among others. His research interests include AIML, Deep Learning, Human Computing, Sof Computing, Cloud Computing, and Data Mining. A prolific innovator, he has filed 25 patents and received 16 software copyrights from the Ministry of Education (formerly MHRD), India. He has actively contributed to numerous national and international conferences as a speaker and participant. Dr. Choudhury holds lifetime membership with IETA (International Engineering & Technology Association) and is affiliated with several professional bodies, including IEEE, IET (UK), and other technical societies. He serves as a Technical Advisor to multiple corporate organizations such as Deetya Sof Pvt. Ltd., Noida, IVRGURU, and Mydigital360. He is also the Honorary Secretary of the Indian Engineering Teachers' Association (IETA-India) and holds Senior Advisory roles in the INDO-UK Confederate on of Science, Technology and Research Ltd., London, UK, and the International Association of Professional and Fellow Engineers, Delaware, USA.

Bhawna Wadhwa is head of the Department of Cyber Security, Noida Institute of Engineering and Technology Greater Noida, has extensive experience of 17.5 years 10 months in teaching. She is pursuing Ph.D (P) (Computer Engineering) from J.C. Bose University of Science and Technology, YMCA, Faridabad. She has completed M.Tech. (Computer Engineering) from Maharshi Dayanand University, (YMCA) in 2009 and B.E. (Computer Science and Engineering) from Maharshi Dayanand University in 2005. Her research profile includes more than 15 papers in reputed journals which include work on cloud computing, Information security, adhoc network, artificial intelligence and machine learning. She has presented 10 papers in National and International conferences and has published her research as various book chapters and she has guided 7 PG research scholars in the same area.

***

Neha Arora A multidisciplinary educator and counsellor who's delivered Creative, Technical, and engaging solutions across Fashion design, Technology, Communication and Management. She is pursuing Ph.D. in Design (specialised in E-commerce Industry) from Banasthali Vidyapith and completed her Master's and Bachelor's in Fashion Technology from Bhagat Phool singh Vishwavidyalaya, she has also done advance diploma course in E commerce in fashion Industry from NIFT Delhi. She has 13 years of academics and Industry experience. She has been teaching at Amity University, Noida since 2017 and also worked with other Engineering Institutes/University. She is a lifetime member of Textile association of India and Trainer & Assessor with Samarth and other programs run by Ministry of Textiles. She was the member of various board of member syllabus and scheme committee and have a great interest in research and publications.

Manju Dagar, an esteemed Assistant Professor in the Department of Management Studies, boasts an impressive academic and professional profile. She completed her Ph.D. on the topic of "Attrition of women employees in the IT Industry". She has 13 years of academic experience and has also served as the head of the department. Dr. Dagar is a valued member of the University Placement Cell and HR association. Her areas of specialization include Human Resource Management and Marketing Management, showcasing well-rounded expertise that spans crucial aspects of organizational dynamics. In terms of research, Dr. Manju Dagar has made significant contributions, and her work has been published in various prestigious platforms. Her research papers have found a place in UGC CARE-recognized journals, international publications, and the esteemed Scopus database, reflecting high quality and global relevance to her scholarly output. Additionally, she has made innovative contributions to patents further highlighting her multifaceted approach to academic and intellectual pursuits. She has also successfully completed various Faculty Development Programs (FDPs) and workshops. Most recently, she completed a diploma in Coaching & Mentoring from Skills and Education Group based in United Kingdom organized by CEAPS SGT University.

Narayanage Jayantha Dewasiri, Ph.D., is a Chartered Manager at the Chartered Management Institute, UK. He conducts research at the Open University of Sri Lanka and the Sri Lanka Institute of Marketing. He is the founding Editor-in-Chief of the South Asian Journal of Marketing. His research interests are in corporate social responsibility, corporate governance, marketing and sales, management, and finance. He is working as the Head of Field Force at the HMD Global Inc., Sri Lanka. He also serves as a senior lecturer at University of Sabaragamuwa, Sri Lanka. He can be reached at dnjay@ou.ac.lk

Damith Gangodawilage is a practicing scholar and professional with extensive experience in finance and accounting, business management, entrepreneurship, and innovation. Currently serving as the Deputy Dean at the Qasim Ibrahim School of Business (QISB) at Villa College, Maldives, he has consistently demonstrated leadership in fostering academic excellence and industry engagement. Damith is also the Director of the Centre for Innovation and Entrepreneurship Development (CIED), where he spearheads initiatives such as the VIgnite Business Incubator, driving innovation among youth in the Maldives. With a strong academic foundation, Damith has made significant contributions to teaching, research, and curriculum development. His expertise spans strategic areas such as design thinking, sustainable business practices, and digital transformation. An active researcher, he has authored numerous studies and presentations on topics like corporate sustainability and tax compliance behavior, establishing himself as a thought leader. Dr. Damith is a dedicated mentor and educator, passionately committed to equipping students with critical thinking and problem-solving skills.

Isha Goley is an Assistant Professor at Prestige Institute of Management and Reseach Gwalior, in the department of Management. She holds a MBA in Human Resource Management. Her areas of interest include human resource maangement, interpersonal effectiveness and behavioural sciences.

Khushboo Jain, Assistant Professor, in Shri Vaishnav Institute of Management & Science, Indore has a qualification of MBA (HR & Marketing) and Bachelor of Engineering (Instrumentation & Control). Currently she is the Program Coordinator of BBA (Hospital Administration). She is Ph.D. Supervisor and guiding students under DAVV University. She has published more than thirty research papers in different National and International Journals. She has also presented various papers in National and International conferences. She has attended many workshops, MDPs and FDP. She is an active member of NAAC and ISO committees of the institution. She is specialized in streams of Human Resource, Quantitative Techniques, Operation Resarch and General Management.

Sneha Kadyan is a Social science researcher with demonstrated experience and performance at all organisations and institutes in academic and professional development across diverse social issues. Knowledge innovator with methodological proficiency in qualitative and quantitative skills and analysis. Unique perspective as a professor, scholar and researcher in global institutional environments. Excellent relationship builder -team worker with collaborative research across the globe. Presently she is working as Assistant Professor, Jindal Global Business School, O.P Jindal Global University, Haryana, India.

Deepa Katiyal is Professor and Head Department of Management (UG) in Shri Vaishnav Institute of Management & Science, Indore. She is member of Board of Studies of Management; Devi Ahilya Vishwavidyalaya .She is also the member of inspection committee of DAVV, Indore. She has a first class career in M.A. (Eng), M.B.A. (HR), PhD. She has been taking classes of Management for more than twenty years. She is PhD Supervisor and guiding students under DAVV University. She has published more than seventy research papers in different National and International Journals and conferences. She has attended and organized many seminars and workshops related to Entrepreneurship, HR and Management. She has taken expert lectures on Personality Development and entrepreneurship in various

Institutes and Colleges. She has been called as eminent speaker and special guest by various organizations.

Akanksha Kathuria is an accomplished academician with over 16 years of teaching and research experience in Digital Marketing and Management. She is currently serving as an Assistant Professor at the School of Business, Manav Rachna University, Faridabad. Currently pursuing her Ph.D. in Management from JC Bose University of Science & Technology, YMCA Faridabad, Ms. Kathuria also holds an MBA, an MA in English Literature, and an MA in Yoga. She is certified in Digital Marketing, POSH, and Yoga Trainer. Ms. Kathuria has published extensively in Scopus- and ABDC-indexed journals and has authored a book titled An Introductory Approach to Research Methodology. Her research spans green marketing, consumer behavior, AI applications, and sustainable mobility. In addition to academics, she has conducted Management Development Programs (MDPs) and corporate training in Digital Marketing and Workplace Wellness for reputed organizations. She also holds three patents, reflecting her innovative approach to applied research. She serves on the editorial board of the Journal of Research in Business Management and is actively engaged in community service and wellness initiatives.

Mohan Kumar has a total of eight years of professional experience in both the corporate sector and academia. He has worked in a Public Sector Bank and a State Government University, focusing on Marketing and Entrepreneurship Development. He completed his PhD on "Motivational Factors and Challenges of Start-Ups in NCR" at the Institute of Management Studies and Research, Maharshi Dayanand University, Rohtak-124001, Haryana. Dr. Mohan specializes in Marketing Management, Human Resource Management, Quantitative Analysis, and Entrepreneurship Development. His research has been published in esteemed national and international journals, and he has actively participated in both national and international conferences, presenting his research findings. In March 2022, he received the "Best Research Paper Award" at an international conference conducted by IITM, GGSIU, Delhi, for his paper titled "Psychological Wellbeing Before & During the COVID-19 Pandemic: A Longitudinal Sample Study of Creative Writers and Directors." Since May 2023, Dr. Mohan has been working at SGT University, Gurugram. He continues to contribute to academia through research, conferences, and workshops, furthering his expertise in his specialized domains.

Thi Minh Ngoc Luu is currently the Head of the Department of Economics and Management at the International School, Vietnam National University, Hanoi. She has over 17 years of experience in teaching and researching human resource management, management control, corporate governance, individual and organizational behavior, leadership, and sustainable development.

Firdous Ahmad Malik was born in India and is currently employed by the University of People as an Assistant Professor of Economics (instructor), Department of Management. Currently he is perceiving Post Doctoral research at Amity Dubai University. Before that, he was a research fellow at the National Institute of Public Finance and Policy (NIPFP) in Delhi, India.

Ali Najeeb is the Vice Rector and Chief Operating Officer (COO) of Villa College (VC), Maldives and CEO of Bristol Institute of Business Management (BIBM), Sri-Lanka. He is also an Adjunct Associate Professor at the School of Hospitality, Tourism and Events, Taylor's University, Malaysia. He holds a PhD in Management and HR strategy at the University of Wollongong, Australia, a Master's Degree in Human Resource Management at the University of Newcastle, Australia, and BA in Business Studies from the University of the South Pacific, Fiji. He is a Chartered Professional Member of Chartered Institute of Personnel and Development (CIPD) and a Certified Professional Member of Australian Human Resource Institute (AHRI). Besides his official roles, he also holds several advisory roles and membership in local and international NGOs and professional bodies. His research interests include higher education leadership, governance and management, industrial relations, employment law and labour issues in small island states, migrant workers, HRM/ER in tourism and hospitality industry. He has published several research papers in refereed international journals, shared his experience in local and international panel discussions and presented his research work in international scholarly conferences. He is a recipient of the Emerald Literati Award for Excellence 2014 – the outstanding paper award. Dr Najeeb is a former school principal and worked as a Director of the Education and Development Centre in the Ministry of Education. He is also a former fulltime member of the Employment Tribunal, Maldives. Besides his administrative, teaching and research work, he is actively engaged in management and corporate consultancy and training.

Ali Najeeb is the Vice Rector and Chief Operating Officer (COO) of Villa College (VC), Maldives and CEO of Bristol Institute of Business Management (BIBM), Sri-Lanka. He is also an Adjunct Associate Professor at the School of Hospitality, Tourism and Events, Taylor's University, Malaysia. He holds a PhD in Management and HR strategy at the University of Wollongong, Australia, a Master's Degree in Human Resource Management at the University of Newcastle, Australia, and BA in Business Studies from the University of the South Pacific, Fiji. He is a Chartered Professional Member of Chartered Institute of Personnel and

Development (CIPD) and a Certified Professional Member of Australian Human Resource Institute (AHRI). Besides his official roles, he also holds several advisory roles and membership in local and international NGOs and professional bodies. His research interests include higher education leadership, governance and management, industrial relations, employment law and labour issues in small island states, migrant workers, HRM/ER in tourism and hospitality industry. He has published several research papers in refereed international journals, shared his experience in local and international panel discussions and presented his research work in international scholarly conferences. He is a recipient of the Emerald Literati Award for Excellence 2014 – the outstanding paper award. Dr Najeeb is a former school principal and worked as a Director of the Education and Development Centre in the Ministry of Education. He is also a former fulltime member of the Employment Tribunal, Maldives. Besides his administrative, teaching and research work, he is actively engaged in management and corporate consultancy and training.

Mananage Shanika Hansini Rathnasiri is a Senior Lecturer attached to the Department of Marketing Management, Faculty of Management Studies (FMS), Sabaragamuwa University of Sri Lanka (SUSL). Currently, she is pursuing her doctoral studies at the College of Society and Professions, London South Bank University, London, United Kingdom. She serves as an Associate Editor cum Editorial Assistant of the South Asian Journal of Marketing, published by Emerald Publishing. She is also serving as an Editor of the book namely "Shaping Entrepreneurial Marketing: A South Asian Perspective," by Emerald Publishing. Furthermore, she contributes as an Editor of the Sri Lankan Marketer Magazine, published by the Sri Lanka Institute of Marketing.

Shimona Shriya is an Assistant Professor at XIME. She is pursuing a PhD in Management from the Indian Institute of Management, Bodh Gaya (The Enlightening IIM). Shimona is in her final PhD stage. She holds a Bachelor of Engineering (B.E.) degree in Electrical and Electronics from BIT Mesra, Ranchi. Beyond academia, Shimona brings valuable industry experience, having worked for two years at Honda Cars India Ltd. and nearly two years at Havells India Ltd. Her corporate roles honed her expertise in operations, supplier relationship management, project management, and quality assurance, bridging practical insights with theoretical frameworks. Her research interests encompass Market Design, Preference Matching, Data Envelopment Analysis, Project Management, and Sports Management. She is equally passionate about teaching, with a keen interest in Operations Management, Project Management, Business Statistics, and Supply Chain Management. Shimona aims to contribute to academia through impactful research and dynamic teaching, fostering a rich learning environment for students

Sandeep Kumar Singh is a faculty in the area of Information Systems and Analytics at Jindal Global Business School, O.P. Jindal Global University, Sonipat. Prior to joining Jindal Global Business School, he worked as an Assistant Professor at JKLU Jaipur for two years. He has earned his Ph.D. in computational physics from Antwerpen University, Belgium. His PhD work focused on mathematical modeling and simulation of two-dimensional materials. Post to his PhD, he worked as a research scientist in Sweden for four years in energy storage systems, bioelectronics. He also did MTech in industrial mathematics and scientific computing from IIT Madras, Chennai. His current research interests are to apply complex mathematical modeling in the business and financial world.

Anugamini Priya Srivastava is a distinguished academic and researcher whose scholarship spans educational supply chain management, entrepreneurship education, digital transformation in higher education, organizational behavior, and sustainable practices in management and education. Her research is characterized by methodological rigor, conceptual innovation, and practical relevance, positioning her as a well-recognized scholar in national and international research communities. With 59+ peer-reviewed publications, she has established a strong academic footprint, including a Scopus-indexed book and numerous book chapters written in collaboration with international and national faculty. She has consistently published in leading ABDC–A and Q1-ranked journals, such as the International Journal of Productivity and Performance Management, Tourism Management Perspectives, International Journal of Manpower, European Journal of Training and Development, Current Psychology, and Journal of Cleaner Production. Her publications, many of which she has led as first author, highlight her intellectual independence and dedication to advancing research across diverse fields. Dr. Srivastava's excellence has been recognized through several prestigious awards: Best Doctoral Research Award at the PAN-IM World Management Conference. Top 20 Research Paper Recognition at the IISc Bangalore Conference. Best Paper Award at the Symbiosis Centre for Behavioral Studies (SCBS). Multiple Best Paper Awards at other national and international conferences. Invited as a nominated faculty and conference chair at international forums, including BITS Pilani and other reputed institutions. In addition to her publications and awards, Dr. Srivastava has made notable contributions to the scholarly community through editorial and reviewing roles. She has served as a Guest Editor for reputed international journals such as the European Journal of Training and Development and the International Journal of Educational Management. She is also a member of the Editorial Advisory Board for the International Journal of Educational Management and worked as an Associate Editor for the South Asian Journal of Marketing. Her extensive service as a peer reviewer has been formally recognized by multiple journals for her high-quality and constructive feedback to authors. She has also advanced research through conceptual model development and scale construction, particularly in areas such as AI skill gaps among higher education faculty and digital globalization readiness. These models and tools offer both theoretical contributions and actionable insights for institutions navigating transformation in globalized education systems. With her record of scholarly publications, leadership in editorial roles, multiple awards, and consistent international recognition, Dr. Srivastava has built a highly credible research profile. Her academic journey demonstrates scholarly leadership, intellectual contribution, and a strong commitment to advancing global knowledge in management and education research.

Jeayaram Subramanian is an Assistant Professor at the Jindal Global Business School, O. P. Jindal Global Unversity, Sonipat, Delhi NCR, India.

Mohit Yadav is an Professor in the area of Human Resource Management at Jindal Global Business School (JGBS). He has a rich blend of work experience from both Academics as well as Industry. Prof. Mohit holds a Ph.D. from Department of Management Studies, Indian Institute of Technology Roorkee (IIT Roorkee) and has completed Master of Human Resource and Organizational Development (MHROD) from prestigious Delhi School of Economics, University of Delhi. He also holds a B.Com (Hons.) degree from University of Delhi and UGC-JRF scholarship. He has published various research papers and book chapters with reputed publishers like Springer, Sage, Emerald, Elsevier, Inderscience etc. and presented research papers in national and International conferences both in India and abroad. He has many best paper awards on his credit too. He is reviewer of various international journals like Computers in Human Behavior, Policing etc. His areas of interest are Organizational Behavior, HRM, Recruitment and Selection, Organizational Citizenship Behavior, Quality of work life and role.

Index

A

Adaptive Learning 1, 2, 3, 4, 5, 8, 10, 11, 12, 27, 291, 333
Adaptive Operations 99, 102
Affective Computing 262, 367, 368, 369, 383
AI 1, 2, 3, 4, 5, 6, 7, 8, 9, 10, 11, 12, 13, 15, 16, 17, 18, 19, 20, 21, 22, 23, 24, 25, 26, 27, 28, 29, 31, 32, 33, 34, 35, 36, 37, 38, 39, 40, 41, 42, 43, 45, 46, 47, 48, 49, 50, 51, 52, 53, 54, 55, 57, 58, 59, 60, 62, 63, 65, 66, 67, 68, 69, 70, 73, 74, 75, 76, 77, 78, 79, 80, 81, 82, 83, 84, 85, 87, 88, 90, 91, 92, 93, 94, 95, 96, 97, 99, 100, 101, 102, 103, 104, 105, 106, 107, 108, 109, 110, 113, 114, 115, 116, 117, 118, 119, 120, 121, 122, 123, 124, 125, 127, 128, 129, 130, 131, 132, 133, 134, 135, 137, 138, 139, 140, 141, 142, 143, 144, 145, 146, 147, 148, 149, 150, 151, 152, 153, 183, 184, 185, 186, 187, 188, 189, 190, 191, 192, 193, 194, 195, 197, 199, 200, 202, 206, 209, 212, 213, 214, 217, 218, 219, 220, 221, 222, 225, 226, 227, 228, 229, 230, 231, 232, 237, 238, 242, 246, 247, 251, 252, 253, 254, 255, 256, 257, 258, 259, 260, 261, 262, 263, 264, 267, 268, 271, 273, 274, 275, 276, 277, 278, 279, 280, 281, 282, 283, 284, 285, 286, 287, 288, 289, 290, 291, 292, 293, 294, 295, 296, 297, 298, 299, 300, 301, 302, 303, 304, 305, 307, 308, 309, 310, 311, 312, 313, 314, 315, 316, 317, 318, 319, 320, 321, 322, 325, 326, 327, 328, 329, 330, 331, 332, 333, 334, 335, 337, 338, 339, 340, 341, 342, 344, 345, 346, 347, 349, 351, 352, 353, 354, 355, 356, 357, 358, 359, 360, 361, 363, 364, 365, 367, 368, 369, 370, 371, 372, 373, 374, 375, 376, 377, 378, 379, 380, 381, 382, 385, 386, 388, 389, 390, 391, 395, 396, 398, 399, 400, 401, 402, 403, 404, 405, 406, 407, 408, 409, 411, 412, 413
AI-Augmented Leadership 73, 75, 79
AI Ethics 25, 36, 57, 59, 65, 70, 124, 301, 304, 319
AI Governance 24, 146, 194, 297, 299, 301, 302, 303, 304, 305, 339, 341, 399
AI-Powered Supply Chains 99, 101, 106
Algorithmic Bias 2, 10, 12, 25, 53, 60, 80, 95, 118, 135, 192, 236, 274, 295, 326, 327, 332, 337, 340, 367, 390
Algorithmic Trust 325, 327, 334, 341
Artificial Intelligence (AI) 1, 2, 3, 6, 10, 12, 16, 17, 18, 24, 26, 29, 31, 32, 34, 35, 36, 37, 38, 39, 42, 43, 45, 46, 48, 54, 55, 56, 57, 58, 61, 65, 70, 71, 73, 74, 75, 82, 84, 85, 87, 88, 90, 95, 99, 100, 109, 111, 113, 114, 116, 124, 125, 127, 128, 133, 137, 138, 140, 152, 183, 193, 195, 199, 200, 202, 204, 206, 212, 214, 217, 218, 219, 224, 238, 249, 252, 256, 261, 264, 268, 277, 279, 280, 281, 285, 286, 288, 290, 291, 293, 294, 295, 296, 305, 307, 309, 311, 322, 323, 325, 326, 327, 328, 329, 340, 343, 344, 345, 346, 347, 350, 351, 352, 355, 362, 363, 364, 365, 368, 395, 396, 399, 400, 403, 405, 406, 412
Automated Decision-Making 23, 274
Automating Intelligence 140, 197, 217, 349
Automation 3, 6, 8, 16, 18, 22, 24, 27, 28, 31, 32, 39, 41, 45, 48, 57, 58, 61, 64, 69, 70, 74, 75, 84, 85, 91, 94, 97, 100, 103, 107, 108, 110, 114, 116, 120, 125, 128, 129, 131, 132, 133, 139, 140, 141, 142, 144, 184, 185, 186, 202, 203, 204, 206, 235, 236, 237, 238, 239, 240, 241, 242, 243, 244, 245, 246, 247, 248, 249, 251, 252, 254, 267, 268, 269, 270, 271, 272, 273, 274, 275, 277, 278, 281, 282, 283, 286, 290, 294, 295, 300, 305, 307,

322, 335, 338, 386, 387, 388, 393, 399, 400, 401, 402, 403, 404
Autonomous Planning 99, 103

B

Behavioral Finance 332, 338, 340
Blueprint 207, 235, 241, 248, 268, 269, 272, 386
Brain Tumor 349, 350, 352, 353, 354, 356, 358, 359, 360, 361, 362, 363, 364
Business Ecosystems 31, 32, 41
Business Strategy 8, 202, 271, 315

C

Cloud Automation 388, 393
Cloud Computing 2, 4, 28, 45, 46, 48, 54, 55, 56, 73, 74, 75, 82, 85, 87, 88, 90, 95, 97, 102, 202, 204, 206, 237, 238, 249, 270, 295, 296, 305, 307, 309, 311, 325, 326, 327, 328, 329, 332, 333, 334, 337, 341, 347, 353, 363, 373, 400, 401, 402, 404, 412
Cloud-Native AI 15, 16, 17, 18, 19, 20, 21, 24, 26, 27, 28, 139, 140, 141, 143, 144, 145, 146, 147, 148, 149, 150, 197, 217, 251, 252, 253, 254, 255, 256, 257, 259, 260, 261, 263, 267, 276, 296, 305, 309, 310, 311, 320, 321
Cloud-Native Architecture 102, 236, 239, 240, 254, 267, 273, 275, 312, 373
Cloud-Native Computing 100, 109, 236
Cloud-Native Intelligence 99, 251, 256, 258, 259
Consumer Behavior 33, 77, 202, 396, 398
Corporate Governance 16, 92, 192, 304
Corporate Strategy 183, 184, 189, 191, 194, 195
Credit Risk 326, 328, 329, 331, 332, 334, 335, 336, 337, 338, 339, 343, 344, 345, 346
Cross-Cultural Management 113, 114, 115, 118, 120, 121
Cultural Intelligence 77, 114, 120

D

Data Analytics 4, 11, 90, 186, 202, 204, 326, 329, 335, 344, 345, 347, 361, 363, 374, 386, 390, 396
Decision-Making 7, 16, 23, 25, 26, 28, 42, 49, 58, 59, 61, 69, 73, 74, 75, 76, 77, 78, 80, 82, 85, 94, 100, 113, 119, 121, 127, 128, 129, 130, 131, 132, 133, 135, 137, 141, 142, 143, 148, 166, 185, 186, 191, 200, 220, 247, 257, 268, 274, 297, 310, 325, 326, 328, 332, 337, 338, 339, 352, 376, 392, 403, 404
Deep Learning 33, 53, 90, 98, 120, 165, 190, 223, 264, 318, 350, 351, 352, 353, 354, 359, 360, 363, 364, 369, 413
Digital Transformation 12, 45, 46, 70, 98, 197, 198, 199, 200, 201, 202, 203, 204, 205, 206, 207, 269, 310, 341, 344, 346, 403, 413
Digital Twin 50, 78, 101, 104, 111, 345

E

Emotion AI 367, 368, 369, 370, 371, 372, 373, 374, 375, 376, 377, 378, 379, 380, 381
Employee Experience 15, 17, 18, 21, 28, 98, 251, 253, 262, 264, 278, 367, 368, 369, 370, 381, 383
Entrepreneurial Creativity 127, 128, 130, 134, 135
Entrepreneurship 128, 129, 130, 135, 137, 217, 279, 346
ESG Reporting 87, 88, 91, 98
Ethical AI 63, 67, 107, 135, 194, 268, 274, 359
Ethical Governance 24, 67, 80, 258, 381
Ethical Implications 42, 53, 67, 80, 122, 192, 246, 303, 317, 378

F

Financial Inclusion 326, 328, 329, 331, 334, 341, 345, 346
FinTech 37, 132, 325, 326, 327, 328, 329,

330, 331, 332, 333, 334, 335, 336, 337, 338, 339, 340, 341, 343, 344, 345, 347
Flipped Learning 155, 156, 157, 158, 159, 160, 162, 163, 164, 165, 166, 168, 169, 170, 171, 172, 173, 174, 175, 176, 177, 179, 180, 181

G

Generative Tools 217
Global Labor 18
Global Operations 235, 240, 243
Global Teams 119, 123, 257
Green Operations 87, 88, 92

H

HR Technology 67, 373, 391
Human-AI Collaboration 27, 55, 123, 184, 190
Human-Centered Design 41
Human-Centric Management 367
Human-in-the-Loop 61, 66, 298, 304, 305, 411
Human Resources 11, 32, 61, 71, 89, 101, 117, 252, 309, 321, 323, 386
Hybrid Intelligence 55, 114, 116, 124, 193, 281
Hyperautomation 100, 107, 108, 238, 239, 275

I

Infrastructural Intelligence 139, 140, 141, 146
Intelligent Automation 16, 235, 236, 237, 238, 239, 241, 242, 243, 248, 267, 268, 269, 270, 271, 273, 275, 399
Internal Mobility 19, 51, 318, 320, 385, 387, 391

L

Learning Ecosystem 2, 3, 5, 8, 9, 11

M

Machine Learning 3, 6, 10, 20, 21, 23, 28, 48, 49, 56, 59, 65, 71, 76, 77, 91, 92, 102, 103, 105, 106, 111, 116, 128, 129, 132, 183, 184, 185, 186, 199, 200, 209, 214, 223, 231, 232, 238, 241, 242, 246, 249, 268, 269, 270, 304, 307, 308, 311, 313, 325, 326, 327, 328, 329, 332, 333, 334, 335, 337, 339, 340, 341, 344, 345, 350, 351, 353, 354, 360, 363, 364, 369, 370, 374, 383, 388, 396, 398, 401, 405, 406, 412
Managerial Decision-Making 73
Managerial Implications 31, 40, 52, 58, 68, 79, 114, 122, 191, 247, 295, 303, 340, 392
Managerial Intelligence 183, 184, 185, 193
Microservices 15, 16, 17, 21, 22, 23, 24, 99, 102, 105, 106, 235, 237, 240, 242, 244, 248, 249, 253, 270, 273, 277, 311, 312, 322, 373
MLOps 15, 16, 18, 23, 24, 52, 106, 249, 295, 297, 300, 301, 305, 330

O

Operational Efficiency 8, 24, 46, 197, 205, 206, 207, 209, 211, 268, 328, 332, 335, 404
Organizational Agility 1, 7, 45, 46, 54, 56, 105, 268, 272, 390

P

People Analytics 57, 58, 59, 60, 62, 63, 65, 66, 67, 68, 69, 316, 319
Perceived Risk 325, 327, 328, 333, 334, 340, 341
Personalized Learning 3, 11, 13, 160, 315, 389
Policy as Code 295, 300, 305
Predictive Analytics 6, 19, 28, 45, 49, 59, 70, 75, 94, 128, 129, 131, 143, 186, 207, 208, 210, 211, 214, 254, 310, 311, 314, 315, 316, 320, 330, 385,

393, 399, 400, 401
Predictive HR Analytics 309, 312, 316
Predictive Maintenance 49, 50, 197, 200, 207, 208, 209, 212, 213, 214, 215, 216, 223, 240, 260

R

Responsible AI 125, 259, 295, 297, 299, 305, 341
Robotic Process Automation 84, 97, 107, 110, 116, 202, 206, 238, 249, 268, 269, 277, 278, 322

S

Scalable Growth 267, 270, 271
Skill Mapping 385, 388, 389
Skills Taxonomy 5, 390
Smart HR Cloud 385, 386, 387, 388, 389, 390, 391, 392, 393
Strategic Foresight 259
Strategic Human Resource 15, 29
Strategic Lever 267, 268, 272
Supply Chain Resilience 110

T

Talent Acquisition 253, 264, 314, 390, 392
Talent Analytics 17, 59, 70
Talent Development 13, 205, 385, 392
Talent Management 50, 122, 309, 310, 312, 314, 319, 322, 379, 386
Transformation 2, 12, 15, 16, 28, 32, 39, 40, 45, 46, 48, 49, 50, 51, 54, 68, 70, 74, 80, 92, 93, 98, 100, 101, 106, 107, 141, 145, 153, 162, 190, 197, 198, 199, 200, 201, 202, 203, 204, 205, 206, 207, 235, 236, 237, 245, 249, 252, 258, 265, 268, 269, 290, 304, 310, 328, 341, 344, 346, 353, 386, 399, 400, 403, 413

U

Upskilling 1, 2, 51, 52, 67, 79, 84, 121, 191, 205, 245, 261, 274, 302, 316, 321, 390, 399

V

Virtual Try-On 410, 411, 412

W

Workforce Planning 15, 16, 17, 18, 19, 28, 385, 386, 387, 389

www.ingramcontent.com/pod-product-compliance
Lightning Source LLC
LaVergne TN
LVHW062333300926
849161LV00007B/1682